PERSPECTIVES ON BIOINORGANIC CHEMISTRY

Volume 3 • 1996

PERSPECTIVES ON BIOINORGANIC CHEMISTRY

Editors: ROBERT W. HAY
School of Chemistry
University of St. Andrews

JON R. DILWORTH
Department of Biological and
Chemical Sciences
University of Essex

KEVIN B. NOLAN
Chemistry Department
Royal College of Surgeons in Ireland
Dublin

VOLUME 3 • 1996

JAI PRESS INC.

Greenwich, Connecticut *London, England*

55 Old Post Road, No. 2
Greenwich, Connecticut 06836

JAI PRESS LTD.
38 Tavistock Street
Covent Garden
London WC2E 7PB
England

ISBN: 1-55938-642-8

Manufactured in the United States of America

CONTENTS

LIST OF CONTRIBUTORS vii

INTRODUCTION TO THE SERIES:
EDITORS' FOREWORD ix

PREFACE
Robert W. Hay xi

STRUCTURE AND FUNCTION OF
MANGANESE-CONTAINING BIOMOLECULES
David C. Weatherburn 1

REPERTORIES OF METAL IONS AS LEWIS ACID
CATALYSTS IN ORGANIC REACTIONS
Junghun Suh 115

THE MULTICOPPER–ENZYME ASCORBATE OXIDASE
Albrecht Messerschmidt 151

THE BIOINORGANIC CHEMISTRY OF ALUMINUM
Tamas Kiss and Etelka Farkas 199

THE ROLE OF NITRIC OXIDE IN
ANIMAL PHYSIOLOGY
Anthony R. Butler, Frederick W. Flitney, and
Peter Rhodes 251

INDEX 279

LIST OF CONTRIBUTORS

Anthony R. Butler — School of Chemistry, University of St. Andrews, St. Andrews, Scotland

Etelka Farkas — Department of Inorganic and Analytical Chemistry, Kossuth University, Debrecen, Hungary

Frederick W. Flitney — School of Biological and Medical Science, University of St. Andrews, St. Andrews, Scotland

Tamas Kiss — Department of Inorganic and Analytical Chemistry, Attila József University, Szeged, Hungary

Albrecht Messerschmidt — Max-Planck-Institut für Biochemie, München, Germany

Peter Rhodes — Ninewells Hospital, Dundee, Scotland

Junghun Suh — Department of Chemistry, Seoul National University, Seoul, Korea

David C. Weatherburn — Department of Chemistry, Victoria University of Wellington, Wellington, New Zealand

INTRODUCTION TO THE SERIES: EDITORS' FOREWORD

The aim of this series is to provide authoritative reviews in the rapidly expanding area of bioinorganic chemistry. The series will present "state of the art" reviews covering the whole field of bioinorganic chemistry. As the subject is, by its very nature, interdisciplinary, the editors feel that there is a need for articles covering the many different aspects of the subject from medicinal chemistry to biophysical studies. Suggestions from readers regarding topics to be covered in subsequent volumes will be welcomed.

R.W.H.
J.R.D.
K.B.N.

PREFACE

The present volume is the third in a series dealing with the field of bioinorganic chemistry. Articles on a wide range of topics are covered in Volume 3 and this approach will be continued in subsequent volumes where contributors will include Janice Bramham (lithium biochemistry); Eiichi Kimura and Tohru Koike (zinc enzymes and model systems); Michael Wilson (cytochromes); Jim Hoeschele (platinum anticancer drugs); Steve Mann and Trevor Douglas (biomineralization); James A. Edwardson (aluminum and Alzheimers disease); Peter F. Lindley (structure of ceruloplasmin); Philip Blower (uses of rhenium and other therapeutic radionucleides in medicine); and Dan Leussing (metal-activated decarboxylases).

Future volumes will be at approximately yearly intervals and are planned through to Volume 6. The style and presentation of each author has been maintained so that the individuality of the author is communicated to the reader.

I would like to thank all of the contributors to the present volume for their enthusiastic cooperation in the preparation of their manuscripts and meeting all the required deadlines. Their cooperation has greatly aided the editors in the preparation of this volume. I would also like to thank Piers Allen of JAI Press for his help and encouragement.

St. Andrews

Robert W. Hay
Series Editor

STRUCTURE AND FUNCTION OF MANGANESE-CONTAINING BIOMOLECULES

David C. Weatherburn

OUTLINE

1. **Introduction** **2**
2. **Manganese Chemistry** **3**
3. **The Importance of Manganese in Biological Systems** **3**
4. **Phosphatases** **10**
    4.1 Serine/Threonine Protein Phosphatase 2A (EC 3.1.3.16) 10
    4.2 Purple Acid Phosphatase (EC 3.1.3.2) 11
5. **Proteins with One Bound Metal** **13**
    5.1 G-Actin 13
    5.2 Dioxygenase 13
    5.3 Elongation Factor Tu 14
    5.4 Isocitrate Dehydrogenase (EC 1.1.1.42) 15
    5.5 Kinases: Adenylate, Guanylate, and Creatine 16
    5.6 α-Lactalbumin 19
    5.7 Lactoferrin 20

Perspectives on Bioinorganic Chemistry
Volume 3, pages 1–113

ISBN: 1-55938-642-8

5.8 Mandelate Racemase (EC 5.1.2.2) 20
5.9 Manganese-Dependent Peroxidase (EC 1.11.1.13) 22
5.10 *cis,cis*-Muconate Cycloisomerase (EC 5.5.1.1) 24
5.11 Ovalbumin 25
5.12 ras p21 Protein 25
5.13 Ribulose 1,5-Bisphosphate Carboxylase (EC 4.1.1.39) 27
5.14 tRNA 29
5.15 Staphylococcal Nuclease (EC 3.1.4.4 or 3.1.4.7) 29
5.16 Superoxide Dismutase (EC 1.15.1.1) 31
5.17 Transketolase (EC 2.2.1.1) 39
6. Proteins with Two Bound Metals 40
6.1 Acyl Carrier Protein 40
6.2 Catalase (EC 1.11.1.6) 40
6.3 DNA Polymerase (EC 2.7.7.7) 41
6.4 Deoxyribonuclease 43
6.5 Fructose-1-6-bisphosphatase (EC 3.1.3.11) 44
6.6 Glutamine Synthetase (EC 6.3.1.2) 45
6.7 Lectin 46
6.8 Leucine Aminopeptidase (EC 3.4.11.1) 49
6.9 6-Phosphofructokinase (EC 2.7.1.11) 50
6.10 3-Phosphoglycerate Kinase (EC 2.7.2.3) 52
6.11 Ribonuclease H (EC 3.1.27.5) 53
6.12 Ribonucleotide Reductase (EC 1.17.4.1) 54
6.13 D-Xylose Isomerase (EC 5.3.1.5) 56
7. Proteins with Three Bound Metals 59
7.1 Alkaline Phosphatase (EC 3.1.3.1) 59
7.2 Enolase (EC 4.2.1.11) 60
7.3 Parvalbumin 62
7.4 Pyruvate Kinase (EC 2.7.1.40) 62
8. Proteins with Four Bound Metals 64
8.1 Inorganic Pyrophosphatase (EC 3.6.1.1) 64
9. Conclusions 68
Acknowledgments 69
Addendum: Update to Middle of 1994 70
References 84

1. INTRODUCTION

This chapter is concerned with the binding of manganese to biological molecules and the nature of the metal binding site in these molecules. There have been previous reviews [1–9] and books [10, 11] on the biochemistry and the bioinorganic chemistry of manganese. Significant progress has been made since the appearance of most of these reviews, particularly in the determination of the structures of manganese-containing proteins. Many of the earlier reviews concentrated on the important water-oxidizing system of Photosystem II. Because that system has been so well covered in a number of recent reviews [12–21] it will not be treated in this chapter. Even after excluding the work on Photosystem II the number of possible references to manganese in biological systems is large and the selection of references has been a problem in compiling this review. Usually only the most recent reference(s) on a particular topic will be given and reference will be made to earlier reviews in particular areas whenever possible.

2. MANGANESE CHEMISTRY

Manganese in biological systems adopts a range of coordination environments and oxidation states. Mn^{2+} is labile, it has no coordination geometry preferences, and it forms relatively weak complexes with ligands (in general stronger than Ca^{2+} and Mg^{2+} but weaker than Fe^{2+}, Co^{2+}, or Zn^{2+}). Consequently it forms exchangeable metal complexes with biological ligands and does not usually form metalloproteins. The ionic radius of Mn^{2+} (0.80 Å) is intermediate between Ca^{2+} (0.99 Å) and Mg^{2+} (0.65 Å) or Zn^{2+} (0.71 Å) so that it is possible for Mn^{2+} to replace Mg^{2+}, Ca^{2+}, and Zn^{2+} in active and/or structural sites of proteins.

The coordination chemistry of Mn^{3+} has not been extensively studied but it would be expected to form stronger complexes than Mn^{2+}. Simple Mn^{3+} complexes tend to disproportionate to Mn^{2+} and Mn^{4+} but can be stabilized by strong field ligands. Mn^{3+} is labile due to the Jahn–Teller distortion caused by the d^4 electron configuration. Mn^{3+}-containing proteins are frequently isolated as metalloproteins with the metal ion bound to the protein, but the lability is reflected in the fact that the manganese content in the isolated protein is often substoichiometric. Consequently a number of dinuclear Mn(III)-containing enzymes have been recognized only recently; there are examples which are controversial and it is probable that there are others that have not yet been recognized.

The coordination chemistry of Mn^{4+} is sparse but is dominated by a tendency to form polynuclear mixed valence complexes with μ_2-oxo ligands and in most preparative systems there is a pronounced tendency to form MnO_2. The coordination geometry is likely to be octahedral. In the few biological systems which are believed to contain Mn^{4+} similar tendencies are evident. The polynuclear clusters may also be labile because although Mn^{4+} with a d^3 electron configuration would be expected to be inert this may not be the case in mixed-valent clusters.

3. THE IMPORTANCE OF MANGANESE IN BIOLOGICAL SYSTEMS

Plants need between 0.1 and 1.0 mg of Mn/kg (dry weight) to be regarded as manganese sufficient [22]. The concentration of Mn in maple leaves, exposed to the sun and therefore actively photosynthesizing, is 90 μg/g [23]. Plants and other soil microorganisms acquire manganese by exuding organic acids and/or proteins which sequester manganese and other metal ions in the soil, but the nature of these proteins is not known [24–26]. For humans the recommended daily intake of manganese is 2.5 to 5 mg/day with a minimum requirement of 0.74 mg/day [27]. Animals have a higher requirement for manganese. The total manganese content of adult humans is between 12 and 20 mg, approximately half in soft tissue, with the highest concentrations in the liver, thyroid, pituitary, pancreas, and kidney [3]. Typical concentrations in these tissues is 0.4 ng/g [28]. The concentration of manganese in the various components of blood shows considerable variation [29]. Concentrations of Mn^{2+} above normal values have been reported in malignant thyroid tumors [30] and in the spinal cord of patients suffering motor neuron disease [31], while elevated manganese-containing superoxide dismutase levels have been found in the blood stream of patients suffering from leukemia [32]. The clinical significance of these observations is not clear, however. Excess Mn^{2+} in humans leads to impairment of intellectual function (related to Parkinsonism) and a lack of visual discrimination [33]. Insufficient Mn^{2+} in the diet has only been observed in human volunteers fed a deliberately deficient diet and these subjects developed symptoms of dermatitis [34]. In animal studies, manganese deficiency leads to abnormal fetal development, bone deformation, and impaired glucose tolerance among other symptoms [35]. In some bacteria, manganese deficiency alters the membrane composition due apparently to inhibition of DNA synthesis or replication [36].

In humans the proteins to which manganese is bound in various organs of the body is not known with any certainty. In blood, transferrin, the iron(III) transporting protein, is the major Mn^{3+} binding protein [37]. In milk, lactoferrin is the major manganese (Mn^{3+}) binding protein [38]. In the rat pancreas Mn^{2+} is bound to the inactive, immature form of carboxypeptidase B [39]. These observations suggest that manganese in higher animals is opportunistic and binds to whatever protein is available. In eukaryote cells, different manganese concentrations have been observed in different compartments of the cell. In general the concentration of Mn^{2+} in biological systems is so low that the ions Mg^{2+} and Ca^{2+} have a substantial concentration advantage when the activation of a biological molecule is considered. However the total concentration in the organism may bear no relationship to the local concentration at the actual site of the molecule within the organism. The exact concentration of manganese within cells is usually not known but it is generally assumed to be very low; in unicellular eukaryotes the Mn^{2+} concentration has been estimated to be $< 10^{-6}$ mol L^{-1} [40]. A similar figure has been determined in rat heptacyte cells [41, 42]. Tholey et al. have shown [43, 44] that the concentration of manganese in cultured glial cells from chick brain is $\sim 7.5 \times 10^{-5}$ mol L^{-1}, which was 200 times the concentration in the growth medium, while in neuronal cells the concentration is even higher.

In lower organisms, specific active transport systems for Mn^{2+} have been reported in several species of bacteria, yeast, lactic acid bacteria, and fungi [45–47]. A number of microorganisms appear to have special requirements for manganese. Some bacteria are capable of oxidizing Mn(II) to Mn(IV) [48] and others reduce Mn(IV) to Mn(II). This topic has been the subject of a number of recent reviews [49–51]. In no case has a Mn(II) oxidase or a Mn(IV) reductase been identified, but some organisms excrete low molecular weight dialyzable components [52], and others excrete proteins [53] to accomplish both tasks. In mammals, little is known about active manganese transport systems; manganese is absorbed from the diet with 2–10% efficiency via the small intestine where it binds to the brush boarder membrane vesicles. The data suggest that specific binding sites are involved [54]. Manganese may also enter cells via Ca^{2+} channels [55]. Manganese is transported into the brain across the blood brain barrier possibly bound to transferrin [56, 57].

Some organisms accumulate relatively high concentrations of manganese. *Micrococcus radiodurans* is a radiation resistant organism which is believed to have developed an extraordinarily efficient DNA excision/repair system. This organism accumulates Mn^{2+} to an extent that the Mn^{2+} concentration in the cell is approximately 100 times higher than that observed in *E. coli*, and it has been estimated that there is one Mn^{2+} ion per 10 nucleotide bases present in the cell [58]. *Lactobacillus plantarum* accumulates Mn(II) from the medium to an intracellular concentration of 0.025 M. It is believed that in this organism the Mn^{2+} is present as a polyphosphate complex and that this complex may function as a superoxide dismutase [59–62]. In yeast, the superoxide dismutase activity is unaffected by high Mn^{2+} concentrations in the cell [63].

Little attention has been paid to the oxidation state of manganese in biological systems [64, 65]. There are large differences in the concentrations of Mn^{2+} as detected by EPR spectroscopy and the total manganese content of some organisms. This difference may be due in part to the reduced sensitivity of the EPR signal of Mn^{2+} bound to proteins or chelated in other ways compared to $Mn(H_2O)_6^{2+}$ [66], but may also be due to the fact that Mn^{3+} is not normally detectable by EPR due to the very fast relaxation time of this ion. Mn^{3+} is known to be present in a number of enzymes, e.g. superoxide dismutase, transferrin, lactoferrin, catalase, and probably in the water oxidizing system of Photosystem II, and it may well be the stable oxidation state in many other enzymes. Studies with inorganic model systems would suggest that octahedrally coordinated Mn^{2+} bound to nitrogen donor ligands is easily oxidized to Mn(III) or Mn(IV) in aerobic systems.

In addition to its role in enzymes, Mn^{2+} has been shown to play a role in a number of other biological processes which will be outside the scope of this review. It has been shown to cause cell division in aging cultures of *Deinococcus radiodurans* [67], influence cell adhesion [68, 69], and influence the interaction between insulin and the insulin receptor [70].

In addition to the manganese-activated enzymes and manganese metalloenzymes there are number of proteins in which manganese (which may be catalytically competent) can be substituted for the native metal. Some of these enzymes have been structurally characterized, in some cases with manganese bound to the protein,

Table 1. Proteins and Enzymes in which Manganese May be the Native Metal and which are Structurally Uncharacterized

Protein/Enzyme	*EC Number*	*Organism/Remarks*	*Reference*
6-Phosphogluconate dehydrogenase	1/1/143	Mouse liver	[80]
Dimethylmalate dehydrogenase	1.1.1.84	*Pseudomonas P-2*	[81]
Auxin oxidase	1.1.1.110	Plants	[82]
Oxalate oxidase	1.2.3.4	*Pseudomonas sp ox-5-3*	[83]
Cucurbitacin BΔ^{23} reductase	1.3.1.5	*Cucurbita maxima*	[84]
2-Aminophenol oxidase	1.10.3.4	*Pycnoporus coccineus*	[85]
N-Acetyltransferase	2.3.1.5	Animals	[86, 87]
Chitin synthetase	2.4.1.16	*Antamoeba invadens*	[88]
Galactosyl transferase	2.4.1.38	Mammals: two Mn^{2+} binding sites	[89–91]
Galactosyl hydroxylysyl glucosyl transferase	2.4.1.66	Chick embryos	[92]
N-Acetyllactosamine synthetase	2.4.1.90	Mouse mastocytoma	[93]
Hypoxanthineguanidine phosphoribosyltransferase	2.4.2.8	Yeast, liver	[94]
UDP-xylose transferase	2.4.2.26	Chicken	[95]
Tyrosine kinase	2.7.1.112	Human insulin receptor	[96–98]
RNA polymerase II	2.7.7.6	Ubiquitous	[99–101]
Poly(A) polymerase	2.7.7.19	Ubiquitous	[102, 103]
Phosphorylase phosphatase	3.1.3.17	Skeletal muscle	[104–106]
Phosphodiesterase	3.1.4.16	*Alkalophilic bacillus No A-40-2*	[107, 108]
Aminopeptidase	3.4.11.9	*Mycoplasma salivarium*	[109]
L-Tryptophan aminopeptidase	3.4.11.17	*Trichosporon cutaneum*	[110]
β-Alanyl-arginine hydrolase	3.4.13.-	Rat brain	[111]
Prolinase	3.4.13.8	Animals	[112, 113]
Prolidase	3.4.14.9	Ubiquitous	[114]
Dipeptidase	3.4.13.11	*Streptococcus cremoris Wg2*	[115]
Oxalacetase	3.7.1.1	*Streptomyces cattleya* *Aspergillus niger*	[116, 117]
Oxaloacetate carboxylase	4.1.1.32	Animal liver kidney	[118]
α-Isopropylmalate synthetase	4.1.3.12	*Alcaligenes eutrophus*	[119]
Imidazole glycerol phosphate dehydratase	4.2.1.19	Yeast	[120]
Ureidoglycollate lyase	4.3.2.3	Legumes, liver	[121, 122]
D-Arabinose isomerase	5.3.1.3	*Aerobacter aerogenes, E. coli*	[123, 124]
L-Arabinose isomerase	5.3.1.4	*E. coli, Lactobacillus gayonii*	[125, 126]
D-Mannose isomerase	5.3.1.7	*Pseudomonas cepacia, S. cerevisiae*	[127, 128]
L-Rhamnose isomerase	5.3.1.14	*Lactobacillus plantarum, E. coli*	[129]
D-Lyxose isomerase	5.3.1.15	*Aerobacter aerogenes*	[130]
D-Glucose isomerase	5.3.1.18	*Bifidobacterium adolescentis*	[131]
D-Ribose isomerase	5.3.1.30	*Mycobacterium smegmatis*	[132]
Phosphoglycerate phosphomutase	5.4.2.1	*Bacillus subtilus*	[133, 134]

(continued)

Table 1. Continued

Protein/Enzyme	*EC Number*	*Organism/Remarks*	*Reference*
β-Alanylarginine hydrolase		Rat brain	[111]
Arginine specific protein kinase		Rat liver	[135]
Calcineurin phosphatase			[136, 137]
Carbofuran Hydrolase		*Achromobacter spWM111*	[138]
Cyanide degrading enzyme		*Bacillus pumilus*	[139]
Citryl-L-glutamate hydrolase		Rat	[140]
Dehydrosqualene synthetase		Bacteria	[141]
Dichloromuconate cycloisomerase		*Alcaligenes eutrophus JMP134*	[142]
UV Endonuclease β		*Deinococcus radiodurans*	[143]
Endonuclease IV		*E. coli*	[144]
Forminoglutamate hydrolase		*Bacillus subtilis*	[145]
		Aerobacter aerogenes	[146]
Fructanase		*Saccarum officanarum*	[147]
Galactosyl-3-*O*-sulfotransferase		Calf thyroid	[148]
Geranyl diphosphate synthetase		*Pelargonium roseum*	[149]
Galactinol synthetase		*Cucumis sativus L*	[150]
Lactose β(1,3) *N*-acetylglucosaminyl transferase		Mammals	[151]
Malolactic enzyme		Leuconostoc, Lactobacillus Pediococcus species	[152–154]
Mannose 6-phosphate receptor		Mammals	[155–159]
Mannosyl transferase		*S. cerevisciae*	[160]
Organophosphorus esterase		*Triatoma infestans*	[161]
Organophosphate acid anhydrase		*Rangia cuneata*	[162, 163]
Oxalacetate acetylhydrolase		*Streptomyces cattleya*	[116]
Phytoene synthetase		*Capsicum annuum chloroplast*	[164, 165]
Protein peptidase		Mammals	[166]
Protein phosphatase		Ubiquitous	[167, 168]
Protein tyrosine phosphatase		Ubiquitous	[169]
Processing protease		Rat liver	[170, 171]
Sabinene hydrate cyclase (synthase)		*Majorana hortensis*	[172]
Serine kinase		Rat adipocytes	[173]
o-Succinylbenzoic acid synthetase		*E. coli*	[174]
Tripeptidase		*Lactococcus lactis subsp cremoris*	[175]

in other cases with the native metal ion present. These structurally characterized proteins will be covered in this chapter. There are many reports in the literature of *in vitro* manganese-activated systems usually using millimolar concentrations of Mn^{2+}. These systems are almost certainly not manganese activated *in vivo*, and if they have not been structurally characterized they are not considered in this chapter. Mn^{2+} substitution into metal-activated enzymes or metalloenzymes allows spectro-

Table 2. Manganese-Containing Proteins that have been Subjected to Biophysical Studies which have Revealed Information Concerning the Nature of the Metal-Ion Active Site

Protein	*ECNumber*	*Organism*	*Technique*	*Remarks*	*Reference*
Tartrate dehydrogenase	1.1.1.93	*P. putida*	ESEEMS	Tartrate and nitrogen bound to metal: monovalent cation required, bound close to Mn^{2+}	[176]
β-1,4-Galactosyl transferase (lactose synthetase)	2.4.1.22.38			Two Mn^{2+} binding sites	[89, 177–180]
S-Adenosylmethionine synthetase	2.5.1.6	*E. coli*	EPR	Binuclear Mn^{2+} center antiferro-magnetically coupled	[181]
Arginine kinase	2.7.3.3	invertebrate muscles	NMR	Mn–arginine distance, 10.9 Å; metal–P (of ATP) distance, 3.2–4.5 Å	[182, 183]
Pyruvate orthophosphate dikinase	2.7.9.1	plants	EPR	Mn^{2+} binds phosphorylated histidine, two water molecules, and a bidentate oxalate; the sixth ligand is unknown	[184]
Prolidase	3.4.13.9	ubiquitous	chemical modification	Arg and Asp or Glu at active site. Asp-276 important	[114, 185, 186]
Anthranilate synthetase	4.1.3.27	*Salmonella typhimurium*	NMR, EPR	Mn^{2+} six-coordinate, five groups from protein, plus one H_2O	[187]
Thiosulfate oxidase		*Thiobacillus versutus*	EPR	Binuclear Mn^{2+} center antiferro-magnetically coupled	[188]

scopic (particularly EPR and NMR) investigations of the active sites of the enzymes which have spectroscopically silent native metals. Work of this nature has been reviewed by Kalbitzer [71], Mildvan [72, 73] and Villafranca [74, 75] and the results of several such studies will be considered further in this review.

There are also enzymes which have been claimed to be activated by manganese, and for which a crystal structure is available, but the crystals contain no metal ions. Examples are aspartate aminopeptidase [76, 77] and dihydrofolate reductase [78, 79]. Such examples are not included in this chapter.

There are many structurally uncharacterized enzymes and proteins which may be manganese-activated *in vivo*. Examples of such species are included in Table 1 with references to either reviews of the function of the molecule in question or to a report(s) of the manganese-substituted form of the molecule.

There is another group of manganese-containing or activated proteins which have been subjected to a variety of studies such as site-directed mutagenesis, EXAFS, EPR, and ESEMS, which have provided information about the nature of the metal binding site, but the structure of this site is still not known. Molecules of this type are listed in Table 2 and will not be further considered in the chapter.

There is a further group of proteins which are activated by different metal ions in a species-dependent fashion. In one example, phenylotic dioxygenases, the structure of an enzyme with iron present has been determined, but the structure of the manganese-dependent enzyme is not known. This example will be discussed very briefly below. There is yet another group of proteins where a manganese form of the protein has been characterized and for which preliminary X-ray structural

Table 3. Proteins which have been Studied in a Mn(II)-Bound Form and which have been the Subject of Preliminary X-ray Structural Investigations

Protein/Enzyme	*EC Number*	*Organism/Remarks*	*References*
Aequorin	—	*Aequorea victoria*	[189–191]
Malic enzyme	1.1.1.40	*Ascaris suum*: two Mn^{2+} at two tightly bound sites	[192–194]
Catechol *O*-methyl transferase	2.1.1.6	Ubiquitous, Mg^{2+} enzyme: Mn^{2+} equally effective	[195, 196]
Purple acid phosphatase	3.1.3.2	Plants Mn or Fe/Zn enzyme, animals Fe^{2+}/Fe^{3+}	[197]
Arginase	3.5.3.1	Microbial, plants, animals: two Mn^{2+} ions spin coupled; Zn^{2+} plays structural role: crystals of rat enzyme diffract to 2.4 Å resolution	[198–202]
ATPase (chloroplast coupling factor)	3.6.1.3	Three tight and three weak Mn^{2+}-binding sites; metal-binding sites 8 Å apart; tetrahedral Mn^{2+}, three protein ligands one water in enzyme; octahedral Mn^{2+}–P distance of 5.3 (±0.5) Å, Mn^{2+}–O bond of 2.05 (±0.15) Å: in Mn–ATP enzyme complex, Mn–P distance 4.95 (±0.15) Å, suggesting ATP is not in first coordination sphere. ε-subunit *E. coli* enzyme crystallized; Mn^{2+} probably on β subunit	[71, 203–213]
Phosphoenolpyruvate carboxykinase	4.1.1.49	Ubiquitous, arginine at Mn^{2+}-binding site	[214–216]
Isocitrate lyase	4.1.3.1	Bacteria, plants, fungi: crystals from *E. coli* diffract to 2 Å	[217–222]

information on the molecule is available. Such proteins are listed in Table 3 but will not be considered further in this review (with the exception of purple acid phosphatase). Details of the metal-ion coordination of the molecules in Table 3 can be expected to be available in the near future.

4. PHOSPHATASES

The hydrolysis of phosphate esters in biological systems is a crucially important process which is linked to energy metabolism, nervous system functions, signal transduction, and metabolic regulation. There are a wide variety of enzymes which accomplish the hydrolysis and they are currently classified into at least six different categories:

1. alkaline phosphatases,
2. purple acid phosphatases,
3. low molecular weight acid phosphatases,
4. high molecular weight acid phosphatases,
5. serine and threonine protein phosphatases, and
6. tyrosine protein phosphatases.

Within each of these categories there is an apparently diverse group of enzymes, some of which are activated by manganese. The manganese-dependent purple acid phosphatases are discussed below. Mn^{2+}-dependent phosphatases have been reported from pig brain membranes [167], and an extracellular Mn^{2+} dependent alkaline phosphatases from *Bacillus sp RK11* [223], turkey gizzard smooth muscle [224], and mouse liver [225]. No information is currently available concerning the nature of the Mn^{2+} binding site in these enzymes. It has been suggested that all serine/threonine phosphatases may contain dinuclear metal centers [226]. With such an important class of enzymes there have been numerous reviews on the structure and function of the enzymes to which the interested reader is referred [168, 227–232].

4.1 Serine/Threonine Protein Phosphatase 2A (EC 3.1.3.16)

Protein phosphatases may be divided into at least three categories that are specific for phosphoserine, phosphothreonine, and phosphotyrosine residues. These important enzymes have been the subject of many reviews [230]. Serine/threonine protein phosphatases are divided into two types depending on their sensitivity to inhibitors: type 1 phosphatases are sensitive to and type 2 are insensitive to inhibitors. The enzymes are further subdivided according to their divalent metal ion requirements. Type 2A phosphatases do not require divalent cations: type 2B (also known as Calcineurin) require Ca^{2+} and are sensitive to Mn^{2+}; type 2C require Mg^{2+}. Protein phosphatases 1, 2A, and 2B are homologous with each other, whereas protein phosphatase 2C is structurally distinct [233]. PP1 and PP2A show 41% amino acid identity and mammalian and *Drosophilia* PP1 and PP2A show >90% sequence identity, placing these enzymes among the most highly conserved of all observed enzymes [234].

4.2 Purple Acid Phosphatase (EC 3.1.3.2)

Studies on purple acid phosphatases have been the subject of a number of excellent reviews [235–237]. There is controversy as to whether manganese containing purple (violet) acid phosphatases exist *in vivo*. Purple acid phosphatases isolated from mammalian sources contain two iron atoms (the active form is Fe^{2+}/Fe^{3+}) and a similar enzyme isolated from red kidney beans is a Zn^{2+}/Fe^{3+} enzyme [238, 239]. Tartrate-resistant acid phosphatases in the biomedical literature and the dinuclear iron purple acid phosphatases are the same molecular species [237]. The structure of the iron/zinc form of the enzyme is under active investigation [197]. It is known from EXAFS studies that the iron atoms are 3.0 ± 0.1 Å apart and the metal ligands are nitrogen and oxygen donor atoms.

A considerable body of literature suggests that at least some purple acid phosphatases isolated from plants contain manganese. The first such report by Uehara [240] reported the isolation of an enzyme from sweet potato and similar enzymes have since been reported from rice, spinach leaves, and soybean. Reported properties of these enzymes are given in Table 4 together with some of the properties of the diiron and iron/zinc enzymes. The properties of the manganese-containing species are consistent with a bound Mn(III) ion. The controversy arises because there is a claim that the enzyme from sweet potato is an iron enzyme [241]. The claim is based on the isolation, by a slightly modified purification procedure, of a dimeric enzyme from sweet potato with a higher purity, a 14-fold higher activity, and an analysis which shows one iron atom per subunit and no manganese. Kawabe et al. [242] had earlier reported a study of a sweet potato enzyme in which Fe(III) was substituted for Mn(III) and reported that the activity was 50% of the native enzyme. They did not report the activity of reconstituted Mn(III) enzyme, however.

There is no doubt that the enzyme studied by the Japanese workers contains manganese and very little iron [243], while that of the American group contains iron, and no manganese. There are a number of ways of reconciling these observations. There is no evidence one way or another that the sweet potatoes are the same species, so the different forms may reflect species differences. Alternatively, the enzyme might be active with either metal *in vivo* and the form isolated may depend on the exact cultivation and isolation conditions. There is a report that sweet potato contains two forms of purple acid phosphatase [244], and these may contain different metal ions. Clearly more work on the characterization of these enzymes is required.

A (μ-oxo)(μ-phosphato)(μ-carboxylato)diiron active site with additional coordination from two carboxylates, two histidines, and a tyrosine residue has been

Table 4. Properties of Purple or Violet Acid Phosphatases Isolated from Various Sources

Organism	*Metal Cofactor*	*M* (*kDa*)	*Subunits*	λ_{max} (*nm*)	ε (M^{-1} cm^{-1})	*Activity Units* (*per mg*)	pH_{max}	*Reference*
Sweet Potato	1 Mn/subunit	110	2	555				[240, 243, 252]
(Kokei No14)								
(Kintoki)	1 Mn/mol	110	2	515	2460	52.4	5.8	[250]
	1 Fe/mol			525	3000	27	4.8	[242]
	2 Fe/mol	113	2	545	3080	751		[241]
Kidney bean	1 Zn/1 Fe	110.708	2	560	3360			[197, 239]
Spinach leaves	Mn	92	2	530		236	5.5	[253]
Rice plant cultured cell	Mn	65		555				[254]
Rice bran	Mn	68		560			4.5	[255]
Soybean	Mn	240	4	540			5.5	[256]
Soybean root cultures	1 Mn	120	2	556		512		[257]
Uteroferrin	1 Fe(III)			510	4000	400–500		[258]
	1 Fe(II)	40	0	530	2000			
Fe(III)/Zn(II)				550	4000			
Fe(III)/Fe(III)								
Beef spleen	2 Fe			550	2100			[259, 260]
Human	4 Fe	30						[261]

proposed for the active site of the iron-containing enzymes [227, 235, 245]. There is good evidence for Mn–tyrosine and Mn–histidine coordination in the sweet potato enzyme [246–248], and somewhat more doubtful claims for Mn–cysteine coordination [249, 250]. ^{31}P and ^{17}O NMR studies of the Mn(III)-containing acid phosphatase from sweet potato showed that a direct Mn(III)–PO_4^{3-} interaction is observed and phosphate is probably in the first coordination sphere of the metal ion [251]. One question which deserves further attention is the stoichiometry of metal binding. Inspection of Table 4 shows that all the studies which have been performed are consistent and suggest one metal ion (either Mn or Fe) per subunit. The electronic spectra are similar to the binuclear enzymes so that it is possible that the isolated enzyme is substoichiometic in metal. However, the properties are sufficiently different from the dinuclear enzymes, e.g. reduction is not required to form an active species, that the metal ion binding site is probably different.

5. PROTEINS WITH ONE BOUND METAL

5.1 G-Actin

Actin together with myosin forms the contractile protein complex actomyosin which plays a central role in muscle motility. As normally isolated, actin from rabbit muscle has a single polypeptide chain of 375 amino acid residues. In the absence of salt it exists as a globular protein G-actin, but traces of salt change it to a fibrous protein F-actin. G-actin has a high-affinity binding site for ATP and another binding site for a metal ion. ^{31}P NMR spectroscopic studies show that the metal site is <10 Å from the ATP binding site and it is considered likely that the metal ion is bound to the ATP [262]. Actin forms a 1:1 complex with bovine pancreatic DNase and this form of the protein has been crystallized and structurally characterized [263]. The actin molecule consists of two peanut-shaped domains of approximately equal size, lying side by side and connected by two crossovers of the polypeptide backbone chain at one end. The result is a hinged molecule with a deep cleft, and the essential cofactors ATP or ADP and a divalent metal ion (Ca^{2+}) bind within the cleft. The metal ion and the nucleotide make extensive contact with both domains. The Ca^{2+} ion is in a hydrophillic pocket formed by the phosphate groups of the adenine nucleotide and the amino acids Asp-11, Gln-137, and Asp-154. These amino acid residues are too far from the Ca^{2+} to be bonded directly to the metal ion, but the β- and γ-phosphate oxygens of ATP and the β-phosphate oxygen of ADP is bonded to the Ca^{2+}

5.2 Dioxygenase

Dioxygenases catalyze the cleavage of dihydroxybenzene rings to yield dicarboxylic acids with incorporation of the dioxygen molecule, e.g. 3,4-protocatechuate catalyzes the reaction as shown in Scheme 1. Most dioxygenases studied are iron

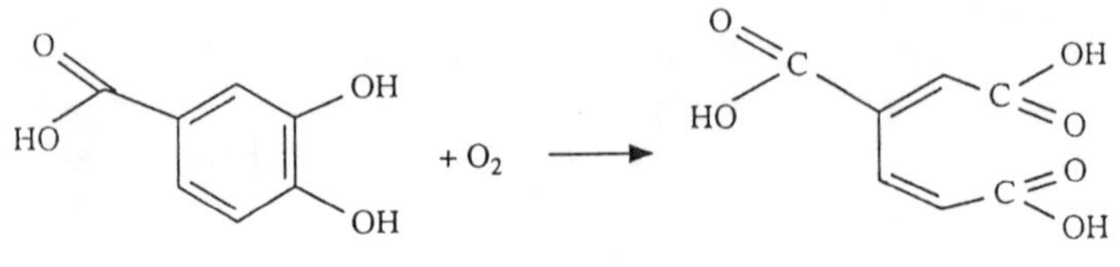

Scheme 1.

enzymes, and the 3,4-protochatechuate dioxygenase from *Pseudomonas aeruginosa* has been characterized crystallographically and shown to contain a five-coordinate iron(III) center. The iron is bound to two tyrosine residues, two histidines, and a water molecule in a trigonal bipyramidal arrangement [264]. An EXFAS study suggests that there are two Fe–N bond lengths of 2.08 Å and three Fe–O bond lengths of 1.90 Å long in this molecule [265]. A manganese-dependent 3,4-dihydroxyphenylacetate 3,4-dioxygenase from *Bacillus brevis* has been isolated, and similar enzymes have been suggested as being present in *B. steareothermophilus* and *B. macerans* but detailed studies of their properties have not been reported [266].

5.3 Elongation Factor Tu

Elongation Factor Tu (EF-Tu) is one of the enzymes essential for bacterial protein biosynthesis and in *E. coli* it is 5% of the total protein [267]. The binding of aminoacyl tRNA to the acceptor site of ribosomes in prokaryotic systems is dependent on EF-Tu and GTP. A ternary complex of aminoacyl–(tRNA)–(EF-Tu)–GTP binds to the ribosome, GTP hydrolysis occurs, and the binary complex (EF-Tu)–GDP is released. The binding of guanosine nucleotides to EF-Tu is markedly affected by metal ions [268]. Mg^{2+} is the physiologically active metal ion but Mn^{2+} imparts a greater activity.

NMR and EPR spectroscopic investigations have been conducted on the Mn^{2+}–(EF-Tu)–GTP complex [71, 268, 269]. The EPR spectra from both *E. coli* and *B. stearothermophilus* are similar and ^{17}O-substitution studies suggest that only the β-phosphate of the GDP is bound to the metal ion. Studies of the $H_2{}^{17}O$ solutions of the (EF-Tu)–Mn and (EF-Tu)–Mn–GDP complex suggests that there are four water molecules bound to the metal in the (EF-Tu)–Mn species and also four water molecules in the coordination sphere when GDP is bound. Five water molecules are bound to the manganese in a (EF-Tu)–Mn–GDP–PO_4^{3-} complex [269]. The coordination pattern of Mn^{2+} in the (EF-Tu)–Mn–GDP complex is pH-dependent, i.e. at pH 8.1 two protein residues are bound to the metal while at pH 6.8 the Mn^{2+} has an additional water bound.

The nucleotide binding site has been subject to intensive studies using site-directed mutagenesis. Drastic alterations of residues Gly-23, Lys-24, Thr-25, and Leu-27 eliminated the ability to bind guanosine nucleotides and Val-20–Gly and

His-84–Gly mutants have very low and low GTPase activity, respectively [270]. The guanine recognition loop has also been tested by site-directed mutagenesis and the Asp-138–Asn mutant has a reduced affinity for GDP; replacement of Lys-136 also leads to reduced affinity.

Studies by three different groups have been reported on the X-ray structure of this enzyme. These studies have been summarized by Kjeldgaard and Nyborg [271] in the most accurate structure of the *E. coli* enzyme reported to date and so the history will not be repeated here. In addition to the studies on the *E. coli* enzyme, studies on the crystals isolated from *Thermus thermophilus* HB8 which diffract at 1.9 Å resolution [272] and from *Thermus aquaticus* which diffract at 2.6 Å resolution [273] have been initiated. The *E. coli* protein contains three domains. The N-terminal α/β domain (residues 1–200) contains the GTP binding site and shares structural and sequence homology with other GDP/GTP binding proteins. The three-dimensional structures of the GTP binding domains of EF-Tu and the ras p21 protein align almost perfectly [274, 275]. Two other domains (residues 209–299 and 300–393) of antiparallel β-barrels constitute the rest of the molecule. The metal ion is located in the middle of a cleft, separating the GTP/GDP binding site from a loop, Asp-80 to His-84. At the best resolution available, only two ligands to the metal ion are observed, an oxygen of the β-phosphate group of the guanosine diphosphate, and the side chain of Thr-25. The other four ligands are probably water molecules. One of the water molecules is replaced by the γ-phosphate oxygen in the (EF-Tu)–GTP complex [271, 277, 278]. An Asp-80–Asn mutant enzyme has a much reduced affinity for MgATP [276].

5.4 Isocitrate Dehydrogenase (EC 1.1.1.42)

Isocitrate dehydrogenase catalyzes the oxidative decarboxylation of isocitrate to α-ketoglutarate and CO_2, one of the reactions of the citric acid cycle. Mammalian tissues have two forms of the enzyme: one is restricted to mitochondria and requires NAD^+ and Mn^{2+} or Mg^{2+} as cofactors; the other form occurs in both the cytosol and the mitochondria and requires $NADP^+$ as a cofactor. The first form of the enzyme is regulated by phosphorylation of a serine residue. There are no large structural changes of the protein following phosphorylation but isocitrate binding is prevented by phosphorylation [279].

Multinuclear NMR studies of the pig heart enzyme in the presence of the ^{113}Cd isocitrate complex suggest that the metal ion has an all-oxygen octahedral coordination sphere. The isocitrate is a bidentate ligand to the metal ion and there are two water molecules bound to the metal [280]. Studies of the Mn^{2+}-substituted form of the enzyme suggest that Mn^{2+} is 7 Å from the 1, 2, and 5 carbons of the product α-ketoglutarate further suggesting the product is not coordinated to Mn^{2+}. NMR studies on this form of the enzyme suggest that the $NADP^+$ is 8–10 Å from the Mn^{2+} [280].

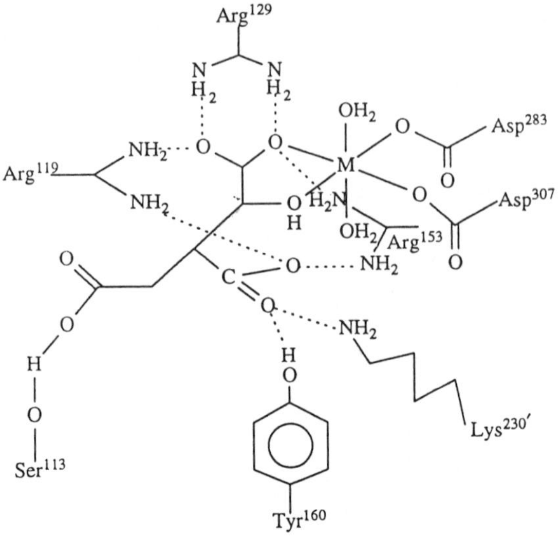

Figure 1. Schematic representation of the metal and isocitrate binding site in isocitrate dehydrogenase.

The crystal structure of the native enzyme from *E. coli* with both Mg^{2+} and Mn^{2+} bound to the metal binding site, and of the phosphorylated native enzyme and the Ser-113–Glu and Ser-113–Asp mutant forms of the enzyme have been reported [281]. There is a large and a smaller domain which are separated by a cleft. The isocitrate is bound to the metal ion as a bidentate ligand in a pocket between the two major domains of the enzyme; both subunits of the dimer participate in forming the metal binding site. Structural changes from the apoenzyme are localized at the metal binding site. Two aspartate residues and two water molecules complete the metal ion coordination sphere. A representation of the metal ion binding site is given in Figure 1.

5.5 Kinases: Adenylate, Guanylate, and Creatine

Kinases are responsible for the transfer of a phosphate group from a nucleotide triphosphate to the acceptor molecule. They are classified according to the type of residue which is phosphorylated [282]. Numerous reviews have been published dealing the properties of kinases [97, 283, 284]. All known kinases have a divalent metal ion requirement; usually Mg^{2+} is the native metal but some kinases, e.g. a serine kinase from rat adipocytes and a protein kinase from *Granulosa* virus [285], are activated by Mn^{2+} alone [173]. The number of divalent metal ions required seems to depend on the particular kinase; all kinases seem to require the metal nucleotide complex as a substrate but a number also require one or more metal ions bound to the enzyme. Reports of protein kinases stimulated specifically by Mn^{2+} are given in Tables 1 and 2. Manganese has been frequently used in association with NMR and EPR to study the nature of the binding site and a number of kinases have been structurally characterized by X-ray crystallography in their Mn^{2+}-bound form.

Adenylate kinase (EC 2.7.4.3) guanylate kinase (E.C. 2.7.4.8), and creatine kinase (EC 2.7.3.2) catalyze the reactions,

$$M^{2+}ATP + AMP \rightarrow 2ADP + M^{2+}$$

$$M^{2+}ATP + GMP \rightarrow ADP + GDP + M^{2+}$$

$$M^{2+}GTP + GMP \rightarrow 2GDP + M^{2+}$$

respectively. Adenylate kinase and guanylate kinase are small monomeric enzymes, M_r of 20,000–28,000, whereas mitochondrial creatine kinase of mammalian heart and brain is an octameric species with M_r of 330,000–340,000 with eight identical subunits [286]. The rabbit muscle creatine kinase is a dimer of identical subunits ($M_r = 41{,}300$). Electron microscopy has revealed that octameric creatine kinase is square shaped with four equal-sized subdomains each of which contains a dimeric unit. *In vivo* the enzyme is bound to the outer side of the inner mitochondrial membrane and it probably exists in a microcompartment between the outer and inner mitochondrial membranes. EPR studies of rabbit muscle creatine kinase suggest that three water molecules are bound to Mn^{2+} in a transition-state analogue complex [287] and that the ATP is bound to the metal as an $\alpha\beta\gamma$ tridentate ligand [288]. ESEEM spectral studies of a creatine kinase–MnADP–NCS^- complex are consistent with N-bonded thiocyanate, and three water molecules bound to the metal ion [287]. The metal binding to the ATP has been investigated using ^{31}P NMR studies of metal nucleotide complexes in the presence of the enzyme. There is no evidence for a direct interaction between the metal ion and the enzyme. The relaxation rates of the Mn^{2+} nucleotide complexes are exchange-limited and are therefore incapable of providing structural information; however with Co^{2+} substitution the relaxation rates are frequency-dependent and suggest a distance of between 2.4 and 4.3 Å from the metal to the P nucleus [289]. This distance which suggests direct coordination is in agreement with the similar distance determined using the EPR spectrum of the Mn^{2+}-substituted enzyme [71].

Little is known about the active site in creatine kinase, but in view of the similarity of the reactions catalyzed by the three enzymes it is probable that the active sites in all three enzymes are very similar. That is certainly the case for adenylate and guanylate kinases which have been structurally characterized.

Guanylate kinases have been isolated from a variety of organisms, the enzyme displays a number of sequence similarities to adenylate kinases particularly in the region of the nucleotide binding site and in a long α-helix covering one side of the enzyme in the adenylate kinase structure. The reasons for the sequence similarity in this region of the protein is not known [290].

Adenylate kinases from *E. coli*, rabbit, pig, and chicken have been studied using ^{1}H, ^{31}P, and ^{17}O NMR and these and other studies (including the structural studies discussed below) have shown that the enzyme has two binding sites, one for AMP and a less specific site for MgATP [71–73, 291–293]. The ATP-binding site is located close to the side chains of Trp-210, His-143, and Tyr-142 using the yeast enzyme amino acid numbering. This site is apparently similar to the site in the *E. coli* enzyme [293] and agrees with the position of one of the two adenosine sites of Ap5A in the crystal structure of adenylate kinase Ap5A–Mg^{2+} complex [294, 295]. Earlier conclusions about the location of this site (close to His-34) are not now tenable [293]. ^{17}O NMR studies on the Mn^{2+}-substituted enzyme suggest that after the Mn^{2+} is bound to the enzyme four water molecules are still bound—it is proposed that one of the β-phosphate oxygens of ATP and one enzyme residue completes the manganese coordination sphere. EPR studies of Mn^{2+} nucleotide complexes agree with this conclusion [71]. This is the only kinase for which a departure in the behavior of the metal ion being bound to all the phosphate groups of ATP or ADP has been detected [183].

Considerable effort has been expended on the identification of these binding sites using NMR and X-ray diffraction studies. Because crystals of adenylate kinase with bound substrates have not been obtained the evidence for the nature of these sites is still indirect. The NMR and X-ray diffraction results are now in accord however [293]. Crystal structures have been determined for the adenylate kinase of porcine muscle enzyme at 3.0 Å resolution [296], porcine cytosolic enzyme at 2.1 Å resolution [297, 298], yeast and *E. coli* enzymes with the inhibitor P^1,P^5-di(adenosine-5′-)pentaphosphate (Ap5A) at 1.9 Å resolution [294, 295] and mutant forms of this enzyme [299], and of guanylate kinase of *S. cerevisiae* at 2.0 Å resolution [300]. Guanylate kinase crystallizes in two temperature-dependent forms, one tetragonal and the other orthorhombic, and the structures of both forms were determined; the protein structure was the same, only the interactions between the molecules in the crystal differ in the two forms [301].

Both adenylate kinase and guanylate kinase contain a deep cleft which divides the molecule into two domains, one of which consists of three helical rods and the other a five-strand pleated sheet. The nucleotide binding sites in this cleft are held by extensive hydrogen-bond contacts with the polypeptide backbone. In the porcine muscle and *E. coli* adenylate kinases, the metal ion (both Mg^{2+} and Mn^{2+} occupy the same site) is also located in this cleft bound to the second and third phosphoryl groups of ATP with no direct contact with the polypeptide [294], but close to the residues identified by the NMR studies. The metal binding site in guanylate kinase has not been identified directly but has been deduced by analogy with the adenylate kinase structures.

5.6 α-Lactalbumin

α-Lactalbumins are milk proteins which together with galactosyltransferases (they form the enzyme complex lactose synthetase) are responsible for the synthesis of the milk sugar lactose from UDP-galactose and glucose [89]. The metal binding properties of α-lactalbumins are among the most comprehensively studied of all the proteins discussed in this chapter. Most of these studies have been comprehensively reviewed by Kronman [302] so this topic will be dealt with only briefly and will include only the studies reported since that review was published. Desmet and co-workers have reported the thermodynamics of metal binding [303–306]. The thermodynamic studies are complicated by the fact that there are two forms of the apoprotein and by their interaction with monovalent cations.

The fact that the binding of Mn^{2+} to the Ca^{2+} binding site in the goat protein is endothermic suggests that the binding site is rather rigid and the bonds to the smaller Mn^{2+} ion are rather weak. There is, however, another possibility. In addition to the calcium binding site there may be a second metal binding site which preferentially binds Mn^{2+} in the native protein. One difficulty with this suggestion for a second metal binding site is that there is no such site obvious in the crystal structure of the enzyme. Arguments for and against Mn^{2+} binding to the Ca^{2+} binding site have been advanced by Kronman [302] and Desmet [306]. The binding of Mn^{2+} to the bovine protein has been investigated by NMR [307] and it was concluded that a low affinity metal binding site occurs about 7.5 Å from the amino terminus of the molecule. A second weak metal binding site for Ca^{2+} in bovine lactalbumin has been reported [308] and has been confirmed recently using ^{43}Ca NMR [309]. The latter study also identified His-68 as an influence in metal binding. Human lactalbumin does not have this histidine and does not bind a second Ca^{2+} ion. Fourier transform infrared studies of the bovine protein have revealed that the protein conformation is different in the Mn^{2+} and Ca^{2+} forms [310]. These studies suggest that the Mn^{2+}-bound form of the protein has a similar configuration to the apoprotein. This may explain at least part of the difference in ΔH of bonding. One would expect that Mn^{2+} would interact with a histidine residue more strongly that Ca^{2+} and so Mn^{2+} may be binding preferentially with the weak Ca^{2+} binding site. A crystal structure of the Mn^{2+}-substituted form of the protein may be necessary to settle this question.

Structures of both baboon α-lactalbumin [311] and human α-lactalbumin [312, 313] have been determined at 1.7 Å resolution with the native metal Ca^{2+} bound to the enzyme. There is no evidence from the structural studies for a second metal binding site. The Ca^{2+} binding site is near the surface of the molecule and is formed by the side chain carboxylates of three aspartate residues (Asp-82, -87, -88), the amide oxygens of two other residues (Lys-79, Asp-84) and two water molecules. The Ca^{2+} is thus seven-coordinate with a distorted pentagonal bipyramidal arrangement of oxygen donors [311]. The amide oxygens are in the axial positions.

5.7 Lactoferrin

Lactoferrin, a Fe(III)-containing protein from milk, binds a large proportion of the manganese in human milk [314]. Spectroscopic studies of the Mn(III)-substituted protein have been reported and the intense absorption band at 400 nm was identified as a Mn(III)–phenolate charge-transfer band [315]. Structures of the apo-, iron(III)-, and copper(II)-containing lactoferrins have been determined [316–318] and an investigation of the manganese(III) species is planned [319]. The iron(III) atom is six-coordinate and is bound to four amino acid side chains, two tyrosine residues, a histidine, and a monodentate aspartate residue. A CO_2 molecule acting as a bidentate ligand completes the coordination sphere. The Cu(II) ion is coordinated to the same amino acid residues.

5.8 Mandelate Racemase (EC 5.1.2.2)

Mandelate racemase catalyzes the reaction shown in Scheme 2. The enzyme is octameric with a subunit mass of 38.4 kDa [320]. It is most active with Mg^{2+}, but Mn^{2+} is able to activate the enzyme [321]. A crystal structure of the enzyme from *Pseudomonas putida* at 2.5 Å resolution (data to 2.0 Å have been collected but the results are not yet available) has established that the polypeptide fold consists of three domains. The N-terminal domain has a β-sheet and an antiparallel four-α-helix bundle. This domain is connected to a modified α/β barrel. The modification is that the eighth α-helix is missing; the space normally occupied by this helix is taken by the third domain which is an extended strand of polypeptide chain. The metal ion binding site is near the mouth of the barrel at the C-terminal end of the β-strands. Mg^{2+}, Mn^{2+}, and Eu^{3+} all bind at the same position. The ligands to the metal are three side chain carboxylate groups from two glutamate (Glu-221 and Glu-247) and one aspartate (Asp-195) residues, two water molecules, plus a sulfate group coordinated at what is probably the mandelate binding site [322].

The mechanism of the racemization reaction has been shown to involve a two-base mechanism and the two acid/base catalyst residues required for the reaction have been identified as Lys-166 for the (*S*)-specific base and His-297 for the (*R*)-specific base [323]. A crystal structure of a His-297–Asn mutant which has

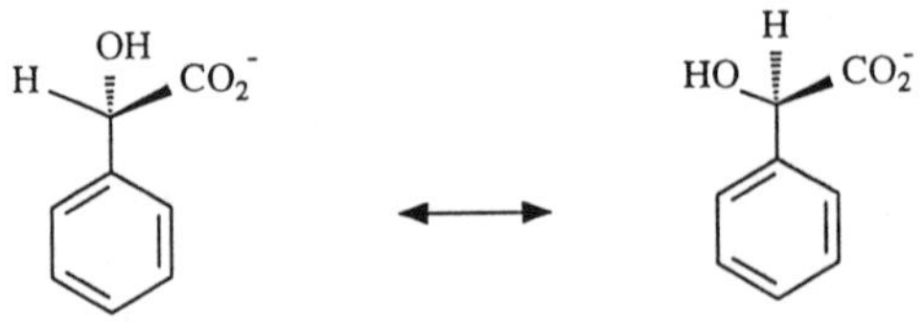

Scheme 2.

Scheme 3.

no mandelate racemase activity has been determined. The structure of the active site is unaffected by the mutation, apart from the substitution itself. The mutant enzyme is able to catalyze the reaction in Scheme 3 at the same rate as the native enzyme (Scheme 3). This suggests that the rate of abstraction of the α-proton from the (*S*) enantiomer is unaffected by the His-297–Asn mutation [324].

The most remarkable feature of the structure of the enzyme is the similarity of the structure to that of the muconate lactonising enzyme from *P. putida* [325]. The C_α backbones of the two molecules are shown in Figure 2. Mn^{2+} is the preferred activating metal ion for muconate lactonizing enzyme but both Mg^{2+} or Mn^{2+} will work with either enzyme. However, despite the structural similarities, mandelate racemase has no muconate lactonizing activity or vice versa.

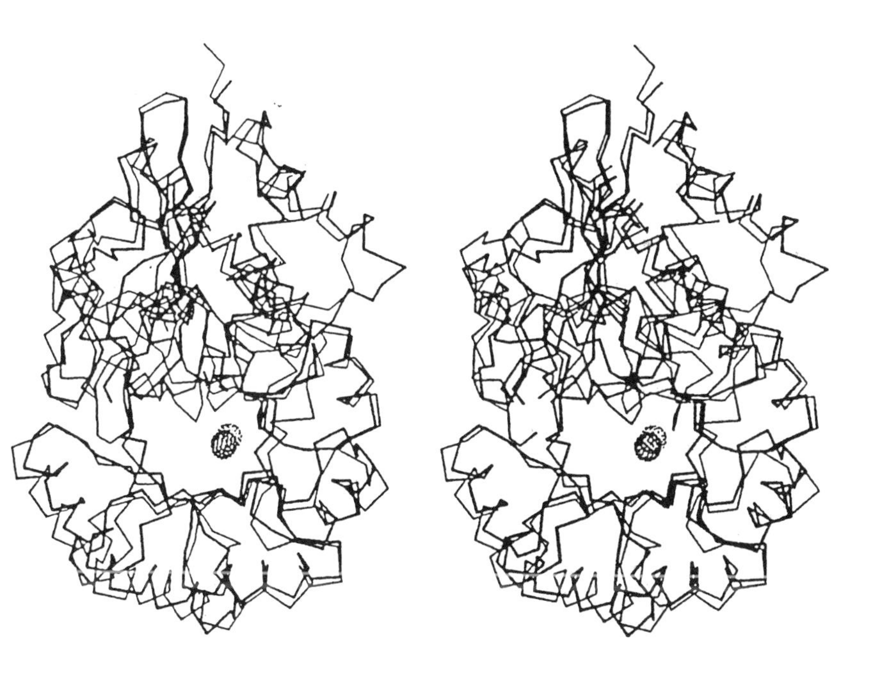

Figure 2. Stereo diagram of the C_α backbone of mandelate racemase and *cis,cis*-muconate cycloisomerase. The backbone of mandelate racemase is shown as a thicker line. Reproduced with permission from *Nature* reference [325], copyright (1990), Macmillan Magazines Limited.

5.9 Manganese-Dependent Peroxidase (EC 1.11.1.13)

Lignin, which constitutes 15–30% of woody plant cell walls, is a heterogeneous aromatic polymer in which monomeric guaiacylpropane units are linked by both ether (**1**) and carbon–carbon (**2**) linkages. Lignin biochemistry has recently been reviewed [326–328]. Wood-degrading (white-rot) fungi produce, usually in response to either carbon or nitrogen depletion, a number of extracellular enzymes which form a synergistic system responsible for the degradation of lignin. Commercial applications of these enzymes have been proposed for the treatment of the effluent of wood pulping plants and also as oxidants of recalcitrant aromatic pollutants [329, 330].

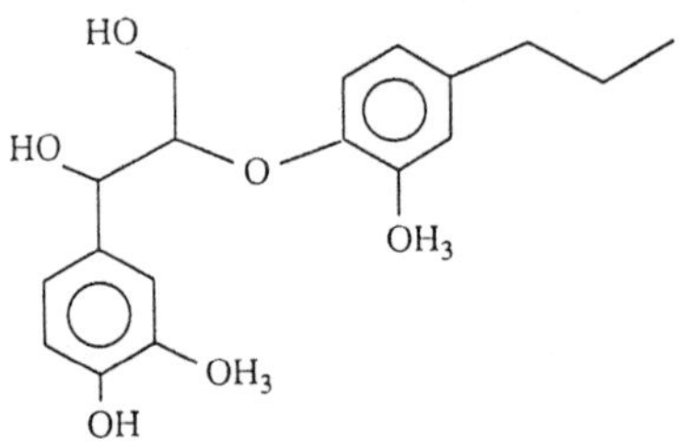

1 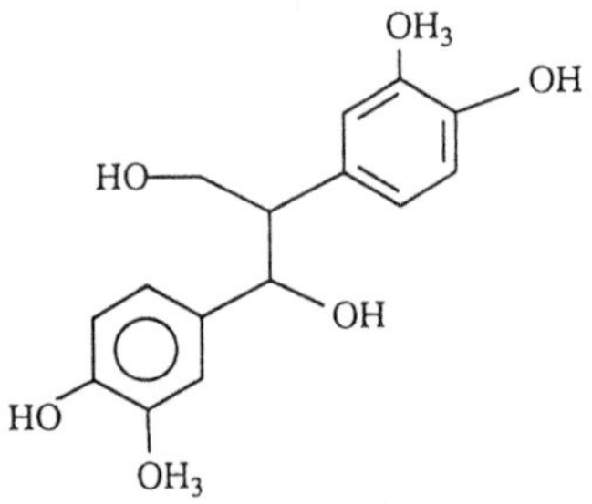 **2**

Manganese-dependent peroxidase, laccase, and lignin peroxidase are examples of the variety of peroxidases excreted by wood-degrading fungi. Some of these fungi, including the most studied and best characterized species *Phanerochaete chrysosporium* [331], produce only lignin peroxidase and manganese-dependent peroxidase. Other organisms such as *Rigidoporus lignosus* and *Lentinus edodes* do not produce a lignin peroxidase but use a combination of manganese-dependent peroxidase and laccase to achieve wood degradation [332–334]. Lignin peroxidase and manganese-dependent peroxidase differ in their catalytic mechanisms. Lignin peroxidase is believed to oxidize non-phenolic aromatic rings and manganese-dependent peroxidase attacks phenolic C_α-hydroxy and C_α-oxo-substructures in lignin. The role of lignin peroxidase in the degradation of lignin has been questioned [326, 335] but manganese-dependent peroxidase does depolymerize lignin [336].

Cultures of *P. chrysosporium* have been shown to contain as many as 21 heme proteins in the extracellular fluid [337]. The lignin and manganese-dependent peroxidases all contain a single ferric protoprophyrin IX (**3**) per molecule and are N- and possibly O-glycosylated [338, 339]. The molecular masses range from 38 to 43 kDa, while the unglycosylated proteins have molecular masses of 37 kDa; the differences between the different forms probably reflect different degrees of gly-

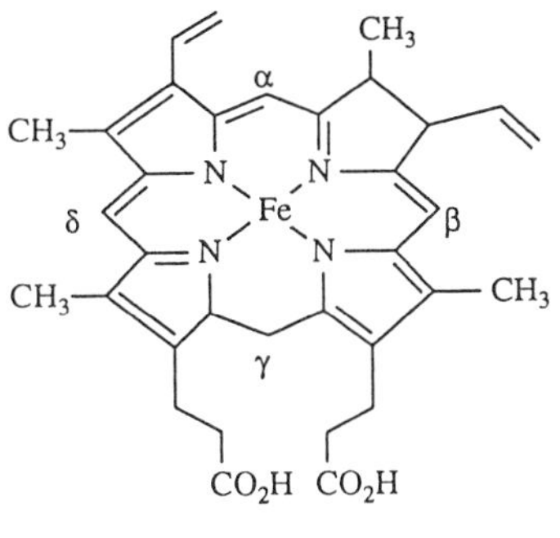

3

cosylation. Some of these proteins are designated as lignin peroxidase because of their ability to directly oxidize veratryl alcohol. Manganese-dependent peroxidase activity was assigned to some of the remaining enzymes because their vanillyl acetone oxidizing ability was shown to be dependent upon the presence of Mn^{2+}, H_2O_2, and an organic ligand such as lactate or tartrate which can stabilize Mn(III) [340]. It has been shown that manganese peroxidase oxidizes substrates via the δ-meso-heme edge, [341] and therefore the Mn^{2+} binding site on the protein is thought to be close to the heme edge; however, Mn^{2+} binding does not block the approach of small molecules to the heme edge. The resulting Mn(III) complexes then oxidize the organic substrate. Manganese-dependent peroxidase has been shown to be present on and within cell corners of Birch wood showing signs of attack by the fungus [342, 343].

The ligands bound to manganese in the *in vivo* systems have not been positively identified. *P. chrysosporium* excretes oxalate and malonate in significant amounts, with glyoxylate and citrate present in trace amounts [344]. In *in vitro* systems, ligands such as lactate, malonate, pyrophosphate, tartrate, combinations of succinate and lactate, and succinate and malate have all been in the buffer system as part of the assay for enzyme activity. Kenten and Mann [345] had shown as early as 1950 that horseradish and turnip peroxidases are capable of generating Mn(III) chelates in the presence of several monophenolic compounds. Recent work has shown that both the lignin peroxidase and the copper-containing oxidase, laccase, together with either an aromatic compound capable of being directly oxidized by the enzyme or malonic or oxalic acids, are also able to oxidize manganese complexes [346–348]. It has also been shown that Mn(III) malonate and lactate complexes oxidize lignin model compounds to give identical products to that of the enzymatic reaction with these substrates [349, 350].

In common with other heme peroxidases, the lignin peroxidase and manganese-dependent peroxidase catalytic cycle begins with the oxidation of the resting iron(III) enzyme by H_2O_2 to yield a Fe(IV) porphyrin cation radical intermediate, compound 1, which is two oxidizing equivalents above the resting state. Compound 1 is reduced by one electron by Mn^{2+}, ferrocyanide, or phenols to yield a Fe(IV)-oxo intermediate, compound 2 [351, 352]. A second single-electron reduction by Mn^{2+}, ferrocyanide, but not phenols returns the enzyme to the resting state. The mechanism of oxidation of lignin by manganese peroxidase has been the subject of numerous investigations mainly using model lignin compounds such as monoaromatic phenols [353], quinones [354], phenolic diarylpropanes [350] methoxybenzenes [355–357], 2,4-dinitrotoluene [358], 2,4-dichlorophenol [359], arylglycerol β-arylethers [349], 2,4-dichlorodibenzo-*p*-dioxin [360], synthetic lignin [336], and chlorolignin [361]. These studies have confirmed that the organic chemistry involves free radical mechanisms.

All three forms of the manganese-dependent peroxidase have been spectroscopically characterized using electronic absorption [351], EPR, resonance Raman spectroscopic, and chemical modification studies. The redox potentials of two forms of the enzyme have also been measured [341, 362].

The structure of manganese-dependent peroxidase has been studied using ^1H NMR [363]. The spectrum of the protein is similar to that of a number of other peroxidases and it is almost identical to the spectrum of lignin peroxidase [364]. The NH group of the proximal histidine residue of both these proteins is accessible to solvent. XAS studies on native lignin peroxidase suggest that the iron is six-coordinate with four pyrrole nitrogen distances at 2.055 Å, an iron-proximal nitrogen distance of 1.93 Å, and an iron distal ligand (thought to be oxygen) distance of 2.17 Å [365]. A crystal structure of lignin peroxidase has been reported [366].

5.10 *cis,cis*-Muconate Cycloisomerase (EC 5.5.1.1)

cis,cis-Muconate cycloisomerase (muconate lactonizing enzyme) catalyzes the conversion of *cis,cis*-muconic acid to muconolactone (see Scheme 4). The enzyme is involved in the bacterial degradation of catechol to β-ketoadipate which is then converted to citric acid cycle intermediates. The enzyme is a hexamer, with a monomer of M_r ~40000, including one Mn^{2+}. Dichloromuconate cycloisomerase which is a different enzyme is also activated by Mn^{2+} [142]. Kinetic studies indicate that the rate-determining step in the reaction is proton transfer from the C5 carbon [367]. Ngai et al. [368] have studied the interaction of the metal ion with the enzyme and its substrates using EPR spectroscopy. These experiments indicate that the

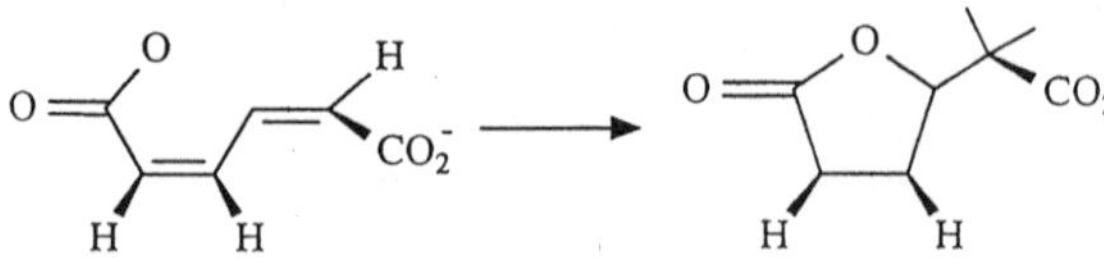

Scheme 4.

Mn^{2+} and the muconolactone are within 5 Å of each other and that the carboxylate rather that the lactone ring of the product is closest to the Mn^{2+}. This distance suggests that the product is not directly coordinated to the Mn^{2+}.

The structure of the enzyme from *Pseudomonas putida* at 3 Å resolution has been determined [369]. The protein contains one large and two smaller domains. The N-terminal domain (residues 1–105) has an α/β polypeptide fold which wraps around the central domain. The C-terminal domain of the molecule does not have a well-defined secondary structure. The structure of the large central domain is a squat α/β barrel about 25 Å long and 40 Å across. The catalytically essential Mn^{2+} ion sits in a hydrophobic pocket about 5 Å from the ends of the α-helices of the central domain. The inhibitor α-ketoglutarate, binds in the hydrophobic pocket near the Mn^{2+}. Details of the coordination to the manganese are not available at the currently reported resolution. The structure of the enzyme is almost identical to that of mandelate racemase, this similarity is discussed in the section on that enzyme.

5.11 Ovalbumin

Ovalbumin is a soluble glycoprotein of 348 amino acids (M_r 42,700) with a single carbohydrate chain linked to asparagine. The protein is heterogeneous due to the variable length of the carbohydrate chain and the degree of phosphorylation of serine side chains. The function of the protein is unclear but it has been suggested that it serves as a metal binding protein and it contains one tight metal binding site. The paramagnetic contribution to the ^{31}P NMR relaxation times due to Mn^{2+} has shown that the metal ion is bound within ~6 Å of two phosphorylated serine residues [370].

The structure of the enzyme has been determined at 1.95 Å resolution and the metal ion in the crystal (probably adventitious) has been shown to be tetrahedrally coordinated to side chains of two adjacent molecules. The ligands to the metal are a phosphoserine, two glutamates, and a water molecule. The chemical identity of the metal ion in the crystals is uncertain. Tetrahedral coordination is unusual for Ca^{2+} or Mg^{2+} and the nature of the coordinating ligands makes Cu^{2+} or Zn^{2+} unlikely. Mn^{2+} is a possibility, however the metal ion was modeled in the structural study as half a Ca^{2+} ion [371].

5.12 ras p21 Protein

The protein, ras p21, is a metal GTP/GDP complex binding protein, the nucleotide binding domain of which shares significant sequence homology with regions of other GTP binding proteins including signal transducting G proteins, cAMP-dependent protein kinases, F1-ATPase, adenylate kinase, myosin, Rec A, and the cystic fibrosis gene product [275, 372, 373]. The binding of GTP to the protein acts as a molecular switch; binding turns the switch on and hydrolysis of GTP turns it off. A structural transition in the protein is propagated from the vicinity of the metal

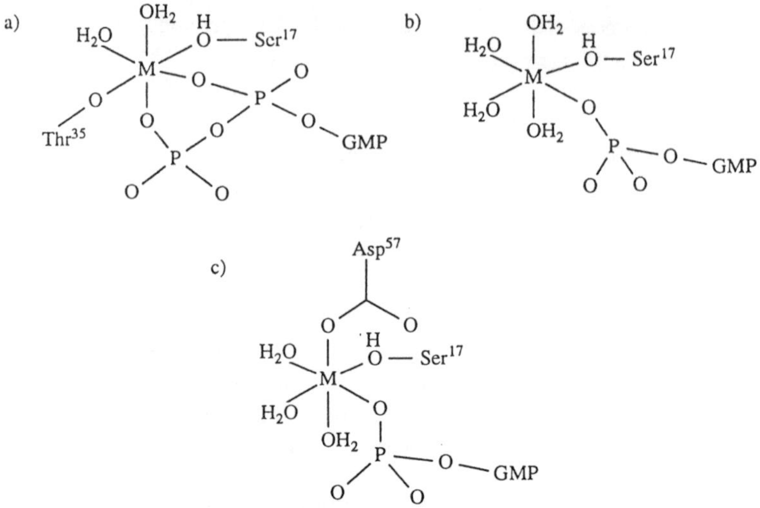

Figure 3. Schematic views of the metal binding sites in the ras p21 protein. (**a**) Metal GTP complex. (**b**) The MgGDP protein complex in crystals obtained from solutions of GDP and enzyme. (**c**) The MgGDP protein complex produced by hydrolysis of GTP within the crystal lattice.

ion some 40 Å. Structures of the active site of the GDP and GTP bound forms of the protein enzyme are shown in Figure 3.

The metal coordination sphere in the enzyme is variable depending on the substrates bound to the protein and on the method of preparation of the crystals. With GTP or GTP analogues bound, the metal has two water molecules, the β- and γ-phosphate of the GTP and the side chains of Ser-17 and Thr-35. The binding is further stabilized by hydrogen-bonding interactions to Asp-57 and Asp-33 and to the α-phosphate [278,374,375]. Upon hydrolysis of the GTP the γ-phosphate is lost from the coordination sphere of the metal ion. The hydroxyl group of Thr-35 is also lost from the coordination sphere and is replaced by a water molecule in the GDP-bound molecule. The coordination of Thr-35 appears to govern the conformation of a loop of polypeptide called loop 2. The change in conformation of loop 2 is believed to trigger the switch. Significant differences between the observed structure in the crystal and the structure of the activated complex have been suggested [376].

One problem unresolved by the crystal structure studies concerns the coordination of Asp-57 to the magnesium. When GTP is hydrolyzed within the crystal the Mg^{2+} is bound to the carboxylate of Asp-57 [377]. Crystals produced from solutions of the protein and Mg–GTP have a water molecule bound to the metal ion in place of the aspartate residue [378, 379]. EPR studies of the Mn–GDP complex in ^{17}O-enriched water show that the manganese has four water molecules, Ser-17, and a phosphate oxygen as ligands, thus supporting the idea that the aspartate is not bound to the metal *in vivo* [380–382]. ESEEM spectra of the Mn^{2+}-substituted protein with ^{15}N enriched serine and glycine residues have been reported [383]. The distance from the metal ion to the Ser-17 amide nitrogen was calculated to be 3.8 ± 0.3 Å, in excellent agreement with the distance in the crystal. The distance from the metal ion to the α-phosphorus of the GDP was found to be 5.3 ± 0.1 Å, in good agreement with the distance in the crystal (5.5 Å). These results confirm that the Mn^{2+} and Mg^{2+} ions bind at the same site in the protein.

5.13 Ribulose 1,5-Bisphosphate Carboxylase (EC 4.1.1.39)

Ribulose 1,5-bisphosphate carboxylase (RuBisCo) catalyzes the crucial carbon fixing step of photosynthesis in which CO_2 is added to the five-carbon species, ribulose 1,5-bisphosphate, and then the C2–C3 bond of the six carbon intermediate is cleaved to yield two molecules of 3-phosphoglycerate (see Scheme 5).

RuBisCo also catalyzes a competing oxygenation reaction, the first step of photorespiration. Mg^{2+} is the native metal but when Mn^{2+} is the metal cofactor, singlet oxygen is produced as a result of the oxygenation reaction [384], and a decrease in the rate of turnover of the enzyme with CO_2 as substrate relative to the Mg^{2+} catalyzed reaction is observed [385]. Two types of RuBisCo are known: type I protein, found in all higher plants and in most photosynthetic bacteria, is a 16-protein aggregate containing 8 large (M_r = 52,000–55,000) and 8 small subunits (M_r = 14,000). Type II protein which is found in certain photosynthetic bacteria, e.g. *Rhodospirillum rubrum*, is a dimer of two identical subunits (M_r = 55,000). The catalytic activity resides on the large subunit of the type I protein. The enzyme must be activated before it is fully active, and the activation involves the carbamalyation of a lysine side chain at the active site (Lys-191 in the *R. rubrum*) enzyme. The carbamate is stabilized by coordination to the required metal ion.

Site-directed mutagenesis studies have suggested that a serine residue, Ser-368 (*R. rubrum* numbering) [386–388], and Asp-188 are involved in substrate binding. EPR and NMR studies of the Mn^{2+}-substituted enzyme complexed with the reaction intermediate analogue, 2-carboxyarabinitol 1,5-bisphosphate (CABP), showed that

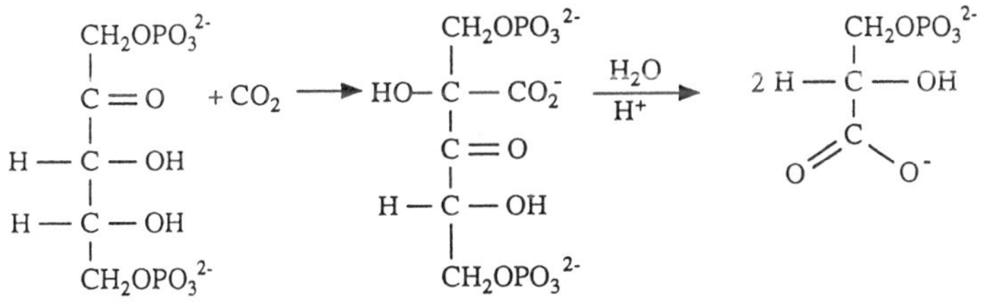

$CH_2OPO_3^{2-}$–C(=O)–CH(OH)–CH(OH)–$CH_2OPO_3^{2-}$ + CO_2 ⟶ $CH_2OPO_3^{2-}$–C(OH)(CO_2^-)–C(=O)–CH(OH)–$CH_2OPO_3^{2-}$ $\xrightarrow[H^+]{H_2O}$ 2 H–C(OH)($CH_2OPO_3^{2-}$)–CO_2^-

Scheme 5.

the metal ion is bound to the carbamate and also interacts with the carboxyl and hydroxyl groups of C2 of CABP [389–392].

Crystal structures of the dimeric form of the enzyme from *R. rubrum* at 1.7 Å resolution [393–396]; a mutant form (Asp-193–Asn) of this enzyme at 2.6 Å resolution [397]; and two plant enzymes, spinach at 2.4 Å resolution [395, 398] and tobacco at 2.0 Å resolution [399, 400], have been determined. Preliminary data on the enzyme from the cyanobacterium *Synechoccus PCC6301* have been reported [401]. The large residue consists of two domains; the active site is at the carboxy end of the C-terminal domain, a β/α barrel in a highly charged region of the protein. Residues from the N-terminal domain of the other subunit are also involved in the active site so that the dimer is the minimal functional subunit. The structural studies show that four amino acid side chains are involved in binding to the metal ion, the carbamylated Lys-191, carboxylate groups of Asp-193 and Glu-194, and an amide oxygen of Asn-111. Two water molecules complete the coordination sphere. In the structure with CABP bound, two oxygen atoms from CABP are coordinated to the metal—the carboxylate group on C2 and the OH on either C2 or C3 [398]. A schematic view of the active site is shown in Figure 4. Site-directed mutagenesis studies show that all three groups are essential for catalytic activity [397].

There is a wealth of chemical, mutagenic, and kinetic evidence suggesting that Lys-166 is the base which mediates the initial enolization step. The crystal structure determinations, however, suggest that this residue is too far from the C3 proton of the substrate to act as the base. A number of other suggestions have been made [388] but no residue has been positively identified as the base in question.

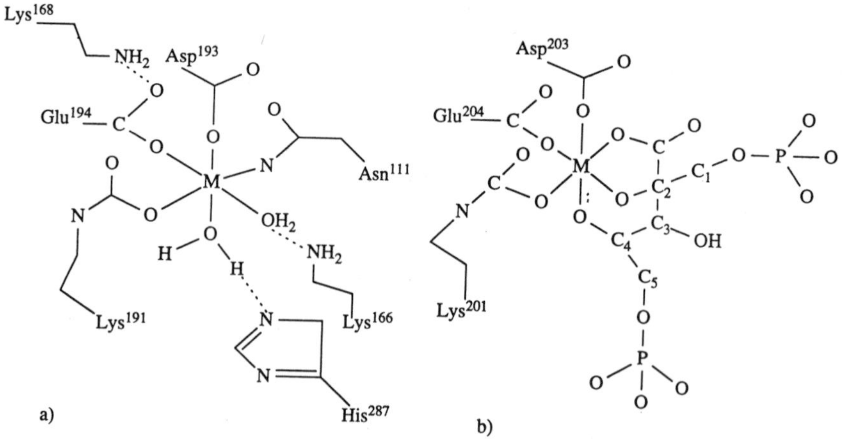

Figure 4. Schematic diagram of the active site in RuBisCo. (**a**) Activated enzyme from *R. rubrum* in absence of substrate. Hydrogen bonds are indicated by the dashed lines. It is not known whether N of Asn-111 is coordinated to the metal ion. (**b**) Activated enzyme from spinach with CABP and substrate CO_2 bound [398].

5.14 tRNA

The binding of cations including Mn^{2+} to yeast phenylalanine transfer RNA has been studied using X-ray crystallography [402]. Metal ions stabilize the tertiary structure of tRNA and the binding is cooperative. Co^{2+} and Mn^{2+} were shown to bind as $M(H_2O)_6^{2+}$ moieties to N7 atoms of guanine residues, whereas Mg^{2+} ions bind to the phosphate oxygens. Mn^{2+} binding to anticoden loop segments of RNA which contain unusual moieties, in this particular case a threonine residue, has been studied by NMR [403]. Mn^{2+} is bound to a phosphate oxygen, the carboxylate of the threonine residue, and to N7 of an adenine ring.

5.15 Staphylococcal Nuclease (EC 3.1.4.4 or 3.1.4.7)

Staphylococcal nuclease is a Ca^{2+}-dependent extracellular enzyme which is a single polypeptide of 144 amino acids (M_r = 16,800) produced by certain strains of *Staphylococcus aurens*. It catalyzes the hydrolysis of both DNA and RNA at the 5′ position of the phosphodiester bond, yielding a free 5′-OH group and a 3′-phosphate monoester (Figure 5). Substitution of Ca^{2+} by Mn^{2+} leads to a decrease in activity of the enzyme but both Mn^{2+} and Co^{2+} bind at the same site as the Ca^{2+} ion. An extracellular endonuclease from the Gram-negative bacterium *Serratia marces-*

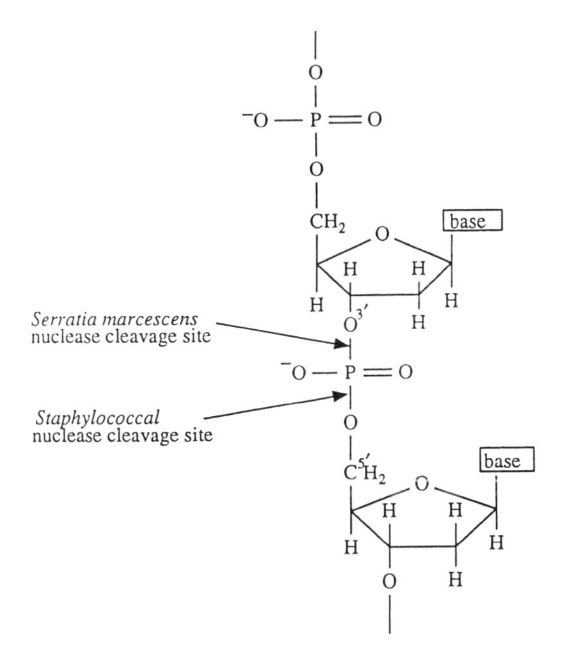

Figure 5. Location of cleavage site of the extracellular nuclease from *Staphyloccus* and *Serratia*.

cens, which cleaves different bonds to those cleaved by staphylococcal nuclease (Figure 5) and uses Mg^{2+} as its native metal, has recently been crystallized and is the subject of a X-ray structure determination [404]. Free energy perturbation calculations of the effects on the activation barrier for the reaction associated with metal-ion substitution in staphylococcal nuclease (~6 kcal mol^{-1}) are in good agreement with the kinetic results [405].

The environment of the active-site metal has been studied using enzyme–Mn^{2+}–nucleotide complexes with both native and mutant enzymes and EPR, NMR, and ESEEM techniques [71, 73, 406, 407]. Pulsed-EPR studies suggest that Mn^{2+} is close to a ^{31}P-nucleus and studies of the Co^{2+}-substituted enzyme suggested distances of 4.4 ± 0.7 Å and 2.7 ± 0.4 Å to the 3′- and 5′-phosphorus atoms of the inhibitor pdTp, respectively. It has been suggested that the 5′-phosphate is coordinated in a bidentate fashion in contrast to the situation observed with the Ca^{2+} complex in the solid state. Two fast-exchanging water ligands to Mn^{2+} were detected by NMR measurements of the enzyme–Mn^{2+} pdTp complex. Similar studies on the enzyme–Mn^{2+} complex showed only one fast-exchanging water molecule bound to Mn^{2+}, and it was suggested that two additional ligands from the protein are bound in this form of the enzyme. EPR studies support this conclusion, and Glu-43 and Asp-19, which are in the second coordination sphere of the metal in the solid state structure, have been suggested as the likely ligands [407]. Studies of the mutant enzymes in which the metal ligands and groups of hydrogen bonded to the phosphate residues are changed show reduced activity in every case. A Glu-43–Ser mutant form of the enzyme has been investigated using kinetic and magnetic resonance techniques with Mn^{2+} as the metal ion. The data suggest that there is one more water molecule bound to the Mn^{2+} ion in the mutant enzyme than is present in the wild-type enzyme. This data supports the suggestion that Glu-43 coordinates the divalent metal ion in the binary enzyme complex but is dissociated from the metal ion, and functions as a general base in the ternary enzyme–metal–DNA complex [408]. NMR studies of staphylococcal nuclease in solution suggest the protein displays considerable heterogeneity in solution [409, 410].

The structure of the enzyme with the inhibitor, deoxythymidine 3′,5′-bisphosphate, was first reported in 1979 [411] and it has recently been redetermined [412, 413]. The structure determinations are in agreement. The inhibitor fits into a cleft with its aromatic ring at the base of the cleft in a hydrophobic pocket. The nucleotide base is hydrogen bonded to water molecules, the phosphate oxygens are hydrogen bonded to the protein, and one of the 5′-phosphate oxygens is coordinated to the Ca^{2+} ion (Figure 6). The Ca^{2+} is also bound to two aspartate residues (Asp-21 and Asp-40) and the carbonyl oxygen of Thr-41.

Mutant enzymes (Val-66–Lys and Glu-43–Asp) have also been the subject of a X-ray structural investigation [414]. Val-66 is located on the hydrophobic side of a α-helix and its side chain is buried in the protein. The side chain of the lysine residue in the mutant enzyme is also fully buried. The structure of the Glu-43–Asp

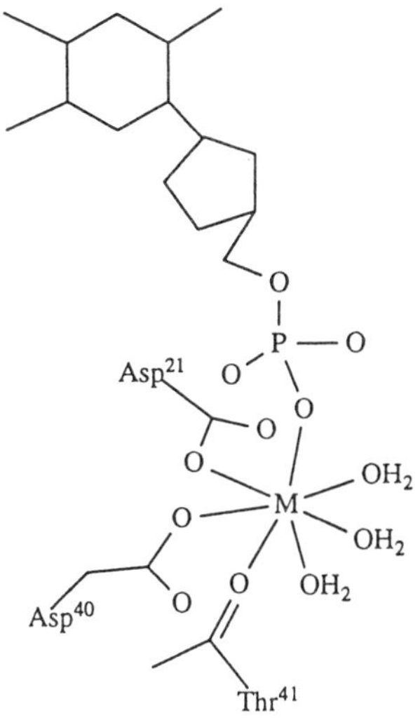

Figure 6. Schematic representation fo the metal binding site with the inhibitor deoxythymidine 3′,5′-bisphosphate in staphylococcal nuclease.

mutant reveals significant changes to the backbone near the active site and to the positions of side chains within the active site and up to 15 Å from that site [415].

5.16 Superoxide Dismutase (EC 1.15.1.1)

Superoxide dismutase (SOD) catalyzes the conversion of O_2^- to O_2 and H_2O_2,

$$2O_2^- + 2H^+ \rightarrow H_2O_2 + O_2$$

and thus contributes to the natural protection of cells from oxygen toxicity [416,417]. Superoxide dismutase has also been shown to participate in the enzymatic formation of the tyrosine radical of ribonucleotide reductase in *E. coli* [418–420], and it plays a physiological role in the inflammatory response of tissue to damage [421] and in various carcinogenic processes [422–424]. Superoxide dismutases have been the subject of numerous studies and a number of excellent reviews have been written [6, 425–430]. There have been considerable advances in our understanding of the enzyme since even the most recent of these reviews, thus the topic is treated in some detail here.

Three different superoxide dismutases are known, each of which has been structurally characterized. Cu/Zn superoxide dismutase [432–434] is found in the cytosols of eukaryotic cells and a few bacteria. The evolutionary unrelated Mn and Fe enzymes are found in prokaryotes and in mitochondria. Strictly anaerobic organisms either have no superoxide dismutase or have the iron form of the enzyme.

Several oxygen tolerant organisms which lack enzymatic superoxide dismutases contain high levels of Mn(II) in their cells [435]. One such organism, *Lactobacillus plantarum* (which is the only known organism not to have an absolute requirement for iron), accumulates Mn(II) from the medium to an intracellular concentration of 0.025 M. This high concentration of Mn(II), which is bound to polyphosphate ions, apparently serves as a functional replacement for SOD [435–438].

Most superoxide dismutases have an absolute metal ion specificity for either manganese or iron and if the Mn(III)–apoenzyme is reconstituted with Fe(III) the molecule is not active (and vice versa) [439–441]. Addition of other metals—e.g. Ni(II), Co(II), and Cu(II)—to the apoenzyme does not impart catalytic activity [442]. *In vivo* it has been demonstrated for *E. coli* that there is competition for the metal binding site by both Mn and Fe [443, 444]. Reasons for the expression of the different forms of the enzyme have been the subject of much attention [445].

There are a number of organisms which have SODs which are active with either Mn or Fe as cofactors (eg. *Bacteroides thetaiotamicron* [446], *Bacteroides fragilis* [447,448], *Propionibacterium shermanii* [449, 450], *Streptococcus mutans* [451], *Porphromonas gingivalis* [452], *Methylomonas J* [453], and *Bacteroides gingivalis* [454–456]). Many organisms which possess both Mn and Fe SODs express the iron form of the enzyme under anaerobic conditions and the manganese form is produced upon exposure to oxygen [455, 457]. The manganese enzyme is dimeric in most bacteria but tetrameric in some mycobacterial and thermophilic species and eukaryotes. The M_r of the subunit of the MnSOD is ~22,500; the redox potential of the manganese enzyme, E° = 0.31 V vs. NHE [458]; and the human Cu/Zn enzyme has a potential of 0.36 V [459, 460].

The spectral properties (visible, EPR, MCD) of the enzyme from *E. coli* and its fluoride and azide complexes in both the Mn(III) and Mn(II) forms have been studied in considerable detail [461]. These studies suggest that binding of F^- and N_3^- occurs at the manganese center. Binding of these anions does not alter the coordination number of the metal but the coordination geometry does change. The *T. thermophilus* enzyme has a reddish purple color, λ_{max} = 480 nm (ε = 950 M^{-1} cm^{-1}), and for the *E. coli* enzyme, λ_{max} = 478 nm (ε = 850 M^{-1} cm^{-1}). There is considerable variation in the reported values of the extinction coefficient of the MnSOD from different organisms. This probably arose because the various workers failed to recognize that some (up to 50%) of the manganese present in the enzyme as isolated is Mn(II) [461, 462]. The absorption spectrum of the Fe-substituted MnSOD from *E. coli* differs significantly from that of the FeSOD from the same organism [443]. This result is surprising in light of the structural studies discussed below which suggest that the environment of the metal ions in both forms of the enzyme should be very similar. The EPR spectrum of native MnSODs lack signals assignable to manganese (consistent with a Mn(III) formulation), although on standing or upon reduction a spectrum typical of Mn(II) is observed. Magnetic studies on the *T. thermophilus* enzyme are consistent with a $S = 2$ spin state (and hence clearly Mn(III)) in the oxidized form of the enzyme, and a $S = 5/2$ state in the reduced form of the enzyme. The zero-field splitting parameter D is +2.44 cm^{-1} in the oxidized form and +0.5 cm^{-1} in the reduced form [462].

The mechanism of the reaction has been the subject of a number of studies [434, 463–466]. Catalysis is believed to involve successive reduction and oxidation of the manganese or iron centers:

$$M^{3+} + O_2^- \rightarrow M^{2+} + O_2$$

$$M^{2+} + 2H^+ + O_2^- \rightarrow M^{3+} + H_2O_2$$

It is usually assumed that coordination of superoxide ion to the metal occurs before reaction occurs, but, apart from the fact that inhibitors are bound to the metal ion, there is little evidence for this assumption. The first step involves proton uptake by the enzyme. This may involve the protonation of a coordinated OH^- group, but if no water or OH^- is bound to the metal a protein residue must be involved. The reaction of superoxide ion with both the iron and manganese forms of the enzyme is close to diffusion controlled ($k_1 = 6.6 \times 10^8$ M^{-1} s^{-1} for FeSOD and 6.8×10^8 M^{-1} s^{-1} for MnSOD) [463, 467]. Studies in D_2O show a decreased rate suggesting that proton transfer makes a significant contribution to the rate-determining step. This rate is remarkable as the entrance to the active site represents only 1% of the total protein surface area.

The complete or almost complete primary structures of thirty nine iron and manganese superoxide dismutases have been determined and these are given in Table 5. Partial sequences have been reported for approximately 10 additional organisms [511, 512]. The amino acid sequences of both the Mn and Fe enzymes from *E. coli* and *Nicotiana plumbaginifolia* have been determined and in both cases there is only ~40% homology between the two forms of the enzyme [499, 504]. Considerable homology, suggesting a common evolutionary origin of the MnSODs and the FeSODs, is evident in the sequences. Fourteen residues (including the four ligands to the metal ions) are invariant throughout and the fact that many of the invariant (or conservatively substituted) residues are involved in subunit–subunit interactions [513] suggests retention of conformation. The positive-charged residue at position 177 has been shown by site-directed mutagenesis studies to be essential for catalytic activity [514, 515]. Sequences of the putative Fe superoxide dismutase from *Tetrahymena pyriformis* and *Aerobacter aerogenes* are clearly anomalous [505, 509] and would be worth further study. The isolated *T. pyriformis* enzyme contains iron, the manganese content was very low, and the enzyme was inactivated by H_2O_2, a characteristic of the iron forms of the enzymes. The apoenzyme was reconstituted with Mn and the Mn form was shown to be inactive. However the reconstituted iron form of the enzyme displayed only low activity also so this may not be a conclusive demonstration that the Mn form is inactive. Inhibition by H_2O_2 and N_3^- are the main criteria for suggesting the *A. aerogenes* enzyme is an iron-containing species [516], but this is not an infallible criteria [494].

Table 5. Comparison of the Amino Acid Sequences of Manganese and Iron Superoxide Dismutases[a,b,c]

	1		10		20		30		40		50		60		70		80		90		100	
		↓		↓↓		*	↓ ↓										*					
1	–PFE	LPALP	YPYDA	LEPHI	DKETM	NIHHT	KHHNT	YVTNL	NAALE	GHPDL	QNKSL	EELLS	NLEAL	PESIR	TAVRN	NGGGH	ANHSL	FWTIL	SP–	NGGGE	PTGEL	ADAIN
2	–PFE	LPALP	YPYDA	LEPHI	DKETM	NIHHT	KHHNT	YVTNL	NAALE	GHPDL	QNKSL	EELLS	NLEAL	PESIR	TAVRN	NGGGH	ANHSL	FWTIL	SP–	NGGGE	PTGEL	AEAIN
3	–SYT	LPSLP	YAYDA	LEPHF	DKQTM	EIHHT	KHHQT	YVNNA	NAALE	SLPEF	ANLPV	EELIT	KLDQL	PADKK	TVLRN	NAGGH	ANHSL	FWKGL	K—	K-GTT	LQGDL	KAAIE
4	-SQHE	LPSLP	YDYDA	LEPHI	SEQVV	TWHHD	THHQS	YVDGL	NSAEE	TLAEN	R—	ET-GD	H-AST	AGAL-	GDVTH	NGCGH	YLHTM	FWEHM	SP–	DGGGE	PSGAL	ADRIA
5	MSEYE	LPPLP	YDYDA	LEPHI	SEQVL	TWHHD	THHQG	YVNGW	NDAEE	TLAEN	R—	ET-GD	H-AST	AGAL-	GDVTH	NGSGH	ILHTL	FWQSM	SP–	AGGDE	PSGAL	ADRIA
6	–TYE	LPKLP	YTYDA	LEPNF	DKETM	EIHYT	KHHNT	YVTKL	NEAVS	GHAEL	ASKPV	EELVA	NLDSV	PEEIR	GAVRN	HGGGH	ANHTL	FWSIL	SP–	NGGGA	PTGNL	KAAIE
7	-MTYE	LPKLP	YTYDA	LEPNF	DKETM	EIHYT	KHHNT	YVTKL	NEAVA	GHPEL	ASKSA	EELVT	NLDSV	PEDIR	GAVRN	HGGGH	AHNTL	FWSIL	SP–	NGGGA	PTGNL	KAAIE
8	PYPFK	LPDLG	YPYEA	LEPHI	DAKTM	EIHHQ	KHHGA	YVTNL	NAALE	KYPYL	HGVEV	EVLLR	HLAAL	PQDIQ	TAVRN	NGGGH	LNHSL	FWRLL	TP–	GGAKE	PVGEL	KKAID
9	-QKHS	LPDLP	YDYGA	LEPHI	NAQIM	QLHHS	KHHAA	YVNNL	NVTEE	KYQEA	LAK–	–GD	VTAQI	–ALQ	PALKF	NGGGH	INHSI	FWTNL	SP–	NGGGE	PKGEL	LEAIK
10	VTTVT	LPDLS	YDFGA	LEPAI	SGEIM	RLHHQ	KHHAT	YVANY	NKALE	QL-ET	AVSK-	–GD	–ASA	VVQLQ	AAIKF	NGGGH	VNHSI	FWKNL	KPISE	GGGEP	PHGKL	GWAID
11	LQTFS	LPDLP	YDYGA	LEPIA	SGDIM	QLHHQ	NHHQT	YVTNY	NKALE	QL-HD	AISK-	–GD	–APT	VAKLH	SAIKF	NGGGH	INHSI	FWKNL	APVRE	GGGEP	PKGSL	GWAID
12	LHVFT	LPDLA	YDYGA	LEPVI	SGEIM	QIHHQ	KHHQT	YITNY	NKALE	QL-HD	AAK–	–AD	-TSTT	VK-LQ	NAIKF	NGGGH	INHSI	FWKNL	APVSE	GGGEP	PKESL	GWAID
13	–KHS	LPDLP	YDYGA	LEPHI	NAQIM	QLHHS	KHHAA	YVNNL	NATEE	KYHEA	LAK–	–GD	VTTQV	–ALQ	PALKF	NGGGH	INHTI	FWTNL	SP–	KGGGV	PKGEL	LEAIK
14	-HKHS	LPDLP	YDYGA	LEPHI	NAQIM	QLHHS	KHHAT	YVNNL	NVTEE	KYHEA	LAK–	–GD	VTTQV	–ALQ	PALKF	NGGGH	INHSI	FWTNL	SP–	KGGGE	PKGEL	LEAIK
15	.....	.....	.....	...HI	NAQIM	QLHHS	EHHAA	YVNNL	NVVEE	KYQEA	LKK–	–GD	VTAQV	–ALQ	PALKF	NGGGH	INHSI	FWTNL	SP–	NGGGE	PKGEL	LEAIK
16	.....	.....	.....	. .HI	SANIM	QLHHS	KHHAT	YVNNL	NVAEQ	KLAEA	VAK–	–GD	VTAEI	–ALQ	PAIKF	NGGGH	INHSI	FWTNL	SP–	NGGGA	PTGDL	QKAIE
17	.....	.....	.....	...HI	SAEIM	QLHHS	KHHAA	YVNNV	NVVEE	KLAEA	ALG–	K–GD	VNTQV	–SLQ	PAFRF	NGGGH	INHSI	FWRNL	SP–	SGGGQ	PCGDL	LKAIE
18	.....	.....	.....	...VI	SAEIM	QVHHQ	KHHAT	YVNNL	NAAEE	QLAEA	IHK–	K–QD	VTKMI	–ALQ	SAIKF	NGGGH	INHSI	FWNNL	CP–	SGGGE	PTPPL	AEAIT
19	.....	.....	.....	...VI	IGDIM	ELHHK	KHHAT	YTNNL	NAAEE	KLAAA	HAE–	-GD	IGGMI	–ALQ	PALKF	NGGGF	INHCI	FWTNL	SP–	QGGGV	PEGDL	ADAIN
20	.....	.....	.....	...HI	SEMIM	QIHHT	KHHQA	YINNL	KACTE	KLKQA	EQA–	A-ND	VAAMN	–ALL	PAIKF	NGGGH	INHTI	FWTNM	AP–	SAGGE	PAGPV	ADAIT
21	.....	.....	.....	...II	CREIM	ELHHQ	KHHQT	YVNNL	NAAEE	QLEEA	KSK–	–SD	TTKLI	–QLA	PALRF	NGGGH	INHTI	FWQNL	SP–	N-KTQ	PSDDL	KKAIE
22	–KVT	LPDLK	WDFGA	LEPYI	SGQIN	ELHYT	KHHQT	YVNGF	NTAVD	QFQEL	SDLLA	KEPSP	ANARK	MIAIQ	QNIKF	HGGGH	TNHCL	FWENL	APES-	QGGGE	PTGAL	ADAID
23	–AYT	LPPLD	YAYTA	LEPHI	DAQTM	EIHHT	KHHQT	YINNV	NAALE	GTSFA	NEPEV	EALLQ	KLDSL	PENLR	GPVRN	NGGGH	ANHSL	FWKVL	TP–	NGGGE	PKGAL	ADAIK
24	KKFYE	LPELP	YPYDA	LEPHI	SREQL	TIHHQ	KHHQA	YVDGA	NALLR	KLDEA	R—	E–SD	TDVDI	KAAL-	KELSF	HVGGY	VLHLF	FWGNM	GPAD-	ECGGE	PSGKL	AEYIE
25	VAEYT	LPDLD	WDYAA	LEPHI	SGEIN	EIHHT	KHHAA	YVKGV	NDALA	KLDEA	RAK–	—DD	HSAIF	LNEK-	NL-AF	HLGGH	VNHSI	WWKNL	SP—	NGGDK	PTGGL	ATDID
26	VAEYT	LPDLD	WDYGA	LEPHI	SGQIN	ELHHS	KHHAT	YVKGA	NDAVA	KLEEA	RAK–	E—D	HSAIL	LNEK-	NL-AF	NLAGH	VNHTI	WWKNL	SP—	NGGDK	PTGEL	AAAIA
27	–SFE	LPALP	YAKDA	LAPHI	SAETI	EYHYG	KHHQT	YVTNL	NN-LI	KGTAF	——	–GKS	LEEII	RRSSE	GGVFN	NAAQV	WNHWF	YWNCL	AP—	NAGGG	PTGKV	AEAIA
28	–AFE	LPPLP	YAHDA	LQPHI	SKETL	EYHHD	KHHNT	YVVNL	NNLVP	GTPEF	——	E-GKT	LEEIV	KSSS-	GGIFN	NAAQV	WNHTF	YWNCL	SP—	DGGGQ	PTGAL	ADAIN
29	–AFE	LPALP	FAMNA	LEPHI	SQETL	EYHYG	KHHNT	TVVKL	NG-LV	EGTEL	A—	E–KS	LEEII	KTST-	GGVFN	NAAQV	WNHTF	YWNCL	AP—	NAGGE	PTGEV	AAAIE
30	-MSYE	LPALP	FDYTA	LAPYI	TKETK	EFHHD	KHHAA	YVNNY	NN-AV	KDTDL	DGQPI	EAVIK	AIAG-	DASK-	AGLFN	NAAQA	WNHSF	YWNSI	KP—	NGGGA	PTGAL	ADKIA

31	NAKFE	LKPPP	YPLNG	LEPVM	SQQTL	EFHWG	KHHKT	YVENL	KKQVV	GTELD	DGKSL	EEIIV	TSYNK	GDIL-	PA-FN	NAAQV	WNHDF	FWECM	KP—	GGGGK	PSGEL	LELIE
32	–KFE	LQPPP	YPMDA	LEPHM	SSRTF	EFHWG	KHHRA	YVDNL	NKQID	GTELD	GKTL-	EDIIL	VTYNK	GAPL-	PA-FN	NAAQV	WNHDF	FWECM	KP—	NGGGE	PSGEL	LEIIN
33	–NYV	LKPPP	FALDA	LEPHM	SKQTL	EFHWG	KHHRA	YVDNL	KKQVL	GTELE	GKPL-	EHIIH	STYNN	GDLL-	PA-FN	NAAQA	WNHEF	FWESM	KP—	GGGGK	PSGEL	LALLE
34	.....	—LN	YEYSD	LEPVL	SAHLL	SFHHG	KHHQA	YVNNL	NATYE	QIAAA	TK—	E–ND	AHKIA	TLQS-	AL-RF	NLGGH	YNHWI	YWDNL	APVK-	SGGGVL	PDEHS	PLTAIK
35	-MTHE	LISLP	YAVDA	LAPVI	SKETV	EFHHG	KHLKT	YVDNL	NKLII	GT-EF	—	E-NAD	LNTIV	QKSE	FFIGN	NAGQT	LNHNL	YFTQF	RP—	GKGGA	PKGKL	GEAID
36	-MAFE	LPDLP	YKLNA	LEPHI	SQETL	EYHHG	KHHRA	YVNKL	NKLIE	GTPFE	K—	EP—	LEEII	RKSD-	GGIFN	NAAQH	WNHTF	YWHCM	SP—	DGGGD	PSGEL	ASAID
37	-MSFQ	LPQLP	YAYNA	LEPHI	SKETL	EFHHD	KHHAT	YVNKL	NGLVK	GT-EQ	EHKT-	—	LEELI	KQKPT	QAIYN	NAAQA	WNHAF	YWKCM	C—	GGGVK	PSEQL	IAKLT
38	–AYE	LPQLP	YAYDA	LEPHI	DAKT?	EIHHS	KHHNT	YVTNL	NAAVE	GT-EF	ADKDI	NDLIA	MLDAL	PADKQ	TAVRN	NGGGH	ANHTL	FWEVI	AP—	GGSNT	PVGEV	AKAID
39	–AYK	LPQLP	YAYNA	LEPHI	DKETM	TIHTS	KHHNT	YVTNL	NKAIE	GS-AL	AEKSV	DELVA	DLNAV	PEDIR	TAVRN	NGGGH	ANHSL	FWTLL	SP—	NGGGE	PTGEL	AEEIK

	110		120		130		140		150				160		170		180		190		200
						↓				↓				* ↓	* ↓						
1	KKFGS	FTAFK	DEFSK	AAA-G	RFGSG	WAWLV	VN–N	GELE-	ITSTP	NQDS-	PIME-	EGK–	TPILG	LDVWE	HAYYL	KYQNR	RPEYI	AAFWN	IVNWD	EVAKR	YSEA
2	KKFGS	FTAFK	DEFSK	AAA-G	RFGSG	WAWLV	VN–N	GELE-	ITSTP	NQDS-	PIME-	EGK–	TPILG	LDVWE	HAYYL	KYQNR	RPEYI	AAFWN	IVNWD	EVAKR	YSEA
3	RDFGS	VDNFK	AEFEK	AAA-S	RFGSG	WAWLV	LK–G	DKLA-	VVSTA	NQDS-	PLMGEAI	SGASG	FPILG	LDVWE	HAYYL	KFQNR	RPDYI	KEFWN	VVNWD	EAAAR	FAAK
4	ADFGS	YENWR	AEFE-	VAA-G	A-ASG	WALLV	YDPVA	KQLR-	NVAVD	NHDEG	AL—	WGS–	HPILA	LDVWE	HSYYY	DYGPD	RGSFV	DAFFE	VIDWD	PIAAN	YDDVVSLFE
5	ADFGS	YENWR	AEFE-	AAA-S	A-ASG	WALLV	YDSHS	NTLR-	NVAVD	NHDEG	AL—	WGS–	HPILA	LDVWE	HSYYY	DYGPD	RGSFV	DAFFE	VVDWD	EPTER	FEQAAERFE
6	SEFGT	FDEFK	EKFNA	AAA-A	RFGSG	WAWLV	VN–N	GKLE-	IVSTA	NQDS-	PLS–	EGK–	TPVLG	LDVWE	HAYYL	KFQNR	RPEYI	DTFWN	VINWD	ERNKR	FDAAK
7	SEFGT	FDEFK	EKFNA	AAA-A	RFGSG	WAWLV	VN–D	GKLE-	IVSTA	NQDS-	PLS–	DGK–	TPVLG	LDVWE	HAYYL	KFQNR	REYI	ETFWN	VIVWD	EANKR	FDAAK
8	EQFGG	FQALK	EKLTQ	AAM-G	RFGSG	WAWLV	KDP-F	GKLH-	VLSTP	NQDN-	PVM–	EGF–	TPIVG	IDVWE	HAYYL	KYQNR	RADYL	QAIWN	VLNWD	VAEEF	FKKA
9	RDFGS	FDKFK	EKLT-	AASVG	VQGSG	WGWLG	FNKER	GHLQ-	IAACP	NQD–	PLQGT	TGL–	IPLLG	IDVWE	HAYYL	QYKNV	RPDYL	KAIWN	VINWE	NVTER	YMACLL
10	EDFGS	FEALV	KKMN-	AEGAA	LQGSG	WVWLA	LDKEA	KKVS-	VETTA	NQD–	PLV–	KGASL	VPLLG	IDVWE	HAYYL	QYKNV	RPDYL	NNIWK	VMNWK	YAGEV	YENV
11	TNFGS	LEALV	QKMN-	AEGAA	LQGSG	WVWLG	VDKEL	KRLV-	IETTA	NQD–	PLV–	KGANL	VPLLG	IDVWE	HAYYL	QYKNV	RPDYL	KNIWK	WMNWK	YANEV	YEKECP
12	TNFGS	LEALI	QKIN-	AEGAA	LQASG	WVWLG	LDKDL	LRLV-	VETTA	NQD–	PLV–	KGASL	VPLLW	IDVWE	HAYYL	QYKNV	RPDYL	KNIWK	VINWK	GASEV	YEKESS
13	RDFGS	FEKFK	EKLT-	AMSVG	VQGSG	WGWLG	FNKEQ	GRLQ-	IACSS	NQD–	PLQGT	TGL–	IPLLG	IDVWE	HAYYL	QYKNV	RPDYL	KAIWN	VINWE	NVTER	YTAC
14	RDFGS	FEKFK	EKLT-	AVSVG	VQGSG	WGWLG	FNKEQ	GRLQ-	IAACS	NHD–	PLQGT	TGL–	IPLLG	IDVWE	HAYLL	QYKNV	RPDYL	KAIWN	VINWE	NVSQR	YIVCLL
15	RDFGS	FEKFK	EKLT-	AVSVG	VQGSG	WGWLG	FNKEQ	GRLQI	IAACS	NQD-	PLQGT	TGL–	VPLLG	IDV..	.....	.....	.....	.....	.....	.....	.....
16	TDFGS	FTKLQ	EKMS-	AVSVA	VQGSG	WGWLG	YDKET	GRLR-	IAACA	NQD–	PLQAT	TGL–	IPLLG	IDV..	.....	.....	.....	.....	.....	.....	.....
17	NDFGS	VDKLR	EKLV-	AAAVG	VQGSG	WAWLG	FNKES	KRLQ-	IATCA	NQD-	PLQGT	TGL–	FPLLG	IDV..	.....	.....	.....	.....	.....	.....	......
18	RDFGS	FEAFK	EKMT-	AATVA	VQGSG	WGWLG	LDPTS	KKLR-	IVACP	NQD–	PLEGT	TGL–	KPLLG	IDV..	.....	.....	.....	.....	.....	.....	.....
19	RDFGS	FDFSK	TTLT-	AATVA	IQGSG	WGWLG	FDPKT	HHLK-	IATCV	NQD–	PLQAT	TGM–	VPLFG	IDV..	.....	.....	.....	.....	.....	.....	.....
20	KEFGS	FQAFK	DKFS-	TASVG	VKGSG	WGWLG	YCPKN	DKLA-	VATCQ	NQD–	PLQLT	HG–	VPLFG	IDV..	.....	.....	.....	.....	.....	.....	.....
21	SQWKS	LEEFK	KEVT-	TLTVA	VQGSG	WGWLG	FNKKS	GKLQ-	LAALP	NQD–	PLEAS	TGL–	IPLFG	IDV..	.....	.....	.....	.....	.....	.....	.....
22	EQFGS	LDELI	KLTN-	TKLAG	VQGSG	WAFIV	KNLSNG	GKLD-	VVQTY	NQDT-	V—	TGPL-	VPLVA	IDAWE	HAYYL	QYQNK	KADYF	KAIWN	V-NWK	EASRR	FDAG
23	SDIGG	LDTFK	EAFTK	AAL-T	RFGSG	WAWLS	VTPE-	KKLVV	EST-G	NQDS-	PLS–	TGN–	TPILG	LDVWE	HAYYL	KYQNR	RPEYI	GAFFN	VVNWD	EVSRR	YQEALA
24	KDFGS	FERFR	KEFSQ	AAISA	E-GSG	WAVLT	YCQRT	DRLFI	MQVE-	KHNVN	VIPHF	——	RILLV	LDVWE	HAYYI	DYRNV	RPDYV	EAFWN	IVNWK	EVEKR	FEDIL

Table 5. (Continued)

	110		120		130		140			150			160		170		180		190		200
					↓					↓				* Å *	↓						
25	ETFGS	FDKFR	AQFS-	AAANG	LQGSG	WAVLG	YDTLG	KLTT-	TFQLY	DQQAN	VN—	LGI–	IPPLLQ	VDMWE	HAFYL	QYKNV	KADYV	KAFWN	VVNWA	DVQSR	YMAATSKTQGLIFD
26	DAFGS	FDKFR	AQFH-	AAATT	VQGSG	WAALG	WDTLG	NKLLI	FQVYD	HQTNF	P—	LGI–	VPLLL	LDMWE	HAFYL	QYKNV	KVDFA	KAFWN	VVNWA	DVQSR	YAAATSQTKGLIFG
27	ASFGS	FADFK	AQFTD	AAI-K	NFGSG	WTWLV	KNSD-	GKLAI	VSTS-	NAGT-	PLT–	TDA–	TPLLT	VDVWE	HAYYI	DYRNA	RPGYL	EHFWAL	V-NWE	FVAKN	LAA
28	AAFGS	FDKFK	EEFTK	TSV-G	TFGSG	WAWLV	KAD–	GSLAL	CSTI-	GAGA-	PLT–	SGD–	TPLLT	CDVWE	HAYYI	DYRNL	RPKYV	EAFWNL	V-NWA	FVAEE	GKTFKA
29	KAFGS	FAEFK	AKFTD	SAI-N	NFGSS	WTWLV	KNA-	NGSLAI	VNTS-	NAGC-	PITEE	EGV–	TPLLT	VDLWE	HAYYI	DYRNL	RPSYM	DGFWAL	V-NWD	FVSKN	LAA
30	ADFGS	FENFV	TEFKQ	AAA-T	QFGSG	WAWLV	LID-N	GTLKI	KTT-G	NADT-	PIA–	HGQ–	TPLLT	IDVWE	HAYYL	DYQNR	RPDYI	STFEK	LANWD	FASAN	YAAIA
31	RDFGS	FVKFL	DEFKA	AAA-T	QFGSG	WAWLA	YRARK	FDGENVANPPSPDEDNKLV	VLKSP	NAVN-	PLV–	VWGGY	YPLLT	IDVWE	HAYYL	DFQNR	RPDYI	SVFMDK	LVSWD	AVSSR	LEQAKALISTA
32	RDFGS	YDAFV	KEFKA	AAA-T	QFGSG	WAWLA	YKPEE	KKLA-	LVKTP	NAEN-	PLV–	LGY–	TPLLT	IDVWE	HAYYL	DFQNR	RPDYI	SIFMEK	LVSWE	AVSSR	LKAATA
33	RDFTS	YEKFY	EEFNA	AAA-T	QFGAG	WAWLA	Y–SN	EKLK-	VVKTP	NAVN-	PLV–	LGS–	FPLLT	IDVWE	HAYYL	DFQNR	RPDYI	KTFMTN	LVSWE	AVSAR	LEAAKAASA
34	EKWGS	YENFI	TLFNT	RTAAI	Q-GSG	WGWLG	YDTVS	KSLRL	FEL-G	NQDM-	PE—	WSSI-	VPLLT	IDVWE	HAYYL	DYQNL	RPKYL	TEVWKI	V-NWR	EVEKR	YLQAIE
35	KQFGS	FEKFK	EEFNT	AGT-T	LFGSG	WVWLA	SDA-N	GKLS-	IEKEP	NAGN-	PVR–	LGK–	NPLLG	FDVWE	HAYYL	TYQNR	RADHL	KDLWSI	V-DWD	IVESR	Y
36	KTFGS	LEKFK	ALFTD	SAN-N	HFGSG	WAWLV	KDN-N	GKLEV	LSTV-	NARN-	PMT–	EGK–	KPLMT	CDVWE	HAYYI	DTRND	RPKYV	NNFWQ	VVNWD	FVMKN	FKS
37	AAFGG	LEEFK	KKFTE	KAV-G	HFGSG	WCWLV	EHD–	GKLEI	IDTH-	DAVN-	PMT–	NGM–	KPLLT	CDVWE	HAYYI	DTRNN	RAAYL	EHWWN	VVNWK	FVEEQ	L
38	AKFGS	FDAFK	EEFAK	AAT-T	RFGSG	WAWLI	VD–G	DSVAV	TSTPN	QDS–	PVME-	GK—	TPVLG	LDVWE	HAYY.	.....	.....	.....	.....	.....	...
39	STFGS	FDQFK	EKFAA	AAAG-	RFGSG	WAWLV	VN–N	GKLEI	TSTPN	QDS–	PLSE-	GK—	TPVLG	LDVWE	HAYY.	.....	.....	.....	.....	.....	...

Note: [a]Gaps have been introduced into the sequence to maximize homology.

[b]Ligands to the metal atoms are indicated by *; residues involved in subunit-subunit interactions in the *Bacillus steareothermophilus* protein are indicated by ↓.

[c]Key and references: 1. *Bacillus stearothermophilus* (Mn), [468]; 2. *Bacillus caldotenax* (Mn), [469]; 3. *Escherichia coli* (Mn), [470, 471]; 4. *Halobacterium halobium* (Mn), [472–474]; 5. *Halobacterium cutirbrum* (Mn), [475, 476]; 6. *Listeria ivanovii* (Mn), [477]; 7. *Listeria monocytogenes* [478]; 8. *Thermus thermophilus* (Mn), [479]; 9. *Homo sapien* (Mn), [480–485]; 10. *Zea mays* (maize) (Mn), [486]; 11. *Nicotiana plumba*ginifolia (Mn), [487]; 12. *Pisum sativum* (Mn), [488]; 13. mouse (Mn), [489]; 14. rat (Mn), [490]; 15. *Sus scrofa* (Mn) (pig), [491]; 16. *Petromyzon marinus* (Mn) (lamprey), [491]; 17. *Eptatretus stouti* (Mn) (hagfish), [491]; 18. B*ranchiostoma floridae* (Mn) (amphioxus), [491]; 19. *Parastichopus californicus* (Mn) (sea cucumber), [491]; 20. *Palinurus vulgaris* (Mn) (lobster), [491]; 21. *Drosophila melanogaster* (Mn) , 491]; 22. *Saccharomyces cerevisiae* (Mn), [476, 492]; 23. *Methylomonas J.* (Fe/Mn), [453]; 24. *Methanobacterium thermoautotrophicum* (Fe), [493,494]; 25. *Mycobacterium leprae* (Fe/Mn), [495, 496]; 26. *Mycobacterium tuberculosis* (Fe), [497]; 27. *Escherichia coli* (Fe), [498, 499]; 28. *Pseudomonas ovalis* (Fe), [500]; 29. *Photobacterium leiognathi* (Fe), [501]; 30. *Anacystis nidulans* (Fe), [502]; 31. *Glycine max* (Fe), [503]; 32. *Nicotiana plumbinifolia* (Fe), [503]; 33. *Arabidopsis thaliana* (Fe), [504]; 34. *Tetrahymena pyriformis* (Fe), [505]; 35. *Porphyromonas gingavalis* (Fe/Mn), [452, 454, 456, 506]; 36. *Coxiella burnetii* (Fe), [507, 508]; 37. *Entamoeba histoglytica* (Fe), [509]; 38. *Aerobacter aerogenes* (Fe), [510]; 39. *Bacillus circulans* (Mn), [510].

Attempts have been made to understand the origin of the metal-ion specificity of the enzyme (or lack thereof) in terms of the amino acid sequence data and the observed structures [453, 504, 505, 517, 518]. These attempts have not been successful; most of the suggested explanations have been invalidated by subsequent sequence determinations. If the metals in the enzymes with the anomalous sequences described above are correct, then the explanation cannot be sequence-dependent. Ignoring those exceptions, two residues distinguish FeSODs from MnSODs: residue 76 is glycine in MnSODs and glutamine in FeSODs, and residue 149 is alanine in all FeSODs and either glutamine or histidine in the MnSOD's. The structural studies discussed below show that the residues 149 and 76 are in close proximity in the active site and that the glutamine residue interacts with the strictly conserved Tyr-34. Tyr-34 is believed to be very important for the function of the enzyme—its OH group is pointed towards and only 5 Å from the metal. Enzymes from species which are active with either metal have the manganese sequence at these positions so that the explanation for the metal-ion specificity (or lack of it) cannot lie with these residues alone. The explanation for the metal ion specificity, which must be due to very subtle effects must await future structural investigations.

X-ray structural studies have been reported on the manganese enzymes from *Thermus thermophilus HB8* at 1.8 Å resolution [519–521], *B. stearothermophilus* at 2.4 Å resolution [513, 522], and a recombinant enzyme from humans at 2.2 Å resolution [523, 524]. Preliminary studies have been reported for the enzymes from *S. cerevisiae* [525], *E. coli* [525], and human liver [526, 527]. The iron form of the enzyme has been structurally characterized from *P. ovalis* at 2.1Å resolution [528–530] and *E. coli* with and without the inhibitor azide bound at 2.4 Å resolution [498, 519, 520, 528, 529] and there is a preliminary study of the *Thermoplasma acidophilus* enzyme [532]. The iron and manganese enzymes are structurally homologous. The fold of the monomer in all the structurally characterized proteins consists of two distinct domains. The first domain consists of two long α-helices which are linked by a short α-helix or an unstructured loop. The exact length of these helices seems to be important in determining the nature of the subunit–subunit interactions in the enzyme and whether the enzyme is dimeric or tetrameric [524]. The second domain is a three-layer structure with a three-stranded antiparallel β-sheet in the central layer. The outer layer consists of four α-helices and the third layer is an extended piece of polypeptide chain connecting two of the β-sheet strands. Two ligands to the metal are provided by each domain.

The metal coordination spheres in the structurally characterized proteins show some variability. The protein groups bound to the metal are the same in every case, namely three histidines and a monodentate aspartate. In the *B. steareothermophilus* enzyme these are arranged in a distorted tetrahedral arrangement [513]. ^{19}F NMR studies suggest that two additional ligands can bind the metal ion in this form of the enzyme [533]. In the human and *T. thermophilus* enzymes the metal ion is five-coordinate with a coordinated water molecule (or OH^- group) completing the

Table 6. Reported Bond Distances (Å) and Angles (°) Around Manganese and Iron in Superoxide Dismutases

	B. stearothermophilus	*T. thermophilus*		*Human*	*P. Ovalis*
	Mn(III)	*Mn(III)*	*Mn(II)*	*Mn(III)*	*Fe(III)*
Bond distances					
Mn–^{26}N	2.08	2.11, 2.14	2.09, 2.18	2.10	2.60, 3.00
Mn–^{81}N	2.38	2.12, 2.10	2.13, 2.10	2.10	2.30, 2.30
Mn–^{167}N	1.94	2.16, 2.18	2.18, 2.13	2.10	2.40, 2.30
Mn–^{163}O	1.94	1.75, 1.78	1.83, 1.84	1.94	2.30, 2.30
Mn–^{w}O	—	2.07, 2.09	2.22, 2.24	2.00	–, 2.00(N_3)
Bond Angles					
^{28}N–Mn–^{81}N	96	91, 94	93, 94		105, 90
–^{167}N	97	92, 91	92, 88		90, 90
–^{163}O	88	89, 86	90, 85		95, 90
–^{w}O	—	176, 172	172, 166		–, 175
^{83}N–Mn–^{167}N	134	131, 134	130, 133		125, 120
–^{136}O	92	109, 110	109, 114		110, 120
–^{w}O	—	119, 116	120, 113		–, 90
^{170}N–Mn–^{163}O	133				120, 120
–^{w}O	—				–, 90
O–Mn–^{w}O	—				–, 90

Notes: [a]Residues are numbered according to the sequence in *B. stearothermophilus*.
[b]w, water.

coordination sphere. Metal coordination is unchanged upon reduction to Mn(II). The iron enzymes show similar variability; the *E. coli* enzyme has a five-coordinate iron, a water molecule is the fifth ligand, and the *P. ovalis* enzyme iron center is four-coordinate. These differences extend to the structures of the iron enzymes complexed with the inhibitor azide. In the *P. ovalis* structure the iron center is five-coordinate with the four protein donors and the monodentate azide ion [530]. A representation of the metal ion binding site is shown in Figure 7. In contrast the *E. coli* enzyme complexed with azide has a six-coordinate iron center with four protein ligands, a water, and the azide ion in the coordination sphere [519]. Details of the metal-ion bond lengths and angles in both iron and manganese SODs are given in Table 6.

The active site is lined with an array of five strictly conserved aromatic residues in addition to the metal ligands. There is extensive stacking of these residues in the active site. This hydrophobic environment may be important for enzyme activity. The channel to the active site is formed by residues from two subunits and the entrance to this channel is surrounded by a number of positively charged His, Lys, and Arg residues. It has been proposed that this positively charged cluster of amino acids provides a guiding electrostatic field for the superoxide ion [515].

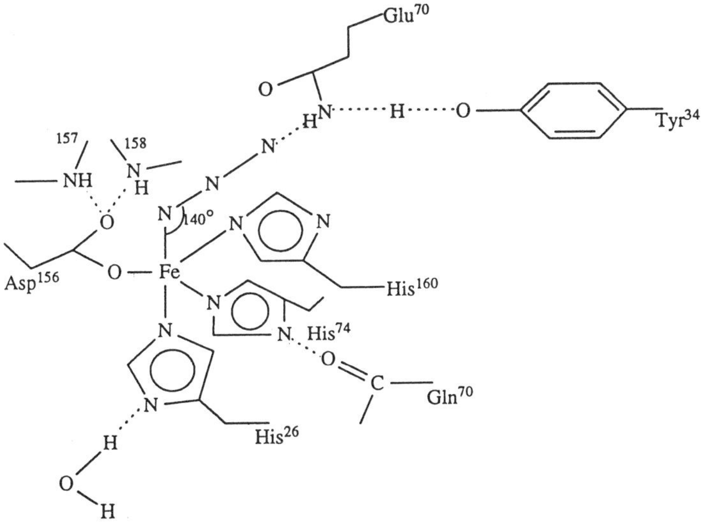

Figure 7. Active site interactions of azide inhibited from superoxide dismutase from *P. ovalis* [530]. The iron geometry is a trigonal bipyramid.

5.17 Transketolase (EC 2.2.1.1)

Transketolase is an important enzyme in the metabolism of glucose-6-phosphate in that it catalyzes the ketol transfer from a ketose to an aldose for a variety of sugar phosphates. The enzyme from *S. cerevisiae* has two identical subunits (M_r = 74,000). It requires thiamine diphosphate as a prosthetic group and Mg^{2+} for

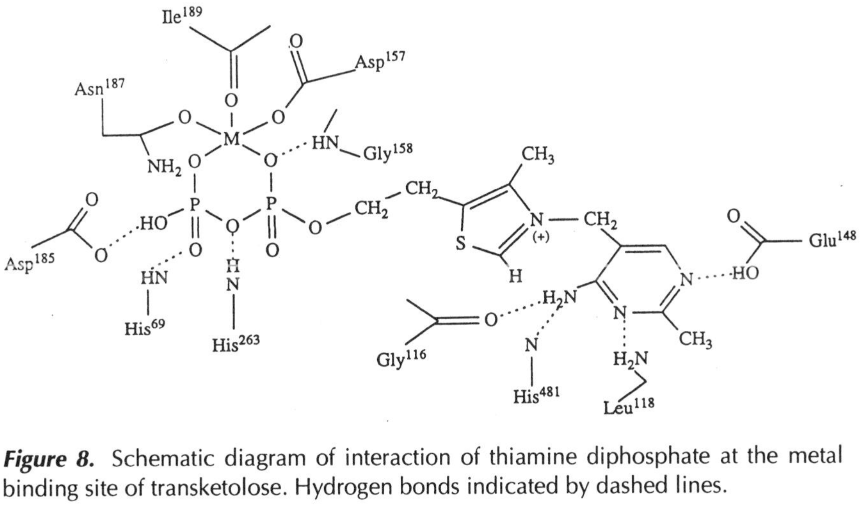

Figure 8. Schematic diagram of interaction of thiamine diphosphate at the metal binding site of transketolose. Hydrogen bonds indicated by dashed lines.

catalytic activity. Mn^{2+}, Ca^{2+}, and other metal ions can replace Mg^{2+} [534]. The crystal structure of *E. coli* transketolase with Ca^{2+} and thiamine diphosphate bound has been determined at 2.5 Å resolution [535]. The coordination to the calcium is five-coordinate, and the side chains of Asp-157 and Asn-187, the peptide oxygen from Ile-189, and two oxygen atoms from the diphosphate are the ligands. A schematic diagram of the metal binding site is shown in Figure 8.

6. PROTEINS WITH TWO BOUND METALS

6.1 Acyl Carrier Protein

Acyl carrier proteins are small, monomeric, highly acidic proteins with a phosphopantetheine prosthetic group which play an essential role in fatty acid biosynthesis in bacteria and plants. The protein probably acts as a shuttle carrying the intermediates which are attached to a 4′-phosphopantetheine group via an acyl thioester linkage from site to site in the fatty acid synthetase system. The 4′-phosphopantetheine group is covalently attached to the protein via a serine (Ser-36) residue. In higher organisms, acyl carrier proteins are associated with a protein complex, but the *E. coli* is a soluble protein. Early work on these proteins has been reviewed [536]. The *E. coli* enzyme (77 amino acids, M_r = 8847) has been the subject of a number of NMR structural determinations [537–542]. At physiological pH, two Mn^{2+} binding sites have been identified from NMR and EPR studies; glutamate or aspartate are suggested as the ligands bound to the metal ion [543, 544]. These studies have identified the metal ion binding sites. One site is around Glu-30, Asp-35, and Asp-38, and the second around Glu-47, Glu-48, Asp-51, Glu-53, and Asp-56.

Crystals suitable for X-ray structural studies were reported some time ago but the structure has not yet been reported [545].

6.2 Catalase (EC 1.11.1.6)

Catalase is responsible for one of the principle pathways of hydrogen peroxide removal from cells:

$$2H_2O_2 \rightarrow 2H_2O + O_2$$

Most catalase enzymes contain the iron protoporphyrin IX prosthetic group and the structures of *Penicillum vitale*, *Micrococcus lysodeikticus* and the beef liver catalase have been determined [546–548]. In addition to the heme-containing catalases, a number of non-heme catalases are known [549, 550] and manganese-containing catalases have been isolated and reasonably characterized from three organisms, *Thermoleophilum album* [551], *Lactobacillus plantarum* [549], and *Thermus thermophilus* [552]. The chemistry and structure of these enzymes has

Table 7. Properties of Manganese-Containing Catalases

Organism	M_r (kDa)	Number of Subunits	Mn per Subunit	λ_{max} (nm)	ε (M^{-1} cm^{-1})	Reference
Lactobacillus plantarum HB8	172	6	1.12	470	—	[549, 555]
Thermus thermophilus	210	6	2	460, 500	1.35×10^3	[552, 556]
Thermoleophilum album	141	4	1.4 (±0.4)	—	—	[551]

been reviewed recently by Penner–Hahn [553] and earlier by Beyer [554], thus will be summarized only briefly here (see Table 7).

The enzymes from all three organisms are very similar. In each case the stoichiometry of the protein is consistent with a binuclear manganese active site. Early reports that the enzyme contains only one manganese are erroneous. EPR characterizations of the *T. album* enzyme by Khangulov and co-workers [557] and similar studies by Penner–Hahn's group on the *L. plantarum* enzyme [552, 553] are consistent with a binuclear active site. The isolated form of the enzyme has a 16-line EPR spectrum together with a much broader signal consistent with Mn(III)/Mn(IV) and Mn(II)/Mn(II) dinuclear sites—i.e. a mixture of oxidation states are present in the isolated enzyme [558]. Upon reduction with NH_2OH the Mn(II)/Mn(III) and Mn(II)/Mn(II) forms of the protein have also been detected by EPR. The Mn(III)/Mn(III) form of the enzyme is EPR silent and there is no evidence for a Mn(IV)/Mn(IV) form of the enzyme [557, 559].

The crystal structure of the protein from *L. plantarum* has been reported at low resolution and the enzyme consists of four antiparallel α-helices [560]. Two regions of enhanced electron density 3.6 ± 0.3 Å apart have been identified as the manganese atoms, but ligands to the manganese have not been identified at the current resolution. Electron spin envelope modulation spectra of the *T. thermophilus* enzyme demonstrate that nitrogen is bonded to Mn and histidine coordination has been suggested [561].

X-ray absorption spectroscopic studies of the *L. plantarum* enzyme [562] suggest that the coordination sphere of the metal ions in the Mn(III)/Mn(IV) form has two oxygens at 1.82 Å and four nitrogen or oxygen ligands at 2.14 Å. The Mn–Mn distance is 2.67 Å. This suggests a di-μ-oxo structure in this form of the enzyme. In the reduced Mn(II)/Mn(II) form of the enzyme the data suggest six ligands at 2.19 Å. The Mn–Mn distance was not determined with certainty but there is a small improvement in the fit if a Mn atom at 3.55 Å is included.

6.3 DNA Polymerase (EC 2.7.7.7)

DNA polymerase catalyzes template directed DNA synthesis and repair [563–566]. Most, if not all organisms contain many different DNA polymerase molecules. DNA polymerases are frequently associated with 3′-5′exonuclease activity, which is involved in editing out mismatched terminal nucleotides of the DNA and a 5′,3′-exonuclease activity which functions in nick translation. DNA polymerase from Bacillus species do not have any exo- or endonuclease activity but are activated by both Mg^{2+} and Mn^{2+} [567]. Manganese is required to obtain maximal DNA polymerase activity in *Drosophila* cells [568] and in C type reteroviruses, but other viruses, e.g. human T-cell leukemia virus, maximal activity is obtained with Mg^{2+}. Bovine DNA polymerase has two strong and one weak binding sites for Mn^{2+} [569]. Ca^{2+} competes for both of these binding sites, while Mg^{2+} competes for only one site. The metal activation of ϕ29 DNA polymerase from *E. coli* has been studied by Esteban et al. [570]. Mn^{2+} and Mg^{2+} activation of the enzyme is almost equally efficient. The polymerase and exonuclease activities of the enzyme are well separated on this enzyme as well. Studies of the metal binding in these systems are complicated because the metal can bind to DNA at multiple sites, the enzyme, and to the nucleotide triphosphate which are all required for reaction. In *E. coli*, Mn^{2+} activated DNA polymerases have been shown to have greater frequency of misincorporation of bases into the molecule. This phenomenon has been shown to be sequence-specific and with some sequences a greater fidelity of incorporation has been demonstrated [571–574].

The DNA polymerase I from *E. coli*, which is active as a single unit, has 928 amino acid residues and three functions: polymerase activity; 3′- to 5′-exonuclease activity; and 5′,3′ exonuclease activity. These three activities take place on three different domains of the protein, each requiring the presence of a divalent metal ion [575]. It is not known which metal ion is active *in vivo*, but Mg^{2+}, which is the most intensively studied ion, Mn^{2+}, Co^{2+}, Ni^{2+}, and Zn^{2+} can all serve as the cofactor. Mn^{2+} has been recommended for use in DNA sequencing studies [576, 577]. NMR studies on DNA polymerase have been reviewed [73]. EPR studies of the binding of Mn^{2+} to the complete enzyme suggest that in the absence nucleotides there is one tight binding site and approximately 20 much weaker binding sites. The affinity of the metal binding sites for Mn^{2+} has been studied quantitatively [575].

Limited proteolysis of the *E. coli* enzyme cleaves the protein into two fragments: the large C-terminal proteolytic fragment, residues 326–928, known as the Klenow fragment; and the N-terminal fragment which contains the 5′,3′-exonuclease activity. The structure of the Klenow fragment has been determined to be complexed with deoxythymidine monophosphate [578], with an eight-base-pair duplex DNA fragment and two single-stranded deoxytetranucleotides [579, 581], and with Asp-424–Ala, Asp-355–Ala, and Glu-357–Ala mutant forms of the fragment [580, 581]. Structures of additional mutant forms are under investigation [581]. The fragment consists of two domains: a small exonuclease domain which is largely β-sheet with α-helix on both sides and which binds two metal ions in a buried hydrophobic pocket; and a carboxy terminal domain which contains a deep cleft about 20–24 Å wide and 25–35 Å deep and which accommodates double-stranded DNA and the DNA polymerase active site. The active sites on the two domains are

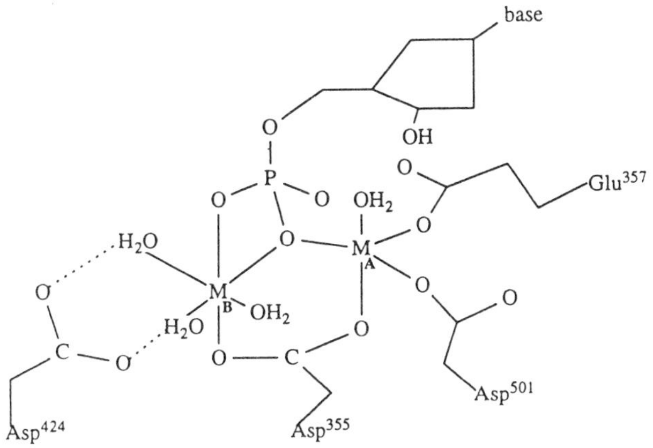

Figure 9. Schematic diagram of the active site of the 3′-5′-exonuclease domain of the *E. coli* DNA polymerase complexed with product. Metal ligand distances are between 1.9 and 2.2 Å. The water or OH^- ion bound to metal A is suggested as the group which attacks the phosphorus atom.

separated by 33 Å [578, 582]. In the absence of substrate or product the exonuclease domain binds only one metal ion (site A), while the deoxynucleoside–monophosphate complex has a second divalent metal ion (site B) 3.9 Å from site A (Figure 9). Zn^{2+}, Mn^{2+}, Sm^{3+}, and Mg^{2+} can all bind to site A which is five-coordinate with Asp-355, Glu-357, Asp-501, the 5′-phosphate of d-TMP, and a water molecule (or OH^-) providing the donor atoms [568, 582]. The site B metal binding site is octahedral. Asp-355, two of the 5′-phosphates of dNMP, and three water molecules comprise the coordination sphere of metal B. Each of these ligands is in turn hydrogen bonded to the protein molecule. Recent work involving Co^{2+} substitution of the Klenow fragment in solution suggests that there may be three metal ion binding sites—site A which is five-coordinate and two octahedral sites [583]. The solid state structures may therefore be misleading.

Mutations of all of the side chains in the active site of the enzyme have revealed that loss of Asp-501 and Asp-355 causes a large decrease in the enzyme activity due to the loss of the metal ions. Both carboxylate oxygens of Asp-424 are necessary to bind metal since the Asp-424–Asn mutant has no activity. Only a single oxygen of the Asp-501 residue is necessary since the Asp-501–Asn mutant has wild-type activity [581].

6.4 Deoxyribonuclease

Deoxyribonuclease (DNase) degrades double-stranded DNA to yield 5′-oligonucleotides in a manner which is neither base- nor sequence-specific. The enzyme requires divalent metal ions for its activity and combinations of Ca^{2+} and Mg^{2+} or

Ca^{2+} and Mn^{2+} have the highest activity. Mn^{2+}-dependent DNases have been claimed [584].

The structures of bovine pancreatic DNase I alone, bound to a short strands of DNA and to actin have been determined [263, 585–588]. Very little conformational change is observed in the enzyme on binding to DNA; the structure of the DNA is, however, significantly altered. A number of different metal binding sites have been identified on the basis of these structural studies. In the free protein, two Ca^{2+} ions are bound at positions remote from the active site. A Pb^{2+} ion was bound at the active site, while Ca^{2+} binding at this site is weak in the absence of nucleotide. Diffusion of Mn^{2+} into crystals containing the bound octanucleotide resulted in another cut in the DNA on an opposite strand from the first cut and 15 Å away from the site of the first cut. This suggests that the DNase has two Mn^{2+}-activated catalytic sites [586]. The structure of the enzyme–(GGUAUACC)$_2$ duplex DNA fragment complex in the presence of Mn^{2+} has also been determined [588]. The metal ion interacts with a glutamate side chain, the scissile phosphate group, and possibly an aspartate residue.

6.5 Fructose-1-6-bisphosphatase (EC 3.1.3.11)

New sugar produced in the photosynthetic cells of plants is diverted toward starch accumulation or sucrose synthesis in the cytosol [589]. In the first step in the latter reaction the dephosphorylation of fructose 1,6-bisphosphate to fructose 6-phosphate is catalyzed by fructose-1,6-bisphosphatase. The enzyme is absent in plants that store starch, e.g. wheat, maize [590]. In other plants the enzyme occurs in two forms: an oxidized form without detectable activity, and an active reduced form [591]. The enzyme also occurs in animals where it is a key factor in the glycogenic pathway. The liver enzymes especially from rabbit and rat have been extensively studied. The enzyme is regulated by the inhibitor, fructose-2,6-bisphosphate, and adenosine monophosphate (AMP).

There are two metal binding sites on each subunit, and the usual activating metals are Mg^{2+} and Mn^{2+}. One of these sites has a high affinity for metal ions and the other is a site of low affinity. There is a report that the enzyme from human fetal brain is stimulated by Zn^{2+}, although Zn^{2+} is usually inhibitory [592, 593]. Mechanistic studies of the reaction of the enzyme have been the subject of many studies [594, 595]. At low metal ion concentrations the rate is much lower with Mn^{2+} than with Mg^{2+}; this difference decreases in the presence of the inhibitor fructose-2,6-bisphosphate and at high concentrations Mn^{2+} gives the faster rate [596].

EPR studies of the binding of fructose 2,6-bisphosphate to the Mn^{2+}-substituted rabbit liver enzyme suggests that the distance from the Mn^{2+} center to the phosphorus atoms is 4.7 ± 0.2 Å for the 2-phosphate and 5.0 ± 0.2 Å for the 6-phosphate [597]. Adenosine 5′-monophosphate (AMP) is an allosteric inhibitor of the enzyme. A ^{31}P NMR study of the binding of AMP and a metal ion indicates that the binding is competitive, which is in agreement with earlier fluorescence and kinetic investigations [598–600]. However this conclusion is at variance with the results of the crystal structure determination discussed below. The solid state structure suggests that the distance from the AMP binding site to the metal is about 30 Å.

Crystallographic studies of the porcine kidney enzyme, which is a tetramer with each subunit containing 335 amino acids, have been performed for the unligated enzyme [601, 602], the enzyme–fructose-6-phosphate–Mg^{2+} complex [603, 605], the enzyme–AMP complex [604], and the inhibitor fructose–2,6-bisphosphate–enzyme complex [602, 606]. The enzyme exists in two conformational states, designated R and T, with the T state observed when AMP is bound to the enzyme. The R and T states are related by a 19° twisting motion of the two subunits about the molecular twofold axis and by a translation of about 1 Å of the AMP domain [604]. Two binding sites for fructose 2,6-bisphosphate have been identified. The major binding site is the same site as the fructose 6-phosphate binding site. The extra phosphate residue interacts via hydrogen bonds with four peptide residues. The metal ion (Mg, Mn, and Zn all occupy the same site) is bound to the side chains of two aspartate residues and a glutamate residue and is ~5 Å from the center of the 2-phosphate residue of the 2,6-bisphosphate inhibitor. If the substrate binds in the same orientation as the inhibitor then an oxygen of the 1-phosphate group would also be bound to the metal ion [602].

6.6 Glutamine Synthetase (EC 6.3.1.2)

Glutamine synthetase catalyzes the reaction of ATP and NH_3 with glutamate to yield glutamine. Since glutamine is the source of nitrogen in the biosynthesis of many metabolites, many of which act as feedback inhibitors, glutamine synthetase serves as the central element in the regulation of cellular nitrogen metabolism [607–609].

Glutamine synthetase is found in bacteria, plants, and in large amounts in mammalian brain where it provides the glutamine precursor of two amino acids involved in nerve transmission, γ-aminobutyric acid, and glutamate [610]. The bacterial enzyme is a dodecomer with 12 identical subunits (M_r = 51,814) arranged in two hexameric rings [611], and the mammalian enzyme is an octamer of apparently identical subunits; the active sites of the different forms are similar [612].

The mammalian enzyme is highly regulated by adenylation/deadenylation of a specific tyrosine residue [613]. The *E. coli* enzyme is regulated by reversible adenylation of Tyr-397 and by the amount of ammonia available. Metal ion specificity is dependent on the adenylation state of the enzyme. If the enzyme is adenylated, Mn^{2+} is required for the biosynthetic reaction; if the enzyme is unadenylated then the Mg^{2+} form is active. The concentration of the metal ions in the cell and product inhibition may also be responsible for control of the bacterial enzyme. The *in vivo* metal is a matter of some controversy. In the brain Mn^{2+} may be the native metal since in at least some cells the concentration of Mn^{2+} is higher than Mg^{2+} and Mn^{2+} binds 200 times more tightly to the enzyme [610, 614]. In other organs Mg^{2+} is apparently the active metal in the unadenylated form of the enzyme, but it is unable to function with the adenylated form of the enzyme, whereas Mn^{2+} is able to do so [615]. The mechanism of the reaction which has been proposed is the formation of a γ-glutamyl phosphate intermediate from ATP and glutamate followed by displacement of the activated phosphate by ammonia. Product release is rate-limiting in the unadenylated form of the enzyme; in the adenylated enzyme phosphoryl transfer is rate-limiting [616].

Studies on the enzyme from *Bacillus cereus* have suggested that the Mg^{2+} and Mn^{2+} ions have different binding sites [617], but this conclusion is not supported by studies on other forms of the enzyme. Many studies including the structure determination have been performed on the manganese form. Studies with lanthanides [618, 619] and other metal ions and metal complexes using EPR and fluorescence techniques [620, 621] have also helped to characterize the active site. Two types of metal binding sites have been distinguished: a high-affinity site, designated n_1, which is near the glutamate binding site [622, 623] and which is thought to keep the enzyme in the right conformation, and the low-affinity n_2 site which is involved in binding the metal ATP complex and may be involved in the phosphoryl transfer reaction [624]. EPR, ESEEM, and NMR studies established that the manganese ions had two bound water molecules and that they were 5.8 Å apart [622, 625, 626].

The crystal structure at 3.5 Å resolution of the enzyme from *Salmonella typhimurium* has been determined [627, 628]. Crystals of the *E. coli* enzyme are isomorphous with the *S. typhimurium* enzyme. The manganese-containing active site is in a cylindrical channel roughly parallel to the six-fold axis of the molecule and open at the top. The metal ions are to one side of the channel which is formed by residues from two subunits. Each manganese ion is coordinated to two water molecules and three protein side chains. The manganese ion, which is believed to bind glutamate, is coordinated to three glutamate residues and there are two glutamates and a histidine ligands for the other manganese ion which is believed to bind ATP. One water (or a OH^-) links the two manganese ions; the Mn–Mn distance is 5.8 Å [628], in good agreement with the NMR studies. Some of the glutamate side chains may act as bidentate ligands but the structure is not sufficiently well defined to be certain. The adenylation site is some 22 Å from the metal ions. Lanthanide ions, which are bound more tightly than either Mn^{2+} or Mg^{2+} are slightly further apart than the manganese ions in the solid state [618, 629].

The thermodynamics of binding of various substrates to the enzyme, and its unfolding have been studied extensively by Ginsburg and co-workers [630, 633].

6.7 Lectin

Lectins are sugar-binding proteins that recognize and bind specific saccharides. Animal lectins are involved in cell surface recognition events and apparently contain only Ca^{2+} as the metal cofactor. Plant lectins which contain both Ca^{2+} and

Mn^{2+} ions as metal cofactors are involved in the recognition of complex carbohydrates [634–636]. The metal-free protein has no carbohydrate binding properties. The binding of the first metal to the lectin induces a conformational change in the protein which results in the formation of a second tight metal binding site. Binding of the second metal causes the saccharide binding site to be formed. In the absence of Ca^{2+}, two molar equivalents of Mn^{2+} will bind Concanavalin A and form the conformation with full saccharide binding capability. Mn^{2+} coordination in lectins has been studied by EPR, NMR, and ESEEM spectroscopy. The dimanganese form of Concanavalin A with bound saccharide has been investigated using EPR and magnetic susceptibility studies [637, 638]. The complex EPR spectrum is consistent with two octahedrally coordinated Mn^{2+} ions antiferromagnetically coupled (J = +1.8 cm^{-1}) with a zero-field splitting of 375 G. The ESEEM study confirmed that two water molecules are coordinated to Mn^{2+} in all the lectins studied [639, 640]. Failure to observe solvent proton relaxation at the Mn^{2+} site in a number of lectins (lentil and pea) which had been reported earlier [641] must therefore be due to slow solvent exchange and not to the absence of the water molecules.

A number of different plant lectins have been structurally characterized: Concanavalin A from Jack Bean (*Canavalia ensiformes*) at 1.75 Å resolution [642–646]; pea lectin at 2.4 Å resolution, but the crystals diffract to 1.2 Å [647–651]; isolectin from *Lathyrus ochrus* at 1.9 Å resolution [652–657]; wheat germ agglutinin isolectin complex with N-acetylneuraminyl-lactose at 2.2 Å resolution [658]; Favin from broad bean *Vicia falsa* at 2.8 Å resolution [659]; a lectin from *Griffonia simplifolia* with a bound tetrasaccharide [660]; a galactose binding lectin from *Erythrina corallodendron* [661]; and a Ca^{2+}-dependent (C-type) mannose binding lectin from rat [662, 663]. Crystals of a *Lens culinaris* lectin have recently been reported as suitable for study [664]. The metals are bound in a shallow depression

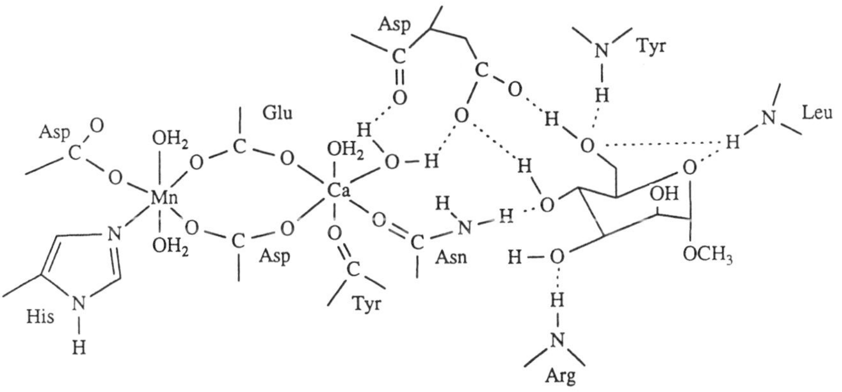

Figure 10. Schematic representation of the metal and saccharide binding sites of Concanavalin A. The saccharide is methyl-α-D-mannopyranoside.

Table 8. Metal Binding Coordination Spheres in Various Lectins

	Con A	*LolI*	*pea*		*Con A*	*LolI*	*pea*
Mn-Asp^{119}	2.26	2.29	2.21	Ca-Asp^{121}	2.44	2.22	2.40
Mn-Asp^{121}	2.06	2.21	1.93	Ca-Asp^{121}	2.76	2.52	2.65
Mn-Asp^{129}	2.30	2.28	2.23	Ca-Phe^{123}	2.33	1.89	2.35
Mn-OH_2	2.15	2.38	2.1	Ca-Asn^{125}	2.45	2.08	2.24
Mn-OH_2	2.13	2.32	2.03	Ca-Asp^{129}	2.33	2.60	2.39
Mn-N	2.45	2.44	2.19	Ca-OH_2	2.26	2.29	2.34
				Ca-OH_2	2.25	2.27	2.39

on the surface of the protein. The metal binding sites of all lectins are highly conserved and the differences in the metal ion positions are small [653, 661]. A schematic representation of the metal binding site is shown in Figure 10. The Mn^{2+} ion is bound to a carboxylate, two carboxylates which serve as a bridge between the metal ions, a histidine, and two water molecules. Ca^{2+} is bound to the two bridging carboxylates, the oxygen of the amido group of an Asn residue and a carbonyl group of a Phe or Tyr, and two water molecules [563, 659]. Metal ligand bond distances as reported for various lectins are listed in Table 8.

The saccharide binding site is in a cleft in the protein surface close to the metal ions. The sugar is hydrogen bonded to protein residues and water molecules and there are also van der Waals interactions with aromatic residues. The protein residues change depending on the particular sugar bound to the protein [655]. Two of the hydrogen-bonding residues are also residues bonded to the Ca^{2+} ion [653]. The most comprehensive structural studies of the sugar binding to these proteins are the studies of the isolectin from *Lathyrus ochrus* by Bourne and co-workers. These studies have involved the apoprotein [654], a bound monosaccharide [653], a trisaccharide,[251] and a biantennary octasaccharide [655].

The metal binding properties of Concanavalin A have been studied in great detail [641, 666–668]. Recent studies using the X-ray absorption fine-structure technique of the Zn^{2+}-substituted form of Concanavalin A have suggested that there are significant structural differences between the structure in solution and in the solid state. The evidence suggests that with Zn^{2+} in the S1 site and Ca^{2+} in the S2 site the zinc ion is hexacoordinate in solution with five oxygen atoms at 2.10 ± 0.02 Å and one N donor at 2.38 ± 0.02 Å. In the solid state the zinc is four-coordinate with four oxygen donors at 2.04 ± 0.02 Å. The geometry change is dependent on the presence of Ca^{2+}; in the absence of this ion the Zn^{2+} is pentacoordinate (four oxygens and 1 nitrogen) in both the crystals and solution. With other metal ions in the S1 site these coordination geometry changes are not observed. These observations are important because although Mn^{2+} does not apparently undergo these changes it has the potential for so doing. They also demonstrate that the crystal structure of a protein may not be the same as the structure in solution [669].

6.8 Leucine Aminopeptidase (EC 3.4.11.1)

This enzyme which occurs in many tissues and organs has a preference for cleavage of leucyl peptides. The specificity is not absolute as substantial rates are observed with other amino acids. The native enzyme from bovine lens, which is the best characterized enzyme, has six identical subunits containing two independent nonidentical binding sites for zinc. Both metal binding sites must be occupied for activity to occur, and the most slowly exchanging site must be occupied by zinc [670]. Manganese can occupy the other zinc binding sites with a significant increase in activity [671]. The distance (determined by NMR) from the manganese ion to the protons of the enzyme inhibitor, N-(leucyl)-*o*-aminobenzene-sulfonate, suggested that the Mn^{2+} ion is close to the carbonyl oxygen of the peptide bond [672]. Leucine aminopeptidase from *E. coli* and *Salmonella typhimurium*, which are identical proteins, have substantial sequence homology with the bovine lens enzyme. This enzyme is activated by Mn^{2+} and inhibited by both Zn^{2+} and EDTA [673]. The amino acids which bind the metal in the bovine lens enzyme are conserved in the *E. coli* enzyme and hence presumably bind Mn^{2+}.

Bovine lens leucine aminopeptidase has been structurally characterized [674–677]. A schematic view of the metal binding site is given in Figure 11 and the legend to the Figure contains additional information about the bond distances. The tight metal binding site is assumed to be Zn(2). Zn(2) is bound to two monodentate

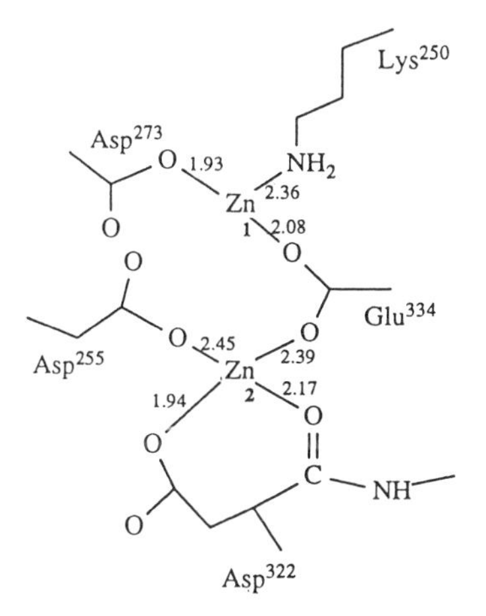

Figure 11. Schematic representation of the metal binding site in leucine aminopeptidase. The zinc ions are 2.9 Å apart. Bond lengths (Å) to the zinc ions are indicated on the bonds.

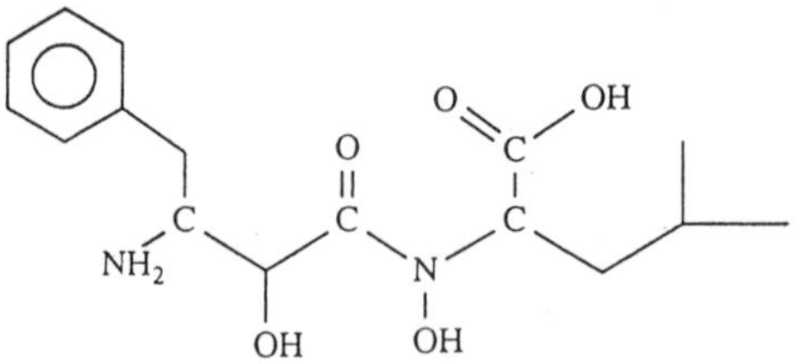

4

aspartate residues, a bridging glutamate carboxylate oxygen, the carbonyl oxygen of an aspartate residue, and weakly by a water molecule. Zn(1) is coordinated by a lysine amino group, the bridging glutamate, and a monodentate aspartate. The Zn–Zn distance is 2.91 Å. One unusual feature of the metal binding is the coordinated lysine residue. The presence of bestatin (**4**), (2*S*,3*R*)-3-amino-2-hydroxy-4-phenylbutanoyl)-Leu-OH) (IV), a slow binding inhibitor of the enzyme, causes no significant conformational changes to the enzyme. The phenyl alanine and leucine side chains of the inhibitor occupy hydrophobic pockets in the active site cavity. The NH_2 and OH groups of the bestatin are bound to Zn(1) and no other changes in the coordination spheres of the metal ions are apparent [677].

6.9 6-Phosphofructokinase (EC 2.7.1.11)

6-Phosphofructokinase (FPK) catalyzes the reversible phosphorylation of fructose-6-phosphate to fructose-1,6-bisphosphate which is a key ATP-producing step in glycolysis. This reaction is a rate-determining reaction in the glycolysis pathway. The enzyme is tetrameric and it displays two different conformational states (R and T) which are in equilibrium.

Enzymes from yeast and mammals and protozoa require two Mg^{2+} (or Mn^{2+}) ions for activity [678]. There are three types of enzymes in humans: PFK-M (muscle), PFK-L (liver), and PFK-P (platelet). Three types of mRNAs are transcribed from a single gene encoding for the enzyme. There are two high affinity Mn^{2+} sites one of which binds the fructose-6-phosphate and the other binds MnATP. The binding constants for Mn^{2+} and Mg^{2+} are the same, but the binding constant of fructose-6-phosphate is lower for Mn^{2+} than for Mg^{2+} [679].

The three-dimensional structure of the enzymes from *E. coli* and *B. stearothermophilus* have been determined [680–684]. The single polypeptide chain is arranged in two distinct domains of approximately equal size joined by a narrow waist region. The N-terminal half is homologous to the carboxy-terminal half. Each subunit has two Mg–ATP binding sites: one is the active site and binds fructose-6-phosphate as well as ATP; the other site is the allosteric effector site and binds the activator MgATP and the inhibitor phosphoenolpyruvate [683]. The active site lies between the two domains of the subunit. In the Mg–ATP–fructose-1,6-bisphosphate–enzyme complex the Mg^{2+} is bound to three water molecules, Asp^{103}, the β-phosphate of ADP, and the 6-phosphate of fructose 1,6-bisphosphate [683]. The structure of the metal binding site and the fructose-6-phosphate binding site is shown in Figure 12. In the effector site the MgATP is bound to the α and β-phosphates of ATP and protein moieties, the carboxylate of Glu-187, and the carbonyl oxygen of Gly-185. The Asp-129 residue which interacts with two water

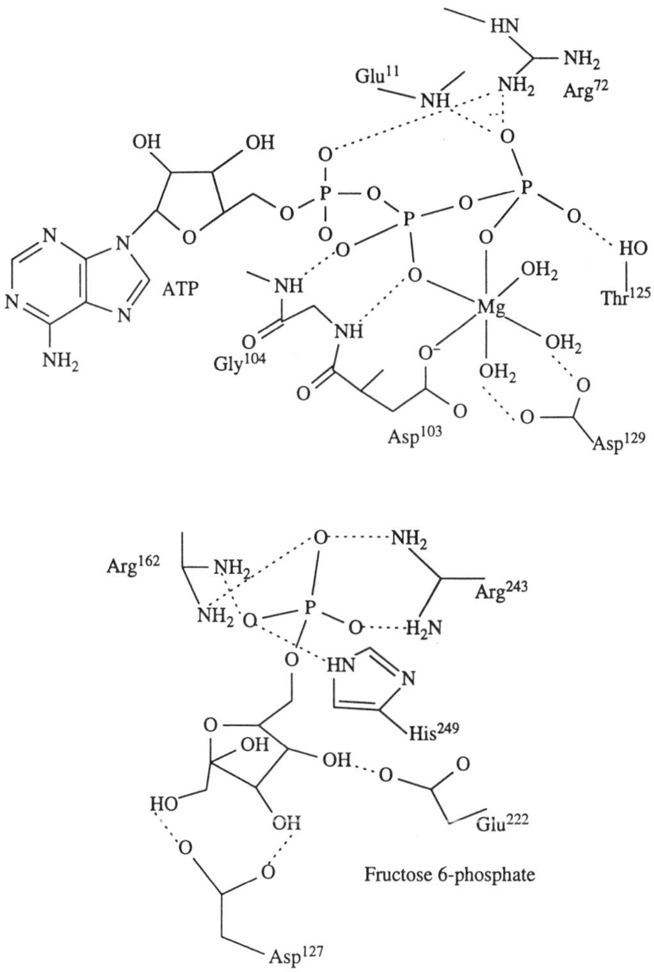

Figure 12. Schematic diagram of the active site residues in *E. coli* 6-phosphofructokinase. Residues Arg-162 and Arg-243 are from an adjacent subunit.

molecules bound to the Mg^{2+} ion has been shown to be important in the catalytic mechanism in both directions [687]. The effects of mutations of active site residues on the kinetics of the reaction have been the subject of a number of studies [688].

6.10 3-Phosphoglycerate Kinase (EC 2.7.2.3)

3-Phosphoglycerate kinase is a monomeric glycolic enzyme responsible for the first ATP generating reaction during glycolysis:

$$\text{1,3-bisphosphoglycerate} + \text{MgADP} \rightarrow \text{3-phosphoglycerate} + \text{MgATP}$$

The structural and spectroscopic properties of the enzyme have been reviewed [71, 689]. The enzyme from a variety of sources is a monomer with a highly conserved sequence and (M_r = 44,000–47,000) [690]. A divalent metal ion is required for activity, and Mg^{2+} is the native metal. Mn^{2+} has been used extensively as a probe for the metal-ion function including some of the structural studies. There are two ATP binding sites on 3-phosphoglycerate kinase and these have very different affinities for the metal ion. Mn^{2+} ion is bound to the enzyme by Asp^{274} and to the α and β-phosphate groups of ADP [691]. Mg^{2+} is bound at the same site. EPR experiments with ^{17}O-substituted ADP analogues show that the results of the solid state structure are valid in solution [692]. The line broadening observed in $H_2{}^{17}O$ is consistent with two or three water molecules bound to the metal ion in the complex. NMR evidence suggests that the molecule is very mobile in solution, particularly in the active site region [693].

Structures of the enzymes from horse muscle [689, 691, 693], pig muscle with bound 3-phosphoglycerate (2.0 Å resolution) [695], and yeast (2.5 Å resolution) [696] have been reported, and the structures of the enzyme from *B. stearothermophilus* and *Thermus thermophilus* are in the process of being determined [697]. The structures of both the horse and the yeast enzymes have been determined with Mn^{2+} bound in the metal binding site. Each form of the enzyme has been shown to have two distinct domains of approximately equal size which are separated from each other by a cleft and which move relative to each other during catalysis. The extent of the hinge bending during catalysis is not known but is being actively investigated using mutant enzymes [685, 698, 699]. Studies of a putative hinge mutant His-338–Gln have shown that this change, which is remote from the active site, has a marked affect on the kinetics of the reaction [685, 686]. Other mutants have been used to define the mechanism of the reaction and to investigate the properties of the active sites [687, 688, 700, 702]. Substrate binding studies have shown that the M^{2+}ATP binding site is on a shallow depression on the inner surface of the C-terminal domain in the region above the cleft. The binding site for the 3-phosphoglycerate in the pig enzyme is on the N-terminal domain in a cluster of basic residues [697]. Earlier crystallographic and NMR studies of the location of this binding site are not in agreement with the binding site observed in the pig enzyme. Most of the crystals used in the X-ray structural studies were isolated from solutions containing high concentrations of sulfate ion. Under these conditions, sulfate is thought to occupy the 3-phosphoglycerate binding site. ^{31}P NMR studies indicate that the sulfate ions affect the structure of the enzyme MgATP complex [703, 704]. NMR and EPR studies of Mn^{2+} and Co^{2+} complexes suggest that the α and β-phosphoryl groups of ADP are bound to the metal [691, 705].

The two substrates are too far apart (>10 Å) to enable direct phosphoryl transfer to occur but in the pig enzyme binding of the 3-phosphoglycerate causes a small hinge bending motion which brings the 3-phosphoglycerate closer to the ATP binding site [695]. The metal ion is bound to the enzyme by only one protein residue and to the ATP by β- and γ-phosphate oxygen atoms. Three water molecules complete the coordination sphere, and an aspartate residue binds the metal ion by hydrogen bonding to two of these water molecules. In the ADP form of the enzyme only a terminal phosphate oxygen of ADP is bound to the metal [691].

6.11 Ribonuclease H (EC 3.1.27.5)

Ribonuclease H is found in a variety of organisms where it hydrolyzes RNA moieties to oligonucleotides only when they are bound to a complementary DNA strand. In *E. coli*, the enzyme consists of a single polypeptide chain of 155 amino acids. HIV reverse transcriptase is a DNA polymerase which can employ either DNA or RNA as a template yielding either a RNA/ DNA hybrid or a duplex DNA product. The HIV reverse transcriptase is produced initially as a polypeptide product (M_r = 66,000) which has both polymerase and a ribonuclease H domain. Subsequent proteolytic cleavage of a homodimer of the subunits with M_r = 66,000 removes the ribonuclease H domain from one subunit leaving a heterodimer with subunits of M_r = 60,000 and 51,000. The heterodimer has one polymerase active site, one Ribonuclease H active site, and one tRNA binding site. Site-directed mutations on viral and *E. coli* ribonuclease H enzymes have shown that Asp-10, Glu-48, Asp-70, and Asp-134 (*E. coli* numbering) are invariant in 26 sequences and are crucial for activity [706, 707].

Jou and Cowan [708] have shown that the inert complex ions $Co(NH_3)_6^{3+}$ and $Co(en)_3^{3+}$ show significant catalytic activity and therefore coordination of the metal to the enzyme is not essential for activity. The metal ion(s) is therefore probably acting as either electrostatic relief for the four carboxylate residues, binding the substrate via hydrogen bonds, or playing a structural role or promoting nucleophilic or base catalysis via an outer sphere mechanism.

Structures of the viral heterodimer at 3.5 Å resolution [709] and the ribonuclease H domain of this dimer at 2.4 Å resolution have been reported [710]. The coordinates of the ribonuclease H domain fit the electron density of the reverse transcriptase very well. Ribonuclease H apo enzyme from *E. coli* at 2.0 Å resolution and the enzyme with bound Ca^{2+}, Mg^{2+}, Ba^{2+}, and Co^{2+} at 1.8 Å resolution [711, 712] have also been structurally characterized. Studies of the enzyme from *Thermus thermophilus* [713] have begun. In the *E. coli* enzyme the metal ion is bound in a depression on the enzyme surface, surrounded by the four invariant carboxylates, Asp-10, Glu-48, Asp-70, and Asp-134, but the last three residues are too far from the metal to be covalently bound to it. Mg^{2+} is bound to the enzyme via the carboxylate of Asp-10 and the carbonyl oxygen of Gly-11. The metal binding to the HIV-1 ribonuclease H was studied using crystals soaked in $MnCl_2$. This structure revealed *two* tightly bound Mn^{2+} ions in the same metal binding site as in the *E. coli* enzyme. The two metal ions are also present in the HIV reverse transcriptase structure. This apparent discrepancy has been explained as a result of crystal packing; in the *E. coli* molecule a lysine side chain from a symmetry-related molecule probably prevents binding of the second metal ion. Arguments against two metal ions in the active site of the *E. coli* enzyme have been advanced [712]. The two Mn^{2+} ions are 4 Å apart with a single carboxylate bridging the metal ions. One metal is bound to two additional carboxylates and the second metal is bound to the fourth carboxylate. The dinuclear metal binding site is thus very similar to the binding site in Klenow fragment of DNA polymerase I (see Section 4.3) [581, 710].

6.12 Ribonucleotide Reductase (EC 1.17.4.1)

Reduction of the four common ribonucleotide diphosphates to the corresponding deoxyribonucleotide diphosphates is accomplished by a variety of enzymes [714–717]. The different forms of the enzyme include a 5′-deoxyadenosylcobalamine-dependent enzyme found in Lactobacillaceae and other predominantly gram-positive bacteria, and a non-heme iron enzyme in *E. coli* and many eukaryotic cells. A manganese-containing enzyme is probably present in a variety of organisms including *Brevibacterium ammoniagenes, Micrococcus leuteus* [718], *Arthrobacter citreus, A. globiformis*, and *A. oxydans* [719]. A fourth form of the enzyme has been reported recently [720].

The best characterized of these molecules is the non-heme iron enzyme from *E. coli* which consists of two nonidentical subunits (R_1 and R_2). R_1 contains two binding sites for the ribonucleotides and the redox active thiols that participate in the reduction of the sugar. The R_2 subunit which consists of two identical polypeptide chains, each of which contains two Fe^{3+} ions and a stable tyrosine radical [721]. The enzyme has been structurally characterized by X-ray crystallography [722] and an EXFAS study of the active site [723]. The EXAFS study showed a short Fe–O bond (1.78 Å) and a Fe–Fe distance of 3.22 Å; the other Fe–ligand distances average 2.06 Å. The bonding arrangement is shown schematically in Figure 13. The structure and EPR spectroscopic properties of the Mn^{2+} substituted form of the R_2 subunit has been reported [724]. The Mn^{2+} form of the enzyme has been suggested as an excellent model for the Fe^{2+} form of the enzyme for which structural data is not yet available. A comparison of the metal-containing sites of the iron and manganese-substituted enzymes is shown in Figure 13. The same amino acids are bound to the metal ions in both forms of the enzyme, but there are significant differences in the coordination spheres. Both Mn^{2+} ions are five-coordinate with

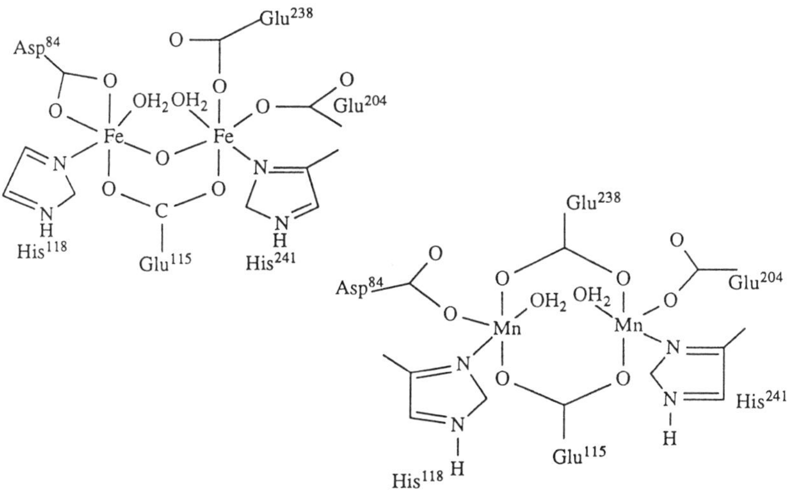

Figure 13. Schematic representations of the diiron and dimanganese centers in *E. coli* ribonucleotide reductase.

trigonal bipyramidal geometry, whereas the Fe^{3+} ions are octahedrally coordinated. There is no evidence for a bridging oxo group in the manganese form of the enzyme. The Mn^{2+}–Mn^{2+} distance is 3.6 Å which is 0.3 Å longer than the distance between the irons in the Fe^{3+} form of the enzyme. The coupling between the manganese centers as revealed by EPR is very weak. Multifield saturation magnetization studies on the oxidized and reduced iron-containing enzymes have been reported [725] and it is suggested that the oxo-bridge is absent in the Fe(II) form of the enzyme with the metal ions linked by two carboxylate residues.

It has been reported that the calf thymus ribonucleotide reductase can be activated by Mn^{2+} *in vitro* [726], but the results reported by Atta et al. [724] for the *E. coli* enzyme suggest that Mn^{2+} is a potent inhibitor of this enzyme.

A manganese requirement for growth and DNA production has been demonstrated in coryneform bacteria [718, 719, 727] and this has been demonstrated to be due to the presence of a manganese-dependent ribonucleotide reductase. Other metals cannot substitute for Mn^{2+} *in vivo*. The enzyme which has been isolated in only small amounts from *Brevibacterium ammoniagenes* has two subunits, a nucleotide binding protein, B1 (M_r = 80,000), and a catalytic protein, B2 (M_r = 100,000) which is apparently a homo dimer. Manganese is bound to the B2 protein, and both B1 and B2 are essential for catalytic activity. The visible spectrum of the protein, a broad absorption in the 350–500 nm region with peaks at 455 and 485 nm is similar to that of *L. plantarum* catalase and to the spectra of a number of dinuclear manganese complexes with bridging oxo and carboxylate groups [5, 728]. This suggests that there may be some similarities to the metal binding site of the *E.*

coli enzyme; however a tyrosine radical has not been detected in the manganese ribonucleotide reductase.

6.13 D-Xylose Isomerase (EC 5.3.1.5)

Sugar isomerases are metal-dependent enzymes [729] and a number of these enzymes display an absolute requirement for Mn^{2+} (see Table 1). D-Xylose isomerase (glucose) is the best characterized of these enzymes. It is an intracellular bacterial enzyme which is widely distributed among Actinomycetaceae and also in Escherichia, Bacillus, Lactobacillus, Clostridium, Thermoanaerobacter, and Thermus aquaticus species. D-Xylose isomerase has considerable commercial importance. Its biological role is to catalyze the isomerization of D-xylose to D-xylulose (Scheme 6) but it also catalyzes the isomerization of D-glucose to D-fructose. Using this latter reactivity, 3 billion kilograms of glucose is converted annually to fructose for use as high-fructose corn syrup in the soft drink industry. D-Xylose isomerase is also used for ethanol production in a process by which xylose is converted to xylulose which is then fermented to alcohol.

Xylose isomerase consists of four identical subunits (M_r = ~173,000) and it has two divalent ion binding sites per subunit [730, 731]. The metal ion requirement (Mg^{2+}, Co^{2+}, and Mn^{2+} all work) is dependent on the organism and the reaction being catalyzed, but both sites must be occupied for activity. For example, in the enzyme from *Thermus aquaticus HB8* Mn^{2+} is more effective at catalyzing the D-xylose isomerase reaction, while Co^{2+} is the preferred ion for the D-glucose reaction [732]. In the soft drink industry the Mg^{2+} form of the enzyme must be used. Metal ion binding decreases in the order $Mn^{2+} > Co^{2+} >> Mg^{2+}$. The dissociation constants for both the metal binding sites may be the same, or may differ by orders of magnitude depending on the species. Binding at one site is independent of metal ion binding at the other site [733]. Kinetic studies of the effects of different metal ions on the reactions of the enzymes have revealed significant differences in the pH of metal binding [734].

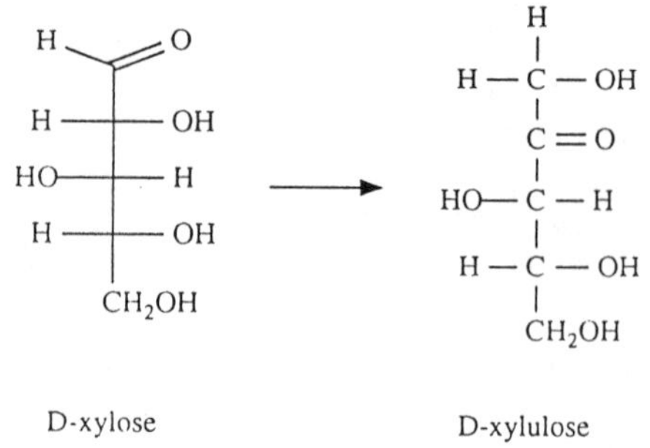

H O
H OH
HO H
H OH
CH_2OH

D-xylose

H
H — C — OH
C = O
HO— C — H
H — C — OH
CH_2OH

D-xylulose

Scheme 6.

Site-directed mutagenic studies have shown that two histidines, His-101 and His-271 (*E. coli* numbering), are essential for enzymatic activity [735]. His-271 is a ligand for one of the metal ions and it was initially suggested that His-101 acted as a catalytic base mediating the reaction [735]. This is not now believed to be the case as the reaction probably proceeds via a hydride transfer mechanism [729, 736, 737]. His-101 is now thought to be important in governing anomeric specificity [738].

The metal ion specificity and pH dependence of the activity are of considerable practical importance. Consequently structural determinations have been reported for xylose isomerase from a number of different organisms, with native and mutant forms of the enzyme, with and without metal ions. The organisms studied include the Mg^{2+}-substituted form of *Arthrobacter strain* B3728 with bound xylitol and D-sorbitol at 2.3 Å resolution [737, 739], more than 20 different mutant and metal-substituted forms of the enzyme from *Actionplanes missouriensis* [740–744], *Streptomyces albus* at 1.65 Å resolution [745, 746], Mg^{2+}-and Mn^{2+}-substituted enzyme from *Streptomyces rubiginosus* complexed with xylose and xylitol at 1.6 Å resolution [736, 747], Mn^{2+}-, Co^{2+}-, Mg^{2+}-, and glucose-substituted enzyme from *Streptomyces olivochromogenes* at 3.0 Å resolution [729, 731], and *Streptomyces violaceoruber* [732].

Three-dimensional structures of all these xylose isomerases are very similar, and the C_α backbone of xylose isomerases from *S. rubiginosus* and *Arthrobacter* have been compared in detail [749]. The molecule has two domains: the N-terminal domain (~300 amino acids) which is an eight-stranded α/β barrel, and the C-terminal domain (50 to 70 amino acids) which is a large loop with little secondary structure. Monomers form tightly coupled dimers by extensive contacts of the core of the first monomer with the C-terminal tail of the second monomer and vice versa. A pair of dimers interact to form the tetramer. The metal ion site is located at the carboxyl terminal end of the β barrel and it is close to the axis which relates the two weakly interacting dimers. The opening to the active site contains ordered solvent molecules, although the entrance is lined with hydrophobic side chains. The active site is an anphipathic pocket with hydrophobic residues on one side and hydrophilic residues on the other side. The latter is involved in binding the divalent metal ions and the sugar substrate.

In the discussion to follow the two different metal ion binding sites will be called site 1 (also called the catalytic site) and site 2 (the structural site), with the sites being 4.9 Å apart. In the dimanganese form of the enzyme the donor atoms of site 2 consist of five protein side chains—Glu-217, Asp-255(bidentate), Asp-257, His-220, and a water molecule. Glu-217 acts as a bridging ligand to the metal at site 1 which is also bound to side chains of Glu-181, Asp-245, Asp-292, and two water molecules. Mg^{2+} in site 1 is five-coordinate. There has been disagreement over which site has the highest affinity for metal ions. It was initially proposed that site 1 was the high-affinity site, but at least for the metal ions VO^{2+}, Co^{2+}, Cd^{2+}, and Pb^{2+} it is probable that site 2 is the site with the highest metal-ion affinity.

Spectroscopic studies on the Co^{2+}-substituted form of the enzyme from *Streptomyces rubiginous* suggest that site 2 is octahedral and that in the absence of substrate site 1 is either tetrahedral or pentacoordinate. Site 1 is octahedral in the presence of substrate [749, 750]. This change in the geometry of the site 1 site upon binding of substrate or inhibitor has been observed in a number of crystal structures, but it appears to be metal-ion dependent. Both metal ions are involved in the binding of D-xylose and the competitive inhibitor xylitol to the protein and are believed to be intimately involved in the catalytic mechanism [741]. The position of the M^{2+} bound at site 1 changes considerably upon sugar binding and a number of different binding sites for this metal ion have been identified [744]. The M^{2+} ion moves approximately 1.0 Å closer to the sugar from its position in the apoenzyme. Bond lengths to the metal ion increase significantly and the coordination number and the groups bound to the metal ion change. There is virtually no change to the coordination of the site 2 metal ion upon sugar binding [735]. A view of the metal-ion binding site is shown in Figure 14.

Binding of the sugar to the metal ion may be different in different forms of the enzyme. In the *Streptomyces* structure [747] xylitol is bound to the metal ions via the O3 and O5 atoms of the sugar, whereas in most other structural determinations the binding to the metal takes place via the O2 and O4 atoms. Collyer et al. [738] have reported the structures of the Mg^{2+} and Mn^{2+} forms of the enzyme from *Arthrobacter strain B3728* with the putative substrates 1-deoxynojirimycin and 2,5-dideoxy-2,5-imino-D-glucinol. The Mg^{2+} ion binds to the sugar via O3 and O4, whereas Mn^{2+} binds the sugar via the imino group and O6. Site-directed mutagenesis of Glu-186, a strictly conserved residue which is close to the metal binding site but not to a metal ligand, showed that the Glu-186–Asp and Glu-186–Gln mutants have low activity with Mg^{2+} as the activating cation, but close to wild-type activity

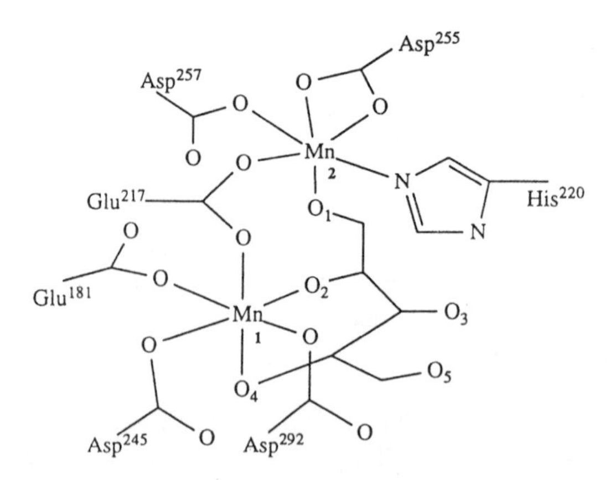

Figure 14. Schematic diagram of xylitol bound to the dimanganese form of xylose isomerase. Manganese atoms at sites 1 and 2 are labeled Mn_1 and Mn_2, respectively.

with Mn^{2+} at neutral pH. At more acidic pHs the Glu-186–Gln mutant has a higher activity than the wild-type enzyme [742]. X-ray structures of these mutant enzymes showed that with Mn^{2+} both metal ion sites are occupied, but with Mg^{2+} only site 1 is occupied. The amide of Gln-186 is hydrogen bonded to the Asp-255 residue, a metal ligand to the metal at site 2. Site-directed mutagenesis studies of most of the active site residues have been reported [742, 743]. Most of the mutations affect the substrate binding and not the catalysis; however, Lys-183, which is hydrogen bonded to O1 of the substrate sugar, is important for catalysis.

7. PROTEINS WITH THREE BOUND METALS

7.1 Alkaline Phosphatase (EC 3.1.3.1)

Alkaline phosphatases are typically dimers of ~94 kDa/subunit and they hydrolyze phosphate monoesters at an optimum pH of ~8. Manganese-substituted forms of this enzyme have been studied by a number of groups [751–755] but the best characterized enzyme is from *E. coli* which contains two Zn^{2+} and one Mg^{2+} ion in the active site [753]. On the basis of EPR and ^{31}P NMR measurements, four Mn^{2+} ions (two per dimer) bind to the enzyme; the first pair added occupy catalytic sites in what was thought to be a tetrahedral environment, and the second pair of Mn^{2+} ions occupy a so-called structural site [752].

The structure of the active site of the $2Zn^{2+}/Mg^{2+}$ enzyme is shown in Figure 15. One zinc ion (M1) is bound to three histidines, a bidentate aspartate, and a phosphate oxygen. The second zinc (M2) is four-coordinate with a histidine, two monodentate aspartates, and a phosphate oxygen in its coordination sphere. The Mg^{2+} (M3) has two aspartates, one threonine, and a glutamate in its coordination

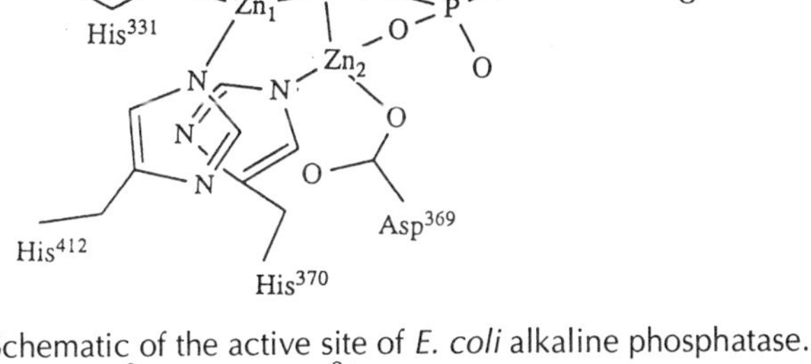

Figure 15. Schematic of the active site of *E. coli* alkaline phosphatase. The Mg^{2+} is 5–7 Å from the two Zn^{2+} ions. The Zn^{2+} ions which are bridged by a phosphate group are 4 Å apart.

sphere [753, 756]. Asp-51 serves as a bridge between the M1 and M3 site. The metal distances are 3.9 Å for M1–M2, 5 Å for M2–M3 and 7 Å for M1–M3. A proposed mechanism of the hydrolysis reaction has been discussed by Kim [753]. EXFAS studies have suggested that the Zn^{2+} in site 1 is five or six coordinate with N or O ligands at an average distance of 2.04(2) Å [757].

7.2 Enolase (EC 4.2.1.11)

Enolase (phosphoenolpyruvate hydratase) catalyzes the reversible dehydration of D-(+)-2-phosphoglycerate (PGA) to phosphoenolpyruvate (see Scheme 7). Enolase from yeast has been shown to be a dimer of identical subunits ($M_r = 93{,}000$) and to exhibit an absolute requirement for divalent metal ions. Mg^{2+} gives the highest activity, but eight different metal ions have been shown to be effective. Mn^{2+} exhibits about 40% of the activity of Mg^{2+} [758]. The role of the metal ions in the binding of substrates and in the catalytic process has been the subject of considerable controversy. Enolase was long thought to bind two metals per monomeric unit, but recent work using Mn^{2+}-substituted enolase and EPR and NMR has shown that each monomer of the apoenzyme has a single metal binding site. However in the presence of substrate, metal ions can bind at three thermodynamically distinct sites (I, II and III) on each monomer [759]. These sites have respectively, the highest to lowest affinity for manganese.

Much of the earlier confusion in the literature is now resolved because it has been shown that Mg^{2+} binds preferentially at site II and Mn^{2+}, Zn^{2+}, and Cu^{2+} bind preferentially at site I [759, 760]. The site I metal ion must be bound before substrate binding can occur. Sites I and II are close together as indicated by spin–spin coupling of the Mn^{2+} ions in the complex with ^{17}O-labeled forms of the inhibitor phosphonoacetohydroxamate, and with two Mn^{2+} ions per enolase monomer [761]. The Mn–Mn distance is 3.2 to 3.8 Å. This is not consistent with a number of earlier studies, but the latest evidence is convincing. Studies with Mn^{2+} in site I and Mg^{2+} in site II reveal that the hydroxamate oxygen is bound to the Mn^{2+} ion. The Mn^{2+} at site I loses a coordinated water molecule when the inhibitor binds. The anisotropic EPR spectrum of this form of the enzyme is consistent with a five-coordinate geometry at this site. With Mg^{2+} in site I and Mn^{2+} in site II the phosphonate oxygen

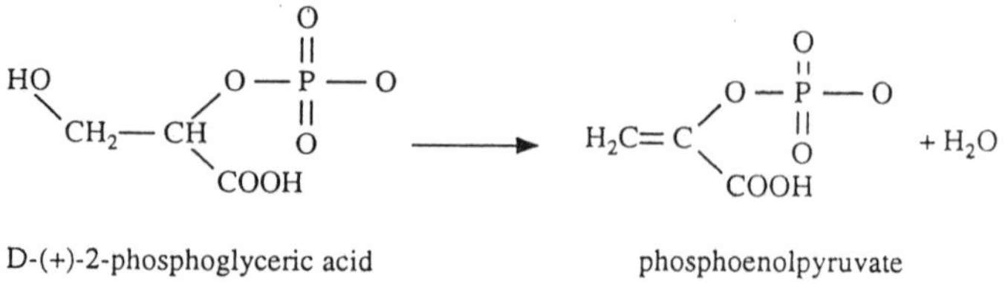

$$HO{-}CH_2{-}CH(COOH){-}O{-}P(=O)_2{-}O \longrightarrow H_2C{=}C(COOH){-}O{-}P(=O)_2{-}O + H_2O$$

D-(+)-2-phosphoglyceric acid phosphoenolpyruvate

Scheme 7.

is bound to the Mn^{2+}. The carbonyl oxygen of the inhibitor is believed to act as the bridge between the two metal ions. EPR data suggest Mn^{2+} at site II is six-coordinate.

In the presence of D-2-phosphoglycerate (PGA) and Mg^{2+}, only a single Mn^{2+} binding site is observed. Mn^{2+} bound in site 1 is bound via the hydroxymethyl group at C3 [760, 762]. Binding to sites I and II is insensitive to pH over the range 5.0–7.5, which suggests that the donor groups to the metal are probably oxygen donors. Binding at site III which is inhibitory, decreases with decreasing pH and it has been suggested that the binding site contains a histidine residue [760]. The binding to site I is tighter by a factor of ~3 in the presence of the substrate. NMR studies have shown that Mn^{2+} bound at site I has two rapidly exchanging water molecules [760]. Water proton relaxation rate studies indicate that with Mg^{2+} occupying site II the distance between Mn^{2+} and the protons of the enzyme inhibitor phosphoglycolate is 5.73 Å. In the absence of Mg^{2+} at site II these distances are slightly longer. These studies suggest that site I is the catalytic site, site II plays a role in influencing the conformation of the substrate, and that site III is inhibitory [763].

The crystal structure of the enzyme in five different forms (apoenzyme, enolase–Mg^{2+}–2-phosphoglycerate, enolase–Ca^{2+}–2-phosphoglycerate, enolase–M^{2+}–phosphonoaceto hydroxamate, and enolase–Zn^{2+}–2-phosphoglycerate) have been reported at 2.2 Å or better resolution. All of these crystals have been grown at acidic pH and in the presence of high concentrations of $(NH_4)_2SO_4$. The polypeptide chain folds into three domains: (1) a N-terminal domain which is wrapped around the (2) main domain which is an α,β-barrel, and (3) a short C-terminal domain. In every structure reported to date only one metal ion is observed bound to the protein [761, 764–778]. This is presumably site I and is in a deep cavity at the carboxylate end of the β-barrel. Zn^{2+} in site I is in an almost regular trigonal bipyramidal coordination with two aspartate carboxylates in axial sites and a glutamate carboxylate and two water molecules in the trigonal plane [768]. One of the water molecules bound to the zinc is replaced by a coordinating group of the substrates PEP or PGA. Nonactivating metals Ca^{2+} and Mg^{2+} have octahedral coordination, the additional site being occupied by another water molecule [764, 766]. Sm^{3+} coordinated to the metal binding site is seven-coordinate with four water molecules. It has been suggested that the Mn^{2+} form may be six-coordinate [769] but the NMR evidence suggests two water molecules are bound to the Mn^{2+} ion so that a five-coordinate structure is likely. If the physiological cofactor Mg^{2+} is bound to enolase then the binding of substrate/product is accompanied by large movements of the two loops of the peptide backbone [766].

One of the heavy metal derivatives has four $PtCl_4^{2-}$ ions bound to the enzyme and it has been suggested that one of these binding sites is the inhibitory site III. The distance from site I is 18 Å which is in agreement with the distance determined by energy transfer measurements on the derivative with Tb^{3+} in site I and Co^{2+} in site III.

7.3 Parvalbumin

Parvalbumins are acidic Ca^{2+} and Mg^{2+} binding proteins of low molecular weight which are found in relatively high amounts in the muscles of vertebrates as well as nonmuscle tissue. It has been proposed that the proteins are involved in the relaxation process of fast muscles. They have a high affinity for Ca^{2+} and a lesser affinity for Mg^{2+}, but under physiological conditions where $[Mg^{2+}] >> [Ca^{2+}]$ both metals are probably bound. The crystal structures of a number of parvalbumins have been determined [770, 771] including a form of pike pI 4.10 parvalbumin with three bound manganese ions, $PaMn_3$ [772]. Comparison of this form of the protein with two Ca^{2+} ions and a NH_4^+ ion, $PaCa_2NH_4$, two Ca^{2+} ions, and a Mg^{2+} ion ($PaCa_2Mg$) and two Mg^{2+} and a Ca^{2+} ions ($PaCaMg_2$) bound has revealed some interesting similarities and differences. At the Ca^{2+} binding site which is present in all forms of the protein, except the $PaMn_3$ form, Ca^{2+} is bound to seven oxygen donors. A glutamate acts as a bidentate ligand and two aspartates, a serine, phenylalanine carbonyl oxygen, and a second glutamate acting as monodentate ligands complete the coordination sphere. The Mn^{2+} which occupies this site in $PaMn_3$ is bound to the same amino acids except that the glutamate which is bidentate in the Ca^{2+} form is a monodentate ligand to the six-coordinate Mn^{2+} ion. The average metal–oxygen bond length of the Mn^{2+} ion (2.22 ± 0.1 Å) is shorter than the average Ca^{2+}–O bond length (2.36 ± 0.1 Å). At the second Ca^{2+} site the Ca^{2+} ion is also seven-coordinate, bound to three aspartate carboxylate groups: a bidentate glutamate carboxylate, a methionine peptide oxygen, and a water molecule. The Mn^{2+} and Mg^{2+} ions bound at this site are six-coordinate; again the reduced coordination number is due to the glutamate acting as a monodentate ligand. The average bond lengths are 2.33 ± 0.1 Å and 2.35 ± 0.12 Å (Ca^{2+}), 2.12 ± 0.1 Å (Mn^{2+}), and 2.11 ± 0.1 Å (Mg^{2+}). The Mn^{2+} ion replaces the Ca^{2+} ion with only a minimal rearrangement of the amino acid side chains, while considerable rearrangement is necessary to accommodate the Mg^{2+} ion.

At the third cation binding site, the ammonium ion is hydrogen bonded to three carboxylate residues and three water molecules. The manganese is bound as a tetrahedral $Mn(H_2O)_4^{2+}$ ion which in turn is hydrogen bonded to the same carboxylate groups as the ammonium ion. In both the Mg^{2+} loaded forms of the protein the Mg^{2+} at this site is octahedrally bound by five water molecules and a carboxylate residue, Asp-53 [772].

7.4 Pyruvate Kinase (EC 2.7.1.40)

The structure and function of pyruvate kinase has been reviewed [773]. Pyruvate kinase has been isolated from many organisms and whether derived from bacterial, plant, or animal sources it consists of four identical subunits with about 500 amino acid residues. It catalyzes the irreversible conversion of phosphoenolpyruvate into

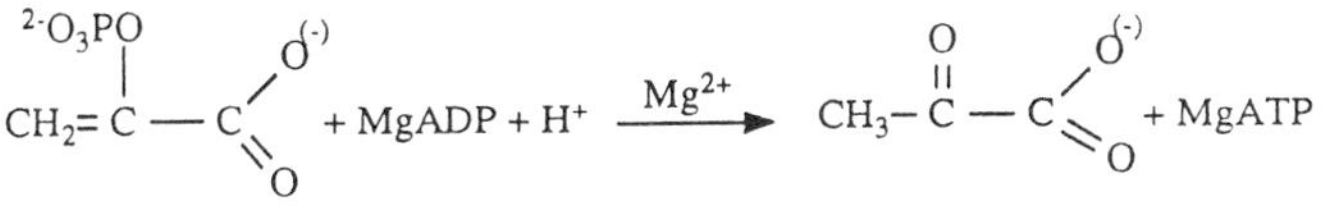

$$^{2-}O_3PO-C(=CH_2)-C(=O)O^{\delta(-)} + MgADP + H^+ \xrightarrow{Mg^{2+}} CH_3-C(=O)-C(=O)O^{\delta(-)} + MgATP$$

Scheme 8.

pyruvate by the addition of a proton and the loss of a phosphate group to ADP (Scheme 8).

$$\text{Phosphoenolpyruvate} + \text{ADP} + \text{H}^+ \rightarrow \text{pyruvate} + \text{ATP}$$

The reaction occurs in two stages: (i) a β-phosphoryl oxygen of ADP attacks the phosphoenolpyruvate phosphorus atom thereby displacing the enolpyruvate and forming ATP, and (ii) the enolpyruvate is converted to pyruvate. The enzyme also catalyzes the metal-ion-dependent decarboxylation oxalacetate [774].

The enzyme has a low nucleotide specificity; guanosine, inosine, uridine, and cytidine diphosphates can all accept the phospho group. Pyruvate kinase differs from other kinases in requiring a divalent metal ion and a monovalent metal ion (K^+), in addition to a metal nucleotide complex. A number of divalent metal ions will activate the enzyme although it is likely that Mg^{2+} is the most common substrate [775]. Synergistic activation of pyruvate kinase by a mixture of Mn^{2+} and Mg^{2+} has been interpreted as due to the enzyme preferring MgATP at the nucleotide binding site, but Mn^{2+} at the metal ion binding site. It has been suggested that in intertidal mollusca a change in metal cofactor from Mg^{2+} to Mn^{2+} may be important in dealing with fluctuations of the organisms pH. Under aerobic conditions (high tide) Mg^{2+} is the best activator, but at low tide the mollusca keep their shells tightly shut and the conditions become anoxic and the pH is decreased. Under these conditions Mn^{2+}-activated pyruvate kinase is the active form [776–778].

Different isoenzymic forms of the enzyme are isolated from mammals depending on the tissue involved. The M1 form is found in skeletal muscle; M2 in lung, adipose tissue, and kidney; L in liver; and R in red blood cells. The bacterial form of the enzyme most closely resembles the M2 mammalian form. One gene encodes the M1 and M2 form of the enzyme but there are two different mRNAs [779]. Similarly the R and L forms of the enzyme have one gene and two mRNAs [780].

EPR and the ESEEM spectra of the Mn^{2+}-substituted enzyme have been the subject of a number of investigations [775, 781–785]. These studies have lead to the suggestion that the ATP-bound metal ion is coordinated to the α, β and γ-phosphate oxygens of ATP, and that the ATP is a bridging ligand via the γ-phosphate to the enzyme-bound metal ion [785]. Oxalate, a structural analogue of pyruvate, is coordinated to the enzyme bound Mn^{2+} as a bidentate ligand, and a single water molecule is also bound to this metal ion [783]. NMR studies with phosphoenolpyruvate and phosphoenolpyruvate analogues suggested that the phos-

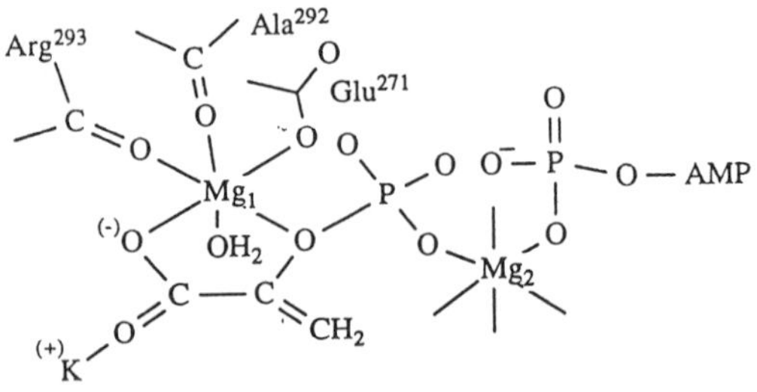

Figure 16. Schematic representation of the metal binding site of pyruvate kinase.

phoenolpyruvate is 5 Å from the metal ion and is therefore not coordinated. This conclusion may not be correct as the NMR results are dependent on the assumption that the rate of ligand exchange is faster than $1/T_1$, which is a difficult condition to satisfy as the ligand approaches the metal ion.

The structure of an M-type enzyme from cat muscle at 2.6 Å resolution shows that Glu-271 and carbonyl oxygens from Ala-292 and Arg-293 coordinate the divalent metal ion in the substrate free enzyme; the remaining three coordination sites are occupied by water molecules. The MgADP complex binds to the enzyme about 3 Å from the enzyme-bound Mg^{2+} ion and phosphoenolpyruvate is bound to the latter Mg^{2+} ion (Figure 16). These results are in good agreement with the spectroscopic conclusions. There is no direct crystallographic evidence for the binding site of the monovalent cation, although it too is thought to be close to the enzyme-bound Mg^{2+} [786]. There are, however, problems with this structure due to the high concentration of $(NH_4)_2SO_4$ in the crystals, thus structural studies of crystals of the enzymes from rabbit muscle complexed with Mn^{2+} and either pyruvate or oxalate, and also from yeast, have been initiated [787, 788].

8. PROTEINS WITH FOUR BOUND METALS

8.1 Inorganic Pyrophosphatase (EC 3.6.1.1)

Inorganic pyrophosphatase is an abundant and widely distributed enzyme in living organisms which catalyzes specifically the hydrolysis of pyrophosphate to orthophosphate. This reaction provides the driving force for protein, DNA, and RNA synthesis. There are two principal types of pyrophosphatases—membrane bound and soluble forms. The membrane bound forms of the enzyme also act as a proton pump [789]. The yeast enzymes are well characterized: one is cytoplasmic having a homodimeric arrangement with M_r = 32,000–35,000; the other form of the enzyme is mitochondrial and is energy-linked. Most bacterial inorganic phosphatases have four or six identical subunits with M_r between 19,000 and 21,000. Comparisons of the sequence of enzymes from bacteria and yeast reveals only a slight degree of homology (see Table 9). The yeast mitochondrial and cytoplasmic enzymes have only 49% sequence identity [796]. The functional residues which bind the metal ions and the pyrophosphate are conserved in all inorganic pyrophosphatases sequenced to date (Table 9) [800].

The number of metal ions required for catalysis has been the subject of studies using different metal ions, and conflicting results were obtained. It is now generally agreed that the *E. coli* enzyme can bind four metal ions [801], two with high affinity, a third with low affinity, and a fourth which binds the substrate for the enzyme, the $MP_2O_7^{2-}$ complex [802]. Mg^{2+} is the native metal, but Mn^{2+}, Co^{2+}, and Zn^{2+} are the active species. Three of the metal ions are absolutely necessary for catalytic activity with the yeast enzyme, and all four are necessary with the *E. coli* enzyme [802, 803]. These metal binding sites are close to each other and close to the $MP_2O_7^{2-}$ binding site [804–807]. Use of complexes such as $Cr(H_2O)_4(P_2O_7)^-$ and $Rh(H_2O)_4$(methylenediphosphonate) have shown that the $M(P_2O_7)^{2-}$ complex binding precedes the binding of the second (or third) metal ion to the enzyme. NMR investigations have revealed two distinct PO_4^{3-} binding sites, and the Mn–^{31}P distance from the tightly bound manganese was calculated to be 6.2 Å [801, 808]. NMR and EPR studies established that two Mn^{2+} ions bind to the free enzyme close enough that they interact magnetically. A distance of 10–14 Å was estimated for this metal–metal distance. In the presence of $[Co(NH_3)_4(P_2O_7)]^-$ the magnetic interactions of the two manganese ions increases; this was attributed to a decrease in the Mn–Mn separation to 7–9 Å. Using $[Cr(NH_3)_4(P_2O_7)]^-$ and $[Cr(H_2O)_4(P_2O_7)]^-$ as substrates, an estimate of the Mn–Cr distance of 5 Å was obtained. These distances are in reasonable agreement with the X-ray structural studies discussed below [805].

There may well be structural differences in the different metal forms of the enzyme since different metals have different specificities. The Mg^{2+}-activated form is specific for pyrophosphate, while the zinc form will hydrolyze a variety of di- and triphosphates as well as $P_2O_7^{4-}$. Reaction of the Mn^{2+}, Co^{2+}, and Zn^{2+} forms of the enzyme are slower than the reaction of the Mg^{2+} form because of slower release of PO_4^{3-} from the enzyme [809]. Chemical modification studies have identified 17 polar residues as essential for binding either the metal ions, or as residues which interact with the pyrophosphate. Using *E. coli* numbering, Tyr-55, Tyr-141, Glu-98, and Lys-104 are important for structural integrity, while Asp-97, Asp-102, Lys-104, His-118, and Arg-77 are important in the catalytic activity of the enzyme [810–815].

The three-dimensional structure of inorganic pyrophosphatase from *S. cerevisiae* has been the subject of a number of different studies. With both the apoenzyme and with Mg^{2+} present the best resolution reported to date is 3.0 Å [816, 817]. A recent more accurate determination of the structure with Mn^{2+} and either two phosphates or a bound pyrophosphate has been reported at 2.35 Å resolution [818]. A structural study of the *E. coli* enzyme is apparently underway. The nature of the active site of the enzyme is clear, it is a hollow about 25 Å in diameter and about 12 Å deep. The metal ions and the phosphates are at the bottom of this hollow and a representation

Table 9. Comparison of the Amino Acid Sequences of Inorganic Pyrophosphatase[a,b,c,d,e]

1	1		10		20		30		40		50
										↓	
1	MSEYTT	REVGA	LNTLD	YQVY-	VEKNG	TPISS	WHDIP	LYANA	EKTIL	NMVVE	IPRWT
2	MS-YTT	RQVGA	KNSLD	YKVY-	IEKDG	KPISA	FHDIP	LYADE	ANGIF	NMVVE	IPRWT
3	SLLNVP	AGKDL	PEDIY	-VVIE	IPANA						
4	AFENKIV	EAFIE	IPTGS								
5	MTYTT	RQIGA	KNTLE	YKVY-	IEKDG	KPVSA	FHDIP	LYADK	ENNIF	NMVVE	IPRWT
6	HRQFST	IQQGS	KYTLG	FKKYL	TLLNG	EVGSF	FHDVP	LDLNE	HEKTV	NMIVE	VPRWT
7	MESFY	HSVPV	GPKPP	—EE	–VYV	-IIVE	IPRGS				
8	RNPNV	TLNER	NFAAF	THRSA	AAHPW	HDLEI	GPEAP	TVFNC	–AVE	ISKGG	
9	-SSFS	SEERA	APFTLE	YRVFL	KNEKG	QYISP	FHDIP	IYADK	–VFH	-MVVE	VPRWS
	↕	↓				↕		*	*		
1	QA-KL	EITKE	ATLNP	IKQDT	KKGKL	RFVRN	CFPHH	GYIWN	YGAFP	QTYED	PNVVH
2	NA-KL	EITKE	EPLNP	IIQDT	KKGKL	RFVRN	CFPHH	GYIHN	YGAFP	QTWED	PNESH
3	DPIKY	EIDKE	S–GA	LFVD-	—	RFMST	AM—	FYPCN	YGYIN	HTLS-	—
4	QN-KY	EFDKE	R—GI	FKLD-	—	RVLYS	PM—	FYPAE	YGYLQ	NTLA-	—
5	NA-KL	EITKE	ETLNP	IIQDT	KKGKL	RFVRN	CFPHH	GYIHN	YGAFP	QTWED	PNVSH
6	TG-KF	EISKE	LRFNP	IVQDT	KNGKL	RFVNN	IFPYH	GYIHN	YGAIP	QTWED	PTIEH
7	RV-KY	EIAKD	F–PG	MLVD-	—	RVLYS	SV—	VYPVD	YGLIP	RTLY-	—
8	KV-KY	ELDKN	S–GL	IKVD-	—	RVLYS	SI—	VYPHN	YGFIP	RTIC-	—
9	NA-KM	EIATK	DPLNP	IKQDV	KKGKL	RYVAN	LFPYK	GYIWN	YGIAP	QTWED	PGHND
		↓ ↓	↓					*	* * ↓	↕	
1	PETKA	KGDSD	PLDVC	EIGEA	RGYTG	QVKQV	KVLGV	MALLD	E-GETD	WKVIV	IDVND
2	PETKA	VGDND	PLDVL	EIGEQ	VAYTG	QVKQV	KVLGV	MALLD	E-GETD	WKVIA	IDIND
3	—	-LDGD	PVDVL	VPTPY	PLQPG	SVIRC	RPVGV	LKMTD	EAGE-D	AKLVA	VP-HS
4	—	-LDGD	PLDIL	VITTN	PPFPG	CVIDT	RVIGY	LNMVD	S-GEED	AKLIG	VP-VE
5	PETKA	VGDND	PIDVL	EIGET	IAYTG	QVKQV	KALGI	MALLD	E-GETD	WKVIA	IDIND
6	KLGKCDVAL	KGDND	PLDCC	EIGSD	VLEMG	SIKKV	KVLGS	LALID	D-GELD	WKVIV	IDVND
7	—	-YDGD	PMDVM	VLISQ	PTFPG	AIMKV	RPIGM	MKMVD	Q-GETD	NKILA	VFDKD
8	—	-EDSD	PMDVL	VLMQE	PVLTG	SFLRA	RAIGL	MPMID	Q-GEKD	DKIIA	VCADD
9	KHTGC	CGDND	PIDVC	EIGSK	VCARG	EIIRV	KVLGI	LAMID	E-GETD	WKVIA	INVED

						↕	↕				
1	PLAPK	LNDIE	DVERH	MPGLI	RA-TN	EWFRI	YK-IP	DGKPE	NSFAFS	GECK	NRKYA
2	PLAPK	LNDIE	DVEKH	LPGLL	RA-TN	EWFRI	YK-IP	DGKPE	NQFAF	SGEAK	NKKYT
3	KLSKE	YDHIK	DVND-	LPELL	KAQIA	HFFEH	YKDLE	KGKWV	KVEGW	ENAEA	AKAEI
4	D–PR	FDEVR	SIED-	LPQHK	LKEIA	HFFER	YKDLQ	-GKRT	EIGTW	EGPEA	AAKLI
5	PLAPK	LNDIE	DVEKY	FPGLL	RA-TN	EWFRI	YK-IP	DGKPE	NQFAF	SGEAK	NKKYA
6	PLSSK	IDDLE	KIEEY	FPGIL	DT-TR	EWFRK	YK-VP	AGKPL	NSFAF	HEQYQ	NSNKT
7	PNVSY	IKDLK	DVNAH	LL—	D-EIA	NFFST	YK-IL	EKKET	KVLGW	EGKEA	ALKEI
8	PEFRH	YRDIK	ELPPH	RL—	A-EIR	RFFED	YK-KN	ENKKV	DVEAF	LPAQA	AIDAI
9	PDAAN	YNDIN	DVKRL	KPGYL	EA-TV	DWFRR	YK-VP	DGKPE	NEFAF	NAEFK	DKNFA
1	EEVVR	ECNEA	WERLI	TGKTD	AKSDF	SLVNV	SVTGS	VANDP	SVSST	IPPAQ	ELAPA
2	LDVIR	ECNEA	WKKLI	SGKSA	DAKKI	DLTNT	TLSDT	ATYSA	EAASA	VPAAN	VLPDE
3	VASFE	RAKNK									
4	DECIA	RYNEQ	K								
5	LDIIK	ETHDS	WKQLI	AGKSS	DSKGI	DLTNV	TLPDT	YSKAA	SDA–	IPPAS	LKADA
6	IQTIK	KCHNS	WKNLI	SGSLQ	EKYDN	LPNTE	RAGNG	VT—	LEDSV	KPPSQ	—
7	EVSIK	MYEEK	YGKKN								
8	KDSMD	LYELT	SK–L	ACNAN	EETS-	—	PFPFL	PVCLD	ITEAA	FYTTC	MLDKI
9	IDIIE	STHDY	WRALV	TKKT-	DGKGI	SCMNT	TVSES	PFQCD	PDAAK	AIVDA	LPPPC
1	PVDPS	VHKWF	YISGS	PL							
2	PIDKS	IDKWF	FISGS	A							
5	PIDKS	IDKWF	FISGS	V							
6	IPPEV	QKWYY	V								
8	SIGAF	NFVM-	LIRKH	C							
9	ESACT	IPTDV	DKWFH	HQKN							

Notes: [a]Gaps have been introduced into the sequence to maximise homology.
[b]Ligands to the metal atoms are indicated by ↓.
[c]Residues involved in binding the pyrophosphate are indicated by ↕.
[d]Other residues in the active site are indicated by *.
[e]Key and references. 1. Schizosaccharomyces pombe, [790]; 2. Kluyveromyces lactis, [791]; 3. Escherichia coli, [792]; 4. Thermophilic bacterium PS-3, [793]; 5. Saccharomyces cerevisiae (cytostolic), [794, 795]; 6. Saccharomyces cerevisiae (mitochondrial), [796]; 7. Thermoplasma acidophilum, [797]; 8. Arabidopsis thaliana, [798]; 9. Bovine retina, [799].

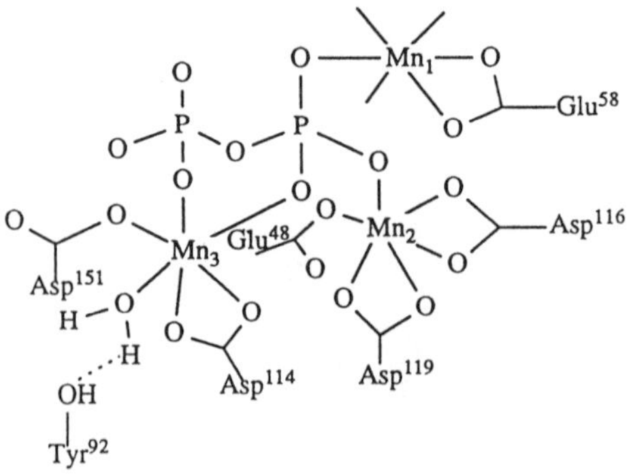

Figure 17. Schematic diagram of the metal binding site in inorganic pyrophosphatase.

of the metal coordination in the active site is given in Figure 17. One phosphate acts as a bridge between all three metal ions present, while the second phosphate is bound to the most deeply buried manganese. The Mn–Mn distances are 3.5, 4.2, and 5.3 Å for Mn_2–Mn_3, Mn_1–Mn_2, and Mn_1–Mn_3, respectively. The P–P distance of the two bound phosphates is 3.6 Å. Mn_3 has strong interactions with the amino acid residues, Asp-114 and Asp-151, and through water with Tyr-92. Mn_2 is coordinated to Asp-116, Asp-119, and Glu-48. The Mn_1 ion is coordinated only to Glu-58 and this ion is probably the Mn^{2+} ion which is bound only in the presence of phosphate or pyrophosphate [819, 820]. Computer modeling studies of the two inorganic pyrophosphatases from yeast and *S. pombe* based on the crystal structure of the yeast enzyme have been reported [821]. These studies suggest that despite the low sequence homology the folding of the proteins is likely to be very similar.

9. CONCLUSIONS

There is no doubt that, despite the considerable quantity and quality of the work described above, much remains to be done. The enzymes and proteins listed in Table 1 have not been properly characterized and basic biochemical characterization of these molecules will be necessary before more detailed structural studies commence. Spectroscopic characterization of the Mn(III) and Mn(IV) centers in proteins along the lines of the work reported by Whittaker for the superoxide dismutase enzyme [461] would be highly desirable. Accurate bond length and angle determinations from EXFAS studies on the proteins which have been structurally characterized are desirable. EXFAS studies on proteins which have not yet been structurally characterized would be useful, particularly if good model complexes become available. This offers a special challenge for Mn^{2+} complexes.

There have been many comments on the similarity of the chemistry of Mn(III) and Fe(III) and how one metal ion may replace the other at the active sites of proteins. Apart from two examples, superoxide dismutase and the more recently characterized ribonucleotide reductase, there are no reports of structurally characterized proteins with both metal ions present. In the case of superoxide dismutase, a structure of an enzyme with the non-native metal present and/or of an enzyme from an organism with a superoxide dismutase functional with either metal, is highly desirable. Ribonucleotide reductase has been structurally characterized in the Fe(III)- and Mn(II)-substituted forms and this has suggested interesting possibilities for the nature of the Fe(II) form of the enzyme.

What have we learned about the nature of the manganese binding sites in proteins? It is clear that Mn(II) prefers oxygen coordination and that properly characterized examples of Mn(II) bound to proteins via a sulfur donor or more than one nitrogen donor atom are unknown. Studies with inorganic model complexes suggest that increasing the number of nitrogen donors in an aerobic environment favors the formation of Mn(III) or Mn(IV) complexes [5]. The coordination number of Mn(II) may be four, five, or six, consistent with the lack of crystal field stabilization of the d^5 electron configuration. It would not be surprising if seven-coordination is observed, although substitution of Mn^{2+} into a seven-coordinate Ca^{2+} site in parvalbumin gave a six-coordinate complex. Mn(III) complexes are less well characterized, but the higher oxidation state is favored by two or more nitrogen (and perhaps tyrosine, cf. lactoferrin) donors, and the coordination number of Mn(III) may be either four, five, or six. Binuclear metal centers are apparently quite common, but the presence of oxo or hydroxo bridges has not yet been demonstrated in a manganese protein. There are strong suggestions of such bonding in catalase. Such structural motifs are extremely common in inorganic model systems and may be confidently expected.

ACKNOWLEDGMENTS

The author is grateful to Professor G. T. Babcock who first aroused his interest in manganese chemistry, and to many other members of the manganese chemistry fraternity for helpful and enlightening discussions. Work on this review began while the author was on study leave at the University of California San Diego working in the laboratory of Professor D. N. Hendrickson. The Leave Committee of Victoria University is thanked for the opportunity to undertake this period of leave. Many helpful discussions were held with Professor Hendrickson and the members of his research group. The many authors who provided both reprints and preprints of their papers are also gratefully acknowledged.

ADDENDUM: UPDATE TO MIDDLE OF 1994

This chapter was written early in 1993, and the following (related to the preceding section numbers) covers the most significant structural work reported during 1993 and early 1994. Reviews on the biological role of manganese in mammalian systems [822] and in human health and disease have appeared [823]. Manganese transport in PC-12 cells has been reviewed [824] and a number of reports have described the influx of Mn^{2+} into cells, across the blood brain barrier and through the intestine [825–833].

A number of new manganese-containing enzymes have been characterized and some will be discussed in Section 5.5. Human carbonic anhydrase, a zinc enzyme in which the zinc is tetrahedrally coordinated by three histidine residues and a water molecule, has been crystallized in metal (Co, Cu, Ni, and Mn) substituted forms. Structures of each of these forms have been determined, the inactive Mn derivative has a hexacoordinated metal ion, three histidines, a water molecule and a bidentate sulfate are coordinated [834]. The bacterium *Pseudomonas diminuta* has a phosphotriesterase which catalyzes the hydrolysis of organophosphate esters. Mn^{2+}, Cd^{2+}, Co^{2+}, or Ni^{2+} can replace the native Zn^{2+} without loss of catalytic activity. The EPR spectrum of the Mn^{2+} substituted enzyme suggests a pair of antiferromagnetically coupled Mn^{2+} ions [835]. ^{113}Cd NMR suggests that three nitrogen and one oxygen donor atoms are bound at one metal binding site, and two nitrogens and two oxygen atoms at the other site [836].

4.2 Purple Acid Phosphatase

A di-iron form of the recombinant human enzyme has been crystallized and is the subject of further studies [837].

5.3 Elongation Factor Tu

Previous structural studies on the *E. coli* Elongation Factor Tu (EF-Tu) were of an inactive form of the protein. An active form of the enzyme from *T. thermophilus* complexed with Mg^{2+} and guanosine-5′-(β,γ-imido)triphosphate (GppNHp) (a GTP analogue) has now been structurally characterized. The residues which form the nucleotide binding domain have the same overall structure as in the *E. coli* enzyme. Mg^{2+} has octahedral coordination with the β and γ-phosphate groups of GppNHp , two OH groups from Thr-25 and Thr-62, and two H_2O molecules bound to the metal [838, 839].

The structure of the (EF-Tu)–GDP complex shows that Tyr-87 in domain I and Tyr-309 in domain III are buried within the protein and are close to each other. Distinct differences in the environment of these residues in this complex compared with the (EF-Tu)–GTP and other complexes has been demonstrated [840].

5.4 Isocitrate Dehydrogenase

Structures of isocitrate dehydrogenase with a bound isocitrate, $NADP^+$, and Ca^{2+}, and with bound α-ketoglutarate, NADPH, and Ca^{2+} have been determined [841, 842]. The first structure shows that the Ca^{2+} is 1 Å from the Mg^{2+} binding site identified in earlier studies. Ca^{2+} is bound to the isocitrate, two water molecules, and the carboxylate residues of Asp-307, Asp-311, and (from an adjacent molecule) Asp-283. The positional change means that the bond to the isocitrate is lengthened to 3.0 Å from the 2.4 Å observed in the Mg^{2+} structure. This increase in bond length is thought to explain the unproductive nature of the Ca^{2+} ternary complex. In the second structure the α-ketoglutarate is bound in the same position and orientation as isocitrate.

5.5 Kinases

A structure of *Myxococcus xanthus* nucleoside diphosphate kinase (EC 2.7.4.6) complexed with a Mg^{2+}–nucleoside complex has been determined and the complex with MnADP has also been characterized, but details have yet to be reported. Mg^{2+} and Mn^{2+} occupy the same position in the active site [843]. In the Mg^{2+}-containing structure, Mg^{2+} does not appear to be coordinated by any ligands; it is 2.9 Å from the α-phosphate of the ADP and even further from protein residues. Such a situation would appear to be extremely unlikely and further details of the Mn^{2+} structure are awaited with interest.

Creatine Kinase and Adenylate Kinase

Creatine kinase has been shown to be specific for the Δ-β-P, exo-α-P, α,β,γ-tridentate $Rh(H_2O)_3ATP$ isomer [845]. The *anti* conformations of the adenosine moiety of MgADP and MgATP bound to rabbit muscle creatine kinase have been determined using NMR [853].

Adenylate kinase specifically catalyzes the reaction of the Δ–β,γ-bidentate $Rh(H_2O)_4ATP$ complex [845]. Crystal structures of two mutants (Pro-9–Leu and Gly-10–Val) of adenylate kinase from *E. coli* complexed with Ap_5A have been determined [854]. There are only slight changes to the overall structure and to the active site of the protein due to these mutations. A mechanism for domain closing in adenylate kinase has been described based on the structural results [855]. Site-directed mutagenic studies of adenylate kinase have concentrated on amino acid residues which are involved in the binding of AMP. Three studies have shown that Thr-23 (conserved among all types of adenylate kinase) interacts directly with ATP during catalysis [856–858]. Substitution of Val-67 indicates that it contributes to AMP binding [859]. An NMR investigation of the chicken muscle enzyme has shown that Thr-39, Leu-43, Gly-64, Leu-66, Val-67, Val-72, and Gln-101 residues are close to the adenosine moiety at the AMP binding site [858]. NMR investigations of single and double mutants of this enzyme indicate that no appreciable change in the conformations of the free enzyme or of the complex with Mg(P-1,P-5-bis(5′-adenosyl)pentaphosphate) occurs for the mutants Asp-140–Ala, Asp-141–Ala and double mutants Arg-138–Met, Asp-140–Ala, Arg-132–Met, and Asp-14–Ala. These results suggest that Asp-140 and Asp-141 are important in assisting Arg-138 and Arg-132 in stabilizing the transition state [860].

Hexokinase (EC 2.7.1.1)

The active site of hexokinase, which catalyzes the ATP-dependent phosphorylation of hexose sugars at the C6–OH group, has been characterized by EPR spectroscopy. A Mn^{II}ATP–nitrate–lyxose complex in this active site has been studied using ^{17}O-enriched ligands [844]. The results suggest that Mn^{2+} is bound to the β-phosphoryl of ADP, the nitrate ion, and four water molecules. Hexokinase has been shown to be specific for the Δ-β, γ-bidentate $Rh(H_2O)_4$ATP isomer [845].

cAMP-Dependent Protein Kinase (EC 2.7.1.37)

Cyclic adenosine monophosphate-dependent protein kinase catalyzes the transfer of the γ-phosphate from ATP to a protein serine residue [846]. Crystal structures of a binary complex of a subunit of the mouse enzyme with an inhibitor peptide, and ternary complexes of the subunit from porcine heart [847] and mouse [848–851] have been determined. The ternary complexes were crystallized together with a peptide inhibitor, Mn^{2+}, and either the ATP analogue adenyliminodiphosphate or ATP [847, 848, 851], or the Mg^{2+}–ATP complex [849]. Detailed kinetic studies with Mg^{2+}, Co^{2+}, and Mn^{2+} as the activating cations have been reported [852]. One metal ion is required to activate the kinase, while a second Mn^{2+} is inhibitory. The catalysis is decreased but not inhibited by a second Mg^{2+}. In the Mn^{2+}-containing structures, one Mn^{2+} is six-coordinate bound to the β and γ-phosphates of the nucleoside, Asp-184 (bidentate), and two water molecules. The other Mn^{2+} is five-coordinate with bonds to the γ- and α-phosphates of the nucleosides, Asn-171 and Asp-184 (long bond), and a water molecule. The metal–metal distance is 4.0 Å.

5.6 α-Lactalbumin

The structure of the human α-lactalbumin–Zn^{2+} complex is of particular interest because it had been proposed that the binding of Zn^{2+} to α-lactalbumin resulted in a different conformation of the protein. The structure shows that the conformation is essentially unchanged by Zn^{2+} binding [861]. Three different metal binding sites on α-lactalbumin were identified using Eu^{III} luminescence spectra [862].

5.8 Mandelate Racemase

The structure of mandelate racemase from *Pseudomonas putida* has been refined at a resolution of 1.9 Å for the Mn^{2+}-substituted enzyme and 2.0 Å for the Mg^{2+}-substituted enzyme [863]. In addition, complexes of the enzyme with Mg^{2+} and (*S*)-atrolactate, and with Mg^{2+} and (*R*)-α-phenylglycidate and a Lys-166–Gln mutant, have been structurally characterized. Mg^{2+} and Mn^{2+} occupy the same binding site. Six oxygen ligands (Asp-195, Glu-221, Glu-247, two water molecules, and an oxygen from a sulfate ion) provide an octahedral coordination sphere. Bonds to the sulfate and to one of the water molecules are relatively long and the structures with the inhibitors show that these ligands are replaced on substrate binding. Binding of the inhibitors to Mg^{2+} occurs via the α-hydroxyl oxygen and a carboxyl oxygen. An active site flap closes over the active site upon substrate binding.

5.9 Manganese-Dependent Peroxidase

Two reviews on the molecular biology of *P. chrysosporium* [866, 867] and one on the biodegradation of lignin have appeared [868]. In addition, the proceedings of a symposium on lignin biodegradation and transformation have appeared [869]. Manganese peroxidase has been isolated from a number of species which were not previously known to produce the enzyme [870–876]. There is continuing interest in the degradation of organic pollutants and Kraft pulp using lignin peroxidase and/or manganese peroxidase [877–889].

^{1}H NMR spectroscopy has been used to study lignin peroxidase and manganese peroxidase containing deuterated histidines. These studies have allowed firm spectral assignments to be made for the protons of the distal and proximal histidines [890]. A second NMR study has shown that $Mn^{2+}_{(aq)}$ binds manganese peroxidase at a number of sites; one site, which is close to the heme edge, has an affinity constant of 10^4 M^{-1}. The other binding sites may be the same as the calcium binding sites identified in lignin peroxidase [891]. Other studies suggest that a Mn^{2+} complex is the physiologically active substrate, and kinetic studies suggest that Mn^{2+} has to be chelated [892, 893]. A variety of ligands (malonate, lactate, oxalate, and gluconic acid) have been suggested [887, 892–894]. It is known that *P. chrysosporium* produces extracellular oxalate under conditions that induce synthesis of the ligninolytic system [895]. However, the NMR study did not observe any effect of added oxalate on the binding of Mn^{2+} to manganese-dependent peroxidase.

Two independent studies have described the structure of lignin peroxidase from *P. chrysosporium* [366, 896, 897]. Structures of a fungal peroxidase of unknown function (but which may degrade lignin) from *Arthromyces ramous* and a peroxidase from *Coprinus cinereus* (which does not) have been determined [898, 899]. These peroxidases and cytochrome *c* peroxidase have similar structures. In each case the iron atom is coordinated by the four heme nitrogens and a histidine, confirming the high-spin five-coordinate iron center deduced spectroscopically. There is a water molecule in the heme pocket above the vacant coordination site but it is too distant from the iron to be considered coordinated. Two calcium ions (one seven-coordinate and the other eight-coordinate) are present in the enzymes. Different sizes to the entrance of the channel to the distal side of the heme are believed to be important in determining substrate specificity of these peroxidases. The peroxidase from *Coprinus cinereus* has a much larger opening to the entrance channel than that observed in lignin peroxidase, suggesting that the latter may use small molecule mediators to oxidize lignin.

Direct evidence that lignin peroxidases are able to catalyze the oxidative cleavage of polymeric lignin *in vivo* has been produced. Small amounts of polymerized lignin were also produced [900].

5.10 *cis,cis*-Muconate Cycloisomerase

Chloromuconate cycloisomerase (EC 5.5.1.7) from *Alcaligenes eutrophus* JMP 134 has been crystallized and a structural study reported [864, 865]. The asymmetric unit consists of two non-identical molecules. Each molecule binds one Mn^{2+} in a site accessible from the surface via a cleft between subunits. In one subunit the Mn^{2+} is coordinated by Asp-194, Glu-221, Glu-245, and probably a chloride ion. The Mn^{2+}–Cl distance is 2.6 Å. In the other subunit the Cl^- is absent and the Mn^{2+} is located in a different position and interacts with the same amino acid residues by rather long bonds (2.7–3.8 Å). It is possible that this is a $Mn^{2+}{}_{(aq)}$ ion. Comparison of the muconate cycloisomerase and chloromuconate cycloisomerase structures reveals differences in the channel leading to the active site.

5.12 ras p21 Protein

Structures of complexes of two mutants, Gly-12–Asp and Gly-12–Pro of ras p21, with Mg^{2+}–guanosine 5′-(β,γ-imido)triphosphate have been determined. The Gly-12–Asp mutant shows the most drastic change of structure in the active site compared to the wild-type protein, although the position of the bound Mg^{2+}–nucleotide complex is similar. In this mutant Mg^{2+} is five-coordinate and one of the coordinated water molecules present in the wild-type is absent. The Gly-12–Pro mutant is similar to the wild-type in the active site [901]. The Gly-12–Asp mutant, and a truncated form of the protein (residues 1–166) have been studied by NMR spectroscopy [902, 903]. Metal-to-ligand distances in Mn^{2+}–GTP and Mn^{2+}–5′guanylylimidodiphosphate (GMPPNP) complexes of the protein have been characterized using ESEEM spectroscopy [904]. In the GDP complex the metal is bound to Ser-17, four water molecules, and the β-phosphate. In the GMPPNP complex Ser-17, the β and γ-phosphates, and three water molecules bind the metal ion.

Site-directed mutagenesis, ^{31}P NMR, and kinetic experiments have been used to study the metal binding site of ras p21 [905]. The mutations studied involve Ser-17, Asp-57, and Thr-35; each of these residues acts as a ligand-to-metal ion in at least one form of the structurally characterized protein. The results, in agreement with the X-ray studies, are interpreted as showing that Thr-35 is only coordinated to the metal in the GTP conformation and the other two residues are important in binding GTP and GDP [906]. Mutations of a number of other important catalytic residues, using unnatural amino acids, have led to important insights. Mutations involving Gln-61, which was thought to be important in γ-phosphate binding, have shown no decrease in GTPase activity. Similar experiments with Gly-60 and Gly-12 have shown that Gly-60 is important in maintaining activity [907]. Kinetic and molecular dynamics simulation studies of Gln-61–Asp and Gly-12–Pro, Gly-12–Arg, Gly-12–Val, and Gln-61–Lys mutants suggest that the side chain of Gln-61 activates a water molecule involved in the GTP hydrolysis reaction [908].

5.13 RuBisCo

Two reviews on the active site of RuBisCo have appeared [909, 910]. A multienzyme complex isolated from spinach chloroplasts, containing ribose-phosphate isomerase, phosphoribulokinase, RuBisCo, phosphoglycerate kinase, and glyceraldehyde-3-phosphate dehydrogenase suggests that the stoichiometry of the subunits is different *in vivo* (two large subunits: four small subunits) from that in the enzyme complex characterized crystallographically (8:8) [911]. The kinetic properties are also different [912].

Crystallization of a complex of RuBisCo with ribulose-1,5-biphosphate promises to yield important information on the structure of the complex in spinach leaves [913]. The crystal structure of unactivated RuBisCo from *Nicotiana tabacum* complexed with 2-carboxy-D-arabinitol 1,5-bisphosphate (CABP) has been determined. CABP binds at the active site in an extended conformation but in the reverse orientation to the same analogue in the activated enzyme. Loop 6 (residues 330–339) remains open and flexible upon binding of the analogue in the unactivated enzyme, in contrast to the closed and ordered loop 6 in the activated enzyme complex [914]. A structure of recombinant RuBisCo from the cyanobacterium *Synechococcus* PCC6301 complexed with CO_2, Mg^{+2}, and CABP has also been reported [915]. Bonding to the Mg^{2+} is octahedral and the carboxylate oxygens of Asp-203 and Glu-204, the carbamate group of Lys-201, the 2′-carboxyl, and the C2 and C3 OH groups of CABP are the metal ligands. Loop 6 is closed over the active site, burying the inhibitor inside the protein. The role of the active site residues in the different stages of the catalytic mechanism has been discussed [915–918].

There have been numerous studies on the characterization of the active site and on altering the CO_2/O_2 specificity of the enzyme using site-directed mutagenesis and chemical modification [919–929]. Carbamylated RuBisCo exhibits chemiluminescence while catalyzing its oxygenase reaction in the presence of Mn^{2+} and this behavior has been the subject of a detailed study [930].

5.15 Staphylococcal Nuclease

Mutant forms (Lys-116–Arg and Lys-116–Gly) of staphylococcal nuclease from *S. aurens* have been studied by X-ray crystallography and NMR. Lys-116 was chosen for mutation because of the single *cis* peptide bond which occurs in the native enzyme between residues 116 and 117. The Lys-116–Arg mutant has the *cis* peptide bond but the Lys-116–Gly mutant contains a predominantly *trans* bond.

The overall structure of the protein is unchanged by these mutations except in the immediate vicinity of the *trans* bond [931].

In the reaction catalyzed by staphylococcal nuclease, Glu-43 had been postulated to act as a general base whose function is to activate the water molecule which attacks the phosphodiester bond. A Glu-43–Asp mutant and a Gly-50–Val, Val-51–Asn double mutant have been studied and the results indicate that this postulate is incorrect [932]. Substitution of Glu-43 with its nitro analogues, (*S*)-4-nitro-2-aminobutyric acid and (*S*)-2-amino-5-hydroxypentanoic acid, and of Arg-35 and Arg-87 by ethylhomocysteine and citrulline have been studied by Judice et al. [933]. In addition to kinetic studies on the other mutants the crystal structure of the Glu-43–(*S*)-2-amino-5-hydroxypentanoic acid mutant was determined.

EPR, NMR, and modeling studies of Lys-49–Ala, Lys-84–Ala, and Tyr-115–Ala mutants of staphylococcal nuclease bound to 3′,5′-pdTp, Ca^{2+}, and Mn^{2+} have suggested that the X-ray structure of the enzyme must be viewed with caution. All three mutants bind metal ions more weakly than the wild-type, and the Lys-49–Ala and Tyr-115–Ala mutations greatly weakened 3′,5′-pdTp binding, whereas the Lys-84–Ala mutation had a smaller effect [934–936]. The NMR results indicate that the metal-coordinated 5′-phosphate of 3′,5′-pdTp is in the same position as in the X-ray structure. However the thymine, deoxyribose, and 3′-phosphate are significantly displaced from their positions in the X-ray structure. The 3′-phosphate is thought to be hydrogen bonded to Lys-49 rather than to Lys-84 and Tyr-85. The repositioned thymine ring is hydrogen bonded to the phenolic hydroxyl of Tyr-115.

5.16 Superoxide Dismutase

Primary structures of manganese and iron superoxide dismutases from cattle (Mn), [937] mouse (Mn, new form) [938], *P. aeruginosa* (Fe and Mn) [939], *Campylobacter jejuni and C. coli* (Fe) [940], *Sulfolobus acidocaldarius* (Fe or Mn) [941], *Haemophilus influenzae Type-b* (Mn) [942], *Hevea brasiliensis* (Mn, two forms) [943], *Chlamydomonas reinhardtii* (Partial sequence) (Fe) [944], *Bordetella pertussiss* (Fe) [945], *Drosophila melanogaster* (Mn) [946], Scots pine (*Pinus sylvestris L.*) (Mn partial sequence) [947], *Helicobacter pylori* (Fe) [948, 949], maize (Mn, a number of forms different from that previously reported) [950] rice *Oryza sativa* [951], *Caenorhabditis elegans* (Mn) [952], and the parasite *Leishmania donovani chagasi* (Fe) [953] have been characterized. A correction has appeared to an earlier report on the sequence of the human enzyme [954].

Wagner et al. reported the structure of tetrameric human SOD at 3.2 Å resolution [955]. The crystals studied were colorless. It is thought that reduction to Mn^{2+} may be responsible for the lack of the usual purple color. Little difference in the structure around the manganese binding site is apparent and the structure is in accord with an earlier report of the structure of the human enzyme. A structural investigation of the iron-containing protein from *Mycobacterium tuberculosis* has been initiated [956]. NMR studies of the Fe(II) form of the enzyme from *E. coli* have shown that the coordination sphere of the Fe(II) is the same as that of Fe(III) in the crystal [957].

Superoxide dismutase from *P. shermanii* is active with either Mn or Fe as the native metal and the Fe form is not inactivated by hydrogen peroxide. This enzyme has been prepared and characterized in the inactive Co^{2+}- and Cu^{2+}-substituted forms [958, 959]. The inactivation of the ferric form of superoxide dismutase by H_2O_2 seems to be due to a tryptophan residue near His-75, a metal ligand. This tryptophan is exchanged for valine in the *P. shermanii* enzyme [960].

5.17 Transketolase

Structures of the complexes of transketolase and the coenzyme analogues, thiamine thiazolone diphosphate, 6′-methyl-, *N*1′-pyridyl-, and *N*3′-pyridyl-thiamin diphosphate have been reported and the complex with thiamine diphosphate has been refined to a resolution of 2 Å [961–963]. All four coenzyme analogues bind to the enzyme in a similar fashion to thiamin diphosphate.

6.1 Acyl Carrier Protein

A FTIR analysis of the structure of acyl carrier protein is in agreement with the structural characteristics determined by other techniques [964].

6.2 Catalase

EPR and ENDOR spectroscopy have been used to characterize the manganese centers in the catalase from *T. thermophilus*. EPR spectra of the oxidized enzyme indicate a high degree of uniformity in the coordination environments of the two manganese ions and simulations of the spectra indicate that the oxidation state of the manganese in the active enzyme is $Mn^{II}Mn^{III}$. The lack of ^{14}N hyperfine couplings in the spectrum of the enzyme suggests either O_6 or O_5N(histidine) ligand donors to each Mn. Comparison with model dimanganese(III,IV) complexes lead to the suggestion that in catalase, manganese is coordinated by carboxylato-type ligands (Asp and Glu) and that there is a di-μ-oxo bridge between Mn ions [965, 966].

6.3 DNA Polymerase

The structure of the DNA polymerase Klenow fragment bound to duplex DNA has been described. This structure suggests that the processing of the DNA is in the opposite direction from that suggested in earlier studies [967]. It was concluded from studies using two mutant enzymes, Arg-682–Ala, and Arg-682–Lys, that the positively charged guanidino group in the side chain of Arg-682 is catalytically important but not absolutely essential for synthesis of DNA [968].

Binary complexes of the Klenow fragment complexed with substrate and products of the reaction, Mg–thymidine 5′-triphosphate, Mn–thymidine 5′-triphosphate, and Mn–thymidine monophosphate and pyrophosphate have been structurally characterized [969]. The nucleoside triphosphate binds within the cleft of the polymerase domain at the position expected from earlier studies, but the precise orientation depends on the exact conditions of crystal growth. The resolution of these structures is such that little can be said about the metal coordination. Unfortunately the Mn and Mg crystals are not isomorphous so that differences in electron scattering is of no use in defining the metal coordination.

6.4 Deoxyribonuclease (DNase)

Crystal structures of inactive mutants (Tyr-76–Ala and His-134–Gln) of bovine pancreatic DNase I have been determined. Tyr-76 stabilizes the wide minor groove and contributes to the binding of DNA; His-134 is believed to function as a proton donor. Disruption of hydrogen bonds between His-134, Asp-78, and Tyr-76 in both mutants leads to an increased mobility and positional shifts in the DNA-binding loop, mainly around Tyr-76. In the Tyr-76–Ala mutant the coordination of the protein to the calcium ion is unchanged from the wild-type enzyme; however the calcium site is unoccupied in the His-134–Gln mutant [970].

6.5 Fructose-1,6-bisphosphatase

The structures of the enzyme binary complex with fructose-1,6-bisphosphate and ternary complexes with the substrate analogues 2,5-anhydro-D-glucitol-1,6-bisphosphate (AhG-1,6-P_2) with Mn^{2+}, Mg^{2+}, and Zn^{2+}, and 2,5-anhydro-D-mannitol-1,6-bisphosphate (AhM-1,6-P_2) with Mn^{2+} and Mg^{2+} have been reported [971]. These substrate analogues represent the α and β anomers of fructose-1,6-bisphosphate, respectively. In the absence of metal ions, fructose-1,6-bisphosphate binds in the active site and is held in place by numerous hydrogen bonds. In the AhG-1,6-P_2 structures the 6-phosphate groups are in the same positions in the active site as in the α-Fru-1,6-P_2 complex. The 1-phosphate group shifts about 1 Å in order to bind to the metal ion(s). Two Zn^{2+} or Mn^{2+} ions (but only one Mg^{2+} ion) are bound in the presence of AgG-1,6-P_2. The $Zn^{2+}$and Mn^{2+} are tetrahedrally coordinated; each metal is bound to three bridging ligands, Glu-97, Asp-118, and the 1-phosphate of the substrate. The carboxylate of Glu-280 and the carbonyl oxygen of Leu-120 complete the coordination spheres. The metal ions are 3.6–3.8 Å apart. In the presence of the β-analogue AhM-1,6-P_2 and in the Mg^{2+}-substituted AhG-1,6-P_2 complex only one five-coordinate metal ion (Mg or Mn) is bound to the protein. Carboxylate oxygens of Glu-97, Asp-118, Asp-121, Glu-280, and the 1-phosphate group of either AhG-1,6-P_2, or AhM-1,6-P_2 are the donor groups. AMP inhibition of fructose-1,6-bisphosphatase is most probably caused by reduced metal-binding affinity due to structural changes of metal ligands (Glu-97, Asp-118, and Asp-121) in the active site [972]. A truncated form of recombinant fructose-2,6-bisphosphatase has been crystallized [973].

6.6 Glutamine Synthetase

Ovine brain glutamine synthetase exhibits a high degree of specificity for binding Mn^{2+} over Mg^{2+} and it has been concluded that Mn^{2+} binds to glutamine synthetase to a significant extent under *in vivo* conditions [974].

Crystal structures of a His-269–Asn mutant (His-269 is a manganese ligand) of glutamine synthetase from *E. coli* and of the oxidized enzyme from *Salmonella typimurium* have been reported [975, 976]. Oxidation converts His-269 to Asn and Arg-344 to γ-glutamyl semialdehyde. The absence of His-269 results in an empty metal binding site in the His-269–Asn mutant. Both binding sites are empty in the oxidized form of the enzyme and the remaining metal ligands are disordered [154]. Four crystal structures of fully adenylated glutamine synthetase from *Salmonella typhimurium,* with Mn^{2+} ions bound, complexed with the substrate Glu and with three feedback inhibitors (L-serine, L-alanine, and glycine) have been determined. Glycine, alanine, and serine inhibit glutamine synthetase by competing with the substrate glutamate for the active site [977].

6.7 Lectin

Four reviews on the chemistry of lectins have appeared [978–981]. There have been six preliminary reports [982–986] and eight reports on completed crystal structures of lectins. The latter include a refined structure of a saccharide-free Cd^{2+}-substituted concanavalin A with a Cd^{2+} ion occupying the Mn^{2+} site [987]; a complex of pea lectin [988]; the human S-Lac lectin, L-14-II, complexed with lactose [989]; a pea lectin–trimannoside complex and lectin IV from *Griffonia simplicifolia* [650, 980, 990]; a peanut (*Arachis hypogaea)* lectin [991]; a *Lathyrus ochrus* isolectin II complexed with both the N2 monoglycosylated fragment of human lactotransferrin and an isolated glycopeptide fragment of human lactotransferrin [992]; and a lentil lectin [993]. This last structure is unusual in that a different mode of binding of the protein to Mn^{2+} has been demonstrated. The normally bridging Asp-121 is rotated 90° with respect to the usual lectin structure with the result that Asp-121 is not coordinated to Mn^{2+} (which is five-coordinate). The pentacoordinate Mn^{2+} is reminiscent of the reduced coordination number for zinc suggested on the basis of EXFAS studies in the Ca^{2+}/Zn^{2+} form of concanavalin A [669]. The thermodynamics of the binding of Mn^{+2}, Co^{+2}, Ni^{+2}, Zn^{+2}, and Cd^{+2} to demetallized concanavalin A and of sugars to concanavalin A, pea lectin, and lentil lectin have been reported [994, 995].

6.8 Leucine Aminopeptidase

The specificity of a Mn^{2+}-substituted leucine aminopeptidase from *Streptomyces griseus*, which contains two zinc ions in its native form, has been investigated [996]. Three reports of the structure of bovine lens leucine aminopeptidase and a review of the structure and function of the enzyme have appeared [997–1000]. The structural studies include a re-refinement of the structure of the complex with bestatin because an incorrect bestatin configuration was used in the original report [997]. The re-refinement results in a different interpretation of the interaction of the Zn^{2+} ions with bestatin. The OH group of bestatin acts as a bridging group between the zinc ions. A structure has also been determined with the inhibitor amastatin. Amastatin binds similarly to bestatin. A mechanism for the leucine aminopeptidase reaction has been proposed based on these structures [998].

Identification of the metal ion site which contains the readily exchangeable metal ion has been reversed. The readily exchanging Zn^{2+} is bonded to Asp-255, Asp-332, Glu-334, and Asp-332. The more tightly binding Zn^{2+} is bonded to Asp-255, Asp-273, Glu-334, and Lys-250 [999].

6.10 Phosphoglycerate Kinase

The structure of the enzyme from *Bacillus stearothermophilus*, which is very similar to that of the yeast enzyme, has been determined [1001, 1002]. Binding of metal ions to the active site has been the subject of two investigations [1003, 1004]. The results suggest that in addition to the interaction of the α-phosphate that is detectable by crystallography [691], the β and/or γ-phosphate(s) of MgATP may also interact with the enzyme.

6.11 Ribonuclease

A crystal structure of an *E. coli*–RNase HI complex with Mg^{2+} has been determined. The crystal, which is not isomorphous with the Mg^{2+}-free crystal previously studied, was grown at a high $MgSO_4$ concentration. The Mg^{2+}-bound and the metal-free enzymes have very similar backbone structures, except for minor regions in the enzyme interface with the DNA/RNA hybrid. The active site contains a single Mg^{2+} atom located at a position almost identical to that previously found by the soaking method; Asn-44, Asp-10, Gly-11, and Glu-48 form coordinate bonds with Mg^{2+} [1005]. This is in contrast to the earlier work of Davies et al. [710] in which two Mn^{2+} ions were bound. Three mutants (Asp-10–Asn, Glu-48–Gln, and Asp-70–Asn) of *E. coli* RNase HI, have been characterized structurally. Asp-10 and Glu-48 are metal ligands and Asp-70 is in the active site, hydrogen bonded to a water molecule which is believed to attack the phosphate. Each of the mutants were inactive as they lacked Mg^{2+} in the metal binding site. Localized conformational changes were observed around the Mg^{2+} binding site. Substitution of a carboxylate residue by an amide group induces the formation of new hydrogen bond networks, presumably due to the cancellation of repulsive forces between the carboxylate residues [1006].

Ribonuclease H from *Thermus thermophilus* HB8 has been crystallized and the structure has been reported at 2.8 Å resolution [1007, 1008]. The active site is very similar to that of the *E. coli* enzyme. Further characterization of the binding site in the *E. coli* enzyme has been achieved using high-field proton and ^{25}Mg NMR, and also by direct titration calorimetry [1009].

6.12 Ribonucleotide Reductase

Our understanding of the structures of the non-heme iron proteins, ribonucleotide reductase [1010] and methane monooxygenase [1011], have been advanced considerably in the period under review. The structure of the Mn^{2+}-substituted form of ribonucleotide reductase had been reported earlier and it is now known that the Fe^{2+} form of the enzyme has the same coordination environment as the Mn^{2+} form, which is different from the Fe^{3+} form of the enzyme. The two iron atoms in methane monoxygenase are linked by three bridging ligands; exogenous μ-hydroxo and μ-acetate groups, and a semibridging glutamate residue (one bond is very long). Octahedral coordination of the iron atoms is completed by two histidines, three monodentate glutamates, and a water molecule. The metal–metal distance is 3.4 Å in good agreement with earlier EXFAS results. In contrast to ribonucleotide reductase, Mn^{2+} bound to methane monooxygenase is EPR silent, which suggests strong coupling between the metals and retention of the μ-hydroxo bridge [1012]. F^- binding to Mn^{2+}-substituted forms of ribonucleotide reductase and methane monooxygenase have been studied using ^{19}F NMR [1013]. In methane monooxygenase, F^- may bind to the manganese at the coordination positions occupied by the exogenous OH^- and acetate groups.

6.13 Xylose Isomerase

The functions of individual amino acid residues in the active site of *Thermoanaerobacterium thermosulfurigenes* D-xylose ketol-isomerase has been studied by site-directed substitution. On the basis of the results, a revised mechanism of D-xylose isomerization has been proposed [1014]. The principal change in this mechanism is that the substrate molecule involved in the hydride-shift step is still in its cyclic form. Alternative mechanisms for an aldose–ketose isomerization have been examined using high level *ab initio* and semiempirical molecular orbital methods. The proton-transfer pathway via an enediol intermediate is shown to be favored in the absence of a metal ion, while the hydride-transfer pathway becomes favored in the presence of a metal ion [1015]. Detailed kinetic studies do not support a proton transport mechanism [1016].

Replacement of His-220, one of the Mn^{2+} ligands, with Ser, Glu, Asn, and Lys has been studied in the enzyme from *Streptomyces rubiginosus*. Crystal structures of the His-220–Ser, His-220–Asn, and His-220–Glu mutant enzymes complexed with Mn^{2+} have been determined. In the His-220–Ser structure, a water molecule replaces the N atom of the imidazole ring and mediates the interaction between Mn^{2+} and Ser-220. A similar water-mediated interaction is observed in the His-220–Asn mutant. Octahedral coordination is maintained for the Mn^{2+} in these mutants, but a pentacoordinate metal is observed in the His-220–Glu mutant [1017]. A crystal structure of the Glu-180–Lys mutant enzyme from *Streptomyces olivochromogenes* has been determined [1018]. This mutant has no isomerase activity and the structural study shows that the lysine residue has replaced the metal at site 1—the metal at site 2 is still present. X and Q band EPR and electronic spectroscopy have been used to study the metal binding sites of this D-xylose isomerase. Mn^{2+} ions have a higher affinity for site 2, the site which contains the histidine ligand. When site 1 is unoccupied, the coordination sphere of Mn^{2+} in site 2 is quite distorted. If both binding sites are occupied by Mn^{2+} the coordination sphere of the site 2 metal ion has higher symmetry. Changes in Mn^{2+} coordination at site 1 caused by occupation of site 2 with Co^{2+}, Cd^{2+}, or Pb^{2+} have also been studied. In the Co^{2+} derivative, the Mn^{2+} has a relatively symmetric ligand environment. With Cd^{2+} or Pb^{2+} in site 2, the coordination environment of site 1 is much more distorted and the enzyme is inactive [1019].

A Tyr-253–Cys mutant of *Arthrobacter* D-xylose isomerase has been structurally characterized with xylitol bound. This mutant showed a changed conformation of Glu-185 and also alternative conformations for Asp-254, a ligand to the site 2 metal ion. It is suggested that electrostatic repulsion from Glu-185 in its new position causes Asp-254 to move when His-219 is unprotonated, thereby preventing M^{2+} binding at site 2 [1020].

Structural studies of the binding of the substrates and substrate analogues D-xylose, D-xylulose, D-sorbitol, xylitol, D-glucose, D-fructose, 1,5-dianhydrosorbitol, L-ascorbic acid, threonate, 3-deoxy-3-fluoromethyleneglucose, ribose and L-sorbose to the enzyme from *Streptomyces rubiginosus* have been reported [1021]. The structure of the enzyme from *Streptomyces murinus* [1022] has been determined and preliminary results reported for the enzymes from *Bacillus coagulans* [1023] and *Thermoanaerobacterium thermosulfurigenes* strain 4B [1024].

7.1 Alkaline Phosphatases

An alkaline phosphatase from *Halobacterium halobium* contains Mn^{2+} and there is evidence that it has a dinuclear metal center [1025].

The carboxylate group of Asp-153 of *E. coli* alkaline phosphatase is hydrogen bonded to two of the three water molecules coordinated to the Mg^{2+}. Mammalian alkaline phosphatases which are an order of magnitude more active than the *E. coli* enzyme, have His in place of Asp-153. An Asp-153–His mutant of the *E. coli* enzyme has been the subject of a number of studies. This mutant exhibits a 3.5-fold decrease in activity at pH 8.0 compared to the wild-type enzyme, and Mg^{2+} is more weakly bound [1026]. A structural study has shown that the octahedral metal binding site in the wild-type enzyme has been converted into a tetrahedral Zn^{2+} binding site in the mutant. The histidine residue is one of the four metal ligands and it replaces three water molecules [1027]. The role of His-372 in *E. coli* alkaline phosphatase has been investigated by studying the His-372–Ala mutant. His-372 is 3.8 Å from the Zn^{2+} and is hydrogen bonded to Asp-327, a bidentate ligand of the Zn^{2+} at the M1 site. The mutant and wild-type enzymes have similar binding affinities for Zn^{2+}, but the kinetic behavior is different. The results suggest that the hydroxyl group coordinated to the Zn^{2+} at the M1 site is a weaker nucleophile in the mutant, thus the interaction between His-372 and Asp-327 may be important for stabilizing the Zn^{2+}–OH^- group responsible for the breakdown of the phosphoserine intermediate [1028].

7.2 Enolase

The structure of the quaternary complex, yeast enolase–Mg^{2+}–F^-–PO_4^{3-}, has been determined. The movable loops Pro-35–Ala-45, Val-153–Phe-169, and Asp-255–Asn-266 are in the closed conformation found previously in the precatalytic substrate enzyme complex. Mg^{2+} is coordinated by F^-, Asp-246, Asp-320, Glu-295, and a water molecule in a trigonal bipyramidal arrangement. Due to movements of the flexible loops the environment of the PO_4^{3-} ion is different from that of the SO_4^{2-} ion in the previously reported structure [1029]. A structure of the enolase–Mn^{2+}–phosphonoacetohydroxamate (PhAH) complex has been determined. This complex contains both structural and catalytic metal ions. The catalytic metal ion is coordinated to two oxygens of the phosphono group of PhAH and to the carbonyl oxygen of Gly-37. Disordered water molecules complete the coordination sphere. The distance between the metals is 8 Å [1030]. The Asp-168–Glu mutant of enolase has been crystallized and a structural study is underway [1031].

7.3 Parvalbumin

The structure of rat parvalbumin has been determined. Two Ca^{2+} binding sites have been characterized and there is some evidence for a third cation binding site which may contain NH_4^+. Both bound Ca^{2+} ions are seven-coordinate with a pentagonal bipyramidal geometry. Side chains from two aspartate, three glutamate, a serine residue, and a phenylalanine carbonyl oxygen are the ligands to one Ca^{2+} ion. Three aspartate and two glutamate side chains, a lysine carbonyl oxygen, and a water molecule are bound to the other Ca^{2+} [1032].

Construction of the Phe-102–Trp mutant of rat parvalbumin introduces a fluorescent label into the protein which enables a study of cation-dependent conformational changes. Trp-102 is confined to a hydrophobic core and conformationally is strongly restricted. Upon Ca^{2+} or Mg^{2+} binding the structural organization of the region around the Trp is hardly affected, but there are significant changes in its electrostatic environment [1033]. The Ca^{2+} and Mg^{2+} forms of parvalbumin have been investigated using two-dimensional ^1H NMR [1034].

7.4 Pyruvate Kinase

The stereochemistry of Mn^{2+} coordinated to the terminal thiophosphoryl group of [γ-^{17}O]adenosine 5′-*O*-(3-thiotriphosphate) in complexes with oxalate and pyruvate kinase from rabbit skeletal muscle has been deduced using EPR [1035]. Mn^{2+} binds to oxalate and the pro-(*R*)-oxygen of the thiophosphoryl group of adenosine 5′-*O*-(3-thiotriphosphate).

Substrate activities and binding affinities of the stereoisomers of the β,γ-bidentate $Rh(H_2O)_4ATP$ and α,β,γ-tridentate $Rh(H_2O)_3ATP$ complexes toward pyruvate kinase have been reported. Pyruvate kinase recognized both Δ-β,γ-bidentate-$Rh(H_2O)_4ATP$ and Δ-β-P,*exo*-α-P α,β,γ-tridentate $Rh(H_2O)_3ATP$ as substrates in the catalyzed phosphorylation of the alternate substrate, glycolate [845].

ENDOR and pulse field-sweep EPR have been used to probe the environment of Mn^{2+} in the oxalate–ATP complex of pyruvate kinase. Well-resolved ENDOR features from ^{31}P γ-phosphate of pyruvate kinase and Mn(II)–nucleotide models were observed [1036, 1037].

8.1 Inorganic Pyrophosphatase

Inorganic pyrophosphatase from *Thermus thermophilus* has been crystallized and is the subject of crystallographic studies [1038]. Refinement of the enzyme from *S. cerevisiae* using an automatic refinement program has been described, but details of the structure are not yet published [1039]. Two independent structure determinations of the *E. coli* enzyme have been reported, but the resolution of the structures is such that details of the metal coordination are not clear at this stage [1040, 1041].

REFERENCES

[1] Hughes, N. P. and Williams, R. J. P., in Graham, R. D. (ed.), *Manganese in Soils and Plants*, Kluwer Academic, 1988, pp. 7–19.

[2] Keen, C. L., Zidenberg-Cherr, S. and Lönnerdal, B., in Kies, C. (ed.), *Nutritional Bioavailability of Manganese*, ACS Symposium Series Vol. 354, American Chemical Society, Washington, DC, 1987, pp. 21–34.

[3] Korc, M., in Prasad, A. S. (ed.), *Essential and Toxic Trace Elements in Human Health and Disease*, Alan R. Liss, New York, 1988, pp. 253–273.

[4] Pecoraro, V. L., Photochem. Photobiol., 48 (1988) 249–264.

[5] Wieghardt, K., Angew. Chem. Int. Ed. Engl., 228 (1989) 1153–1172.

[6] Vincent, J. B. and Christou, G., Adv. Inorg. Chem., 33 (1989) 197–257.

[7] Brudvig, G. W. and Crabtree, R. H., Prog. Inorg. Chem., 37 (1989) 99–142.

[8] Christou, G., Acct. Chem. Res., 22 (1989) 328–335.

[9] Brudvig, G. W., Thorp, H. H. and Crabtree, R. H., Acct. Chem. Res., 24 (1991) 311–316.

[10] Pecoraro, V. L. (ed.), *Manganese Redox Enzymes*, Springer Verlag, New York, 1982.

[11] Schramm, V. L. and Wedler, F. C. (eds.), *Manganese in Metabolism and Enzyme Function*, Academic Press, Orlando, FL, 1986.

[12] Debus, R. J., Biochim. Biophys. Acta, 1102 (1992) 269–352.

[13] Brudvig, G. W., Beck, W. F. and De Paula, J. C., Ann. Rev. Biophys. Biophys. Chem., 18 (1989) 25–46.

[14] Rutherford, A. W., Trends Biochem. Sci., 14 (1989) 227–232.

[15] Babcock, G. T., Barry, B. A., Debus, R. J., Hoganson, C. W., Atamian, M., McIntosh, L., Sithole, I. and Yocum, C. F., Biochemistry, 28 (1989) 9557–9565.

[16] Hansson, O. and Wydrzynski, T., Photosyn. Res., 23 (1990) 131–162.
[17] Ghanotakis, D. F. and Yocum, C. F., Ann. Rev. Plant Physiol. Plant Mol. Biol., 41 (1990) 255–276.
[18] Vermaas, W. F. J. and Ikouchi, M., in Bogorad, L. Vasil, I. K. (eds.), *The Photosynthetic Apparatus Molecular Biology and Operation*, Academic Press, San Diego, 1991, pp. 25–111.
[19] Rutherford, A. W., Zimmermann, J.-L. and Boussac, A., in Barber J. (ed.), *The Photosystems: Structure Function and Molecular Biology*, Elsevier Science Publishers B.V., Amsterdam, 1992, pp. 179–229.
[20] Andersson, B. and Styring, S., in Lee C. P. (ed.), *Current Topics in Bioenergetics*, Vol. 16, Academic Press, San Diego, 1991, pp. 1–81.
[21] Babcock, G. T., in Amesz, J. (ed.), *New Comprehensive Biochemistry*, Vol. 15, Elsevier, Amsterdam, 1987, pp. 125–158.
[22] Son, C. L., Jones, G. P., Graham, R. D. and Zaranas, B. A., Soil Biol. Biochem., 22 (1990) 507–510.
[23] McCain, D. C. and Markley, J. L., Plant Physiol., 90 (1989) 1417–1421.
[24] Przemeck, E. and Haase, N. U., Water Air Soil Pollut., 57–58 (1991) 568–577.
[25] Das, S. C., Mandal, B. and Mandal, L. N., Plant Soil, 138 (1991) 75–84.
[26] Mench, M. and Martin, E., Plant Soil, 132 (1991) 187–196.
[27] Anke, M., Groppel, B., Krause, U., Arnhold, W. and Langer, M., J. Trace Elem. Electrol. Health Dis., 5 (1991) 69–74.
[28] Kihira, T., Mukoyama, M., Kazuo, A., Yoshiro, Y. and Masayuki, Y., J. Neurosci., 98 (1990) 251–258.
[29] Milne, D. B., Sims, R. L. and Ralston, N. V. C., Clin. Chem., 36 (1990) 450–452.
[30] Nishida, M., Sakurai, H., Tezuka, U., Kawada, J., Koyama, M. and Takada, J., Clin. Chim. Acta, 187 (1990) 181–187.
[31] Mitchell, J. D., East, B. W., Harris, I. A. and Pentland, B., Eur. Neurology, 31 (1991) 7–11.
[32] Suzuki, K., Kawaguchi, T., Nakao, H., Kawamura, N., Taniguchi, M., Kanayama, Y., Yonezawa, T. and Taniguchi, N., Cancer Lett., 62 (1992) 211–216.
[33] Hua, M. S. and Huang, C. C., J. Clin. Exp. Neurophysiol., 13 (1991) 495–507.
[34] Friedman, B. J., Freeman-Graves, J. H., Bates, C. W., Behmardi, F., Shorey-Kutschke, R. L., Willis, R. A., Crosby, J. B., Trickett, P. C. and Houston, S. D., J. Nutr., 117 (1987) 133–143.
[35] Linder, M. C., in Linder, M. C. (ed.), *Nutrition and Biochemistry of the Trace Elements*, Elsevier, New York, 1991, pp. 213–276.
[36] Schmid, J. and Auling, G., Agric. Biol. Chem., 53 (1989) 1783–1788.
[37] Aschner, M. and Aschner, J. L., Brain Res. Bull., 24 (1990) 857–860.
[38] Davidson, L. A. and Lönnerdal, B., Am. J. Physiol., 257 (1989) G930–G934.
[39] Kodama, H., Shimojo, N. and Suzuki, K. T., Biochem. J., 278 (1991) 857–862.
[40] Williams, R. J. P., FEBS Lett., 140 (1982) 3–10.
[41] Ash, D. E. and Schramm, V. L., J. Biol. Chem., 257 (1982) 9261–9264.
[42] Nishida, M., Sakurai, H. and Ishizuka, H., Saibo, 22 (1990) 23–27.
[43] Tholey, G., Ledig, M., Mandel, P., Sargentini, L., Frivold, A. H., Leroy, M., Grippo, A. A. and Wedler, F. C., Neurochem. Res., 13 (1988) 45–50.
[44] Tholey, G., Ledig, M., Kopp, P., Sargentini-Maier, L., Leroy, M., Grippo, H. A. and Wedler, F. C., Neurochem. Res., 13 (1988) 1163–1167.
[45] Collins, Y. E. and Stotzky, G., in Beveridge, T. J. Doyle, R. J. (ed.), *Metal Ions and Bacteria*, John Wiley and Sons, New York, 1989, pp. 31–90.
[46] Silver, S. and Walderhaug, M., Microb. Rev., 56 (1992) 195–228.
[47] Archibald, F., CRC Crit. Rev. Microb., 13 (1986) 63–109.
[48] Emerson, D. and Ghiorse, W. C., Appl. Environ. Microbiol., 58 (1992) 4001–4010.
[49] Nealson, K. H., Tebo, B. M. and Rosson, R. A., Adv. Appl. Microb., 33 (1988) 279–318.

[50] Nealson, K. H., Rosson, R. A. and Meyers, C. R., in Beveridge, T. J. Doyle, R. J. (ed.), *Metal Ions and Bacteria*, John Wiley and Sons, New York, 1989, pp. 383–411.
[51] Lovley, D. R., Microb. Rev., 55 (1991) 259–287.
[52] Di-Ruggiero, J. and Gounot, A. M., Microbial Ecology, 20 (1990) 53–63.
[53] Adams, L. F. and Ghiorse, W. C., J. Bacteriol., 169 (1987) 1279–1285.
[54] Kabata, H., Inui, K.-I. and Itokawa, Y., Nutrition Res., 9 (1989) 791–799.
[55] Gavin, C. E., Gunter, K. K. and Gunter, T. E., Analyt. Biochem., 192 (1991) 44–48.
[56] Aschner, M., Gannon, M. and Kimelberg, H. K., J. Neurochem., 58 (1992) 730–735.
[57] Aschner, M. and Aschner, J. L., Neurosci. Biobehav. Rev., 15 (1991) 333–340.
[58] Liebowitz, P. J., Schwartzenberg, L. S. and Bruce, A. K., Photochem. Photobiol., 23 (1976) 45–50.
[59] Archibald, F. S. and Fridovich, I., Arch. Biochem. Biophys., 215 (1982) 589–596.
[60] Archibald, F. S. and Fridovich, I., Arch. Biochem. Biophys., 214 (1982) 452–463.
[61] Archibald, F. S. and Fridovich, I., J. Bacteriol., 146 (1981) 928–936.
[62] Archibald, F. S. and Fridovich, I., J. Bacteriol., 145 (1980) 442–451.
[63] Galiazzo, F., Pedersen, J. Z., Civitareale, P., Schiesser, A. and Rotilio, G., Biol. Met., 2 (1989) 6–10.
[64] Sakurai, H., Nishida, M., Yoshimura, T., Takada, J. and Koyama, M., Biochim. Biophys. Acta, 841 (1985) 208–214.
[65] Ezra, F. S., Lucas, D. S. and Russell, A. F., Biochim. Biophys. Acta, 803 (1984) 90–94.
[66] Mildvan, A. S., Rosevear, P. R., Granot, J., O'Brien, C. A. and Bramson, H. N., Methods Enzymol., 99 (Horm. Act. Part F) (1983) 93–119.
[67] Chou, F. I. and Tan, S. T., J. Bacteriol., 172 (1990) 2029–2035.
[68] Grinnell, F., J. Cell Science, 65 (1984) 61–72.
[69] Yanai, T., Shimo-Oka, T. and Ii, I., Cell Struct. Funct., 16 (1991) 149–156.
[70] Maddox, B. A. and Goldfine, I. D., J. Biol. Chem., 266 (1991) 6731–6736.
[71] Kalbitzer, H. R., in Sigel, H. (ed.), *Metal Ions in Biological Systems*, Vol. 22, Marcel Dekker, New York, 1987, pp. 81–100.
[72] Mildvan, A. S., FASEB J., 3 (1989) 1705–1714.
[73] Mildvan, A. S. and Fry, D. C., Adv. Enzymol., 59 (1987) 241–313.
[74] Villafranca, J. J. and Rauschel, F. M., Adv. Inorg. Biochem., 4 (1982) 289–319.
[75] Villafranca, J. J. and Nowak, T., in Sigman, D. S. (ed.), *The Enzymes* Vol XX, 3rd. Ed., Academic Press, 1992, pp. 63–94.
[76] Ramírez, M., Arechaga, G., Garcia, S., Sanchez, B., Lardelli, P. and de Gandarias, J. M., Brain Res., 522 (1990) 165–167.
[77] Smith, D. L., Almo, S. C., Toney, M. D. and Ringe, D., Biochemistry, 28 (1989) 8161–8167.
[78] McTigue, M. A., Davies, J. F., Kaufman, B. T. and Kraut, J., Biochemistry, 31 (1992) 7264–7273.
[79] Mohammed, A. M. S., Al-Chalabi, K. and Abood, S. A., J. Exp. Bot., 40 (1989) 693–699.
[80] Umazume, K., Nishida, M., Sakurai, H., Yoshimura, Y. and Kawada, J., Biochem. Int., 21 (1990) 85–95.
[81] Magee, P. T. and Snell, E. E., Biochemistry, 5 (1966) 409–416.
[82] Beffa, R., Martin, H. V. and Pilel, P. E., Plant Physiol., 94 (1990) 485–491.
[83] Koyama, H., Agric. Biol. Chem., 52 (1988) 743–748.
[84] Schabort, J. C. and Potgieter, D. J. J., Biochim. Biophys. Acta, 151 (1968) 47–54.
[85] Nair, P. M. and Vining, L. C., Biochim. Biophys. Acta, 96 (1965) 318–327.
[86] Shampengtong, L., Wong, K. P. and Ho, B. C., Insect Biochem., 17 (1987) 111–116.
[87] Mattevi, A., Obmolova, G., Schulze, E., Kalk, K. H., Westphal, A. H., de Kok, A. and Hol, W. G. J., Science, 255 (1992) 1544–1550.
[88] Das, S. and Gillin, F. D., Biochem. J., 280 (1991) 641–647.
[89] Yadav, S. P. and Brew, K., J. Biol. Chem., 266 (1991) 698–703.
[90] Elices, M. J. and Goldstein, I. J., J. Biol. Chem., 263 (1988) 3354–3362.

[91] Witsell, D. L., Casey, C. E. and Neville, M. C., J. Biol. Chem., 265 (1990) 15731–15737.
[92] Myllylä, R., Anttinen, H. and Kivirikko, K. J., Eur. J. Biochem., 101 (1979) 261–269.
[93] Helting, T. and Erbing, B., Biochim. Biophys. Acta, 293 (1973) 94–104.
[94] Sloan, D. L., Ali, L. Z., Picou, D. and Joseph Jr., A., Adv. Exp. Med. Biol., 165 (Purine (1984) Metab. Man-4 Pt. B) 45–50.
[95] McNatt, M. L., Fiser, F. M., Elders, M. J., Kilgore, B. S., Smith, W. G. and Hughes, E. R., Biochem. J., 160 (1976) 211–216.
[96] Wente, S. R., Villalba, M., Schramm, V. L. and Rosen, O. M., Proc. Natl. Acad. Sci. USA, 87 (1990) 2805–2809.
[97] Sefton, B. M. and Hunter, T., Adv. Cyclic Nucleotide Protein Phosphorylation Res., 18 (1984) 195–226.
[98] Koland, J. G. and Cerione, R. A., Biochim. Biophys. Acta, 1052 (1990) 489–498.
[99] Job, C., Briat, J. F., Lescure, A. M. and Job, D., Eur. J. Biochem., 165 (1987) 515–519.
[100] Koren, R. and Mildvan, A. S., Biochemistry, 16 (1977) 241–249.
[101] Gross, G. and Dunn, J. J., Nucleic Acids Res., 15 (1987) 431–442.
[102] Wahle, E., J. Biol. Chem., 266 (1991) 3131–3139.
[103] Lingner, J., Radtke, I., Wahle, E. and Keller, W., J. Biol. Chem., 266 (1991) 8741–8746.
[104] Brautigan, D. L., Ballou, L. M. and Fischer, E. H., Biochemistry, 21 (1982) 1977–1982.
[105] McNall, S. J. and Fischer, E. H., J. Biol. Chem., 263 (1988) 1893–1897.
[106] Stevenson, G. M. M., Arch. Biochem. Biophys., 289 (1991) 324–328.
[107] Ikura, Y. and Horikoshi, K., Agric. Biol. Chem., 54 (1990) 3205–3509.
[108] Pang, X. P., Ross, N. S., Park, M., Juillard, G. J. F., Stanley, T. M. and Hershman, J. M., J. Biol. Chem., 267 (1992) 12826–12830.
[109] Shibata, K. and Watanabe, T., J. Bacteriol., 169 (1987) 3409–3413.
[110] Iwayama, A., Kimura, T., Adachi, O. and Ameyama, M., Agric. Biol. Chem., 47 (1983) 2483–2493.
[111] Kunze, N., Kleinkauf, H. and Bauer, K., Eur. J. Biochem., 160 (1986) 605–613.
[112] Lenney, J. F., Biol. Chem. Hoppe-Seyler, 371 (1990) 433–440.
[113] Lenney, J. F., Biol. Chem. Hoppe-Seyler, 371 (1990) 167–171.
[114] Endo, F. and Matsuda, I., Mol. Biol. Med., 8 (1991) 117–127.
[115] Van Boven, A., Tan, P. S. T. and Konings, W. N., Appl. Environ. Microb., 54 (1988) 43–49.
[116] Houck, D. R. and Inamine, E., Arch. Biochem. Biophys., 259 (1987) 58–65.
[117] Lenz, H., Wunderwald, P. and Eggerer, H., Eur. J. Biochem., 65 (1976) 225–236.
[118] Ash, D. E., Emig, F. A., Chowdhury, S. A., Satoh, Y. and Schramm, V. L., J. Biol. Chem., 265 (1990) 7377–7384.
[119] Wiegel, J., Biochem. Biophys. Res. Commun., 82 (1978) 907–912.
[120] Glaser, R. D. and Houston, L. L., Biochemistry, 13 (1974) 5145–5152.
[121] Takada, Y. and Noguchi, T., Biochem. J., 235 (1986) 391–397.
[122] Wells, X. E. and Lees, E. M., Arch. Biochem. Biophys., 287 (1991) 151–159.
[123] Yamanaka, K. and Izumori, K., Meth. Enzymol., 61 (1975) 465–465.
[124] Boulter, J. R. and Gielow, W. O., J. Bacteriol., 113 (1973) 687–696.
[125] Patrick, J. and Lee, N., Meth. Enzymol., 61 (1975) 453–458.
[126] Yamanaka, K., Meth. Enzymol., 61 (1975) 458–462.
[127] Allenza, P., Morrell, M. J. and Detroy, R. W., Appl. Biochem. Biotechnol., 24–25 (1990) 171–182.
[128] Noltmann, E. A., in Boyer, R. (ed.), *The Enzymes* 3rd. Ed. Vol VI, Academic Press, New York, 1972, pp. 302–314.
[129] Domagk, G. F. and Zech, R., Meth. Enzymol., 9 (1966) 579–582.
[130] Anderson, R. L., Meth. Enzymol., 9 (1966) 593–596.
[131] Hortitsu, H., Kawai, Y., Konishi, H. and Kawai, K., Biosci. Biotech. Biochem., 58 (1992) 165–166.

[132] Elbein, A. D. and Izumori, K., Meth. Enzymol., 89 (1982) 547–550.
[133] Oh, Y. K. and Freese, E., J. Bacteriol., 127 (1976) 739–746.
[134] Kuhn, N. J., Setlow, B. and Setlow, P., Arch. Biochem. Biophys., 306 (1993) 342–349.
[135] Levy-Favatier, F., Depech, M. and Kruth, R., Eur. J. Biochem., 166 (1987) 617–621.
[136] Gupta, R. C., Khandelwal, R. L. and Sulakhe, P. V., FEBS Lett., 196 (1986) 39–43.
[137] Mukai, H., Ito, A., Kishima, K., Kuno, T. and Tanaka, C., J. Biochem., 110 (1991) 402–406.
[138] Karns, J. S. and Tomasek, P. H., J. Agric. Food Chem., 39 (1991) 1004–1008.
[139] Meyers, P. R., Gokool, P., Rawlings, D. E. and Woods, D. R., J. Gen. Microb., 137 (1991) 1397–1400.
[140] Miyake, M., Innami, T. and Kakimoto, Y., Biochim. Biophys. Acta, 760 (1983) 206–214.
[141] Takatsuji, H., Nishino, T., Izui, K. and Katsuki, H., J. Biochem. (Tokyo), 91 (1982) 911–921.
[142] Kuhm, A. E., Schlömann, M., Knackmuss, H.-J. and Pieper, D. H., Biochem. J., 266 (1990) 877–883.
[143] Evans, D. M. and Moseley, B. E. B., Mutat. Res., 145 (1985) 119–128.
[144] Levin, J. D., Shapiro, R. and Demple, B., J. Biol. Chem., 266 (1991) 22893–22898.
[145] Kaminskas, E., Kimhi, Y. and Magasanik, B., J. Biol. Chem., 245 (1968) 3536–3544.
[146] Lund, P. and Magasanik, B., J. Biol. Chem., 240 (1965) 4316–4319.
[147] Legaz, M. E., Martin, L., Pedrosa, M. M., Vincente, C., de Armas, R., Martínez, M., Medina, I. and Rodriquez, C. W., Plant Physiol., 92 (1990) 679–683.
[148] Kato, Y. and Spiro, R. G., J. Biol. Chem., 264 (1989) 3364–3371.
[149] Suga, T. and Endo, T., Phytochem., 30 (1991) 1757–1761.
[150] Handley, L. W. and Pharr, D. M., Z. Pflanzenphysiol., 108 (1982) 447–455.
[151] Takeya, A., Hosomi, O. and Kogure, T., Jpn. J. Med. Sci., 38 (1985) 1–8.
[152] Battermann, G. and Radler, F., Can. J. Microb., 37 (1991) 211–217.
[153] Cox, D. J. and Henick-Kling, T., J. Bacteriol., 171 (1989) 5750–5752.
[154] Lonvaud-Funel, A. and Strasser de Saad, A. M., Appl. Environ. Microb., 43 (1982) 357–361.
[155] Hoflack, B. and Kornfeld, S., J. Biol. Chem., 260 (1985) 12008–12014.
[156] Wendland, M., Waheed, A., Schmidt, B., Hille, A., Nagel, G., von Figura, K. and Pohlmann, R., J. Biol. Chem., 266 (1991) 4598–4604.
[157] Ma, Z., Grubb, J. H. and Sly, W. S., J. Biol. Chem., 266 (1991) 10589–10595.
[158] Klier, H.-J., von Figura, K. and Pohlmann, R., Eur. J. Biochem., 197 (1991) 23–28.
[159] Li, M., Distler, J. J. and Jourdian, G. W., Arch. Biochem. Biophys., 283 (1990) 150–157.
[160] Romero, P. A. and Herscovics, A., J. Biol. Chem., 264 (1989) 1946–1950.
[161] De Malkenson, N. C., Wood, E. J. and Zerba, E. N., Insect Biochem., 14 (1984) 481–486.
[162] Chester, N. A., Comp. Biochem. Physiol. C Comp. Physiol., 94C (1989) 365–371.
[163] Caldwell, S. R. and Raushel, F. M., Biotech. Bioeng., 37 (1991) 103–109.
[164] Graebe, J. E., Phytochem., 7 (1968) 2003–2020.
[165] Dogbo, O., Laferrière, A., D'Harlingue, A. and Camara, B., Proc. Natl. Acad. Sci. USA, 85 (1988) 7054–7058.
[166] Schmidt, B., Wachter, E., Sebald, W. and Neupert, W., Eur. J. Biochem., 144 (1984) 581–588.
[167] Yu, J. S. and Yang, S.-D., J. Protein Chem., 8 (1989) 499–517.
[168] Shenolikar, S. and Nairn, A. C., Adv. Second Messenger Phosphoprotein Res., 23 (1990) 1–121.
[169] Fischer, E. H., Charbonneau, H. and Tonks, N. K., Science, 253 (1991) 401–406.
[170] Ou, W.-J., Ito, A., Okazaki, H. and Omura, T., EMBO J., 8 (1989) 2605–2612.
[171] Hawlitschek, G., Schneider, H., Schmidt, B., Tropschung, M., Hartl, F.-U. and Neupert, W., Cell, 53 (1988) 795–806.
[172] Hallahan, T. W. and Croteau, R., Arch. Biochem. Biophys., 264 (1988) 618–631.
[173] Yu, K.-T., Khalaf, Y. N. and Czech, M. P., J. Biol. Chem., 262 (1987) 16677–16685.
[174] Weische, A., Garvert, W. and Leistner, E., Arch. Biochem. Biophys., 256 (1987) 223–231.
[175] Bosman, B. W., Tan, P. S. T. and Konings, W. N., Appl. Environ. Microb., 56 (1990) 1839–1843.
[176] Tipton, P. A. and Peisach, J., Biochemistry, 30 (1991) 739–744.

[177] Djaballah, H., Staniforth, A. D., Ward, S. and Kuhn, N. J., Biochem. Soc. Trans., 19 (1991) 235S.
[178] Navaratnam, N., Ward, S., Fischer, C., Kuhn, N. J., Keen, J. N. and Findlay, J. B. C., Eur. J. Biochem., 171 (1988) 623–629.
[179] Kuhn, N. J., Ward, S. and Leong, W. S., Eur. J. Biochem., 195 (1991) 243–250.
[180] Aoki, D., Appert, H. E., Johnson, D., Wong, S. S. and Fukuda, M. N., EMBO J., 9, 3171–3178.
[181] Markham, G. D., J. Biol. Chem., 256 (1981) 1903–1909.
[182] Livera, W. C. D. and Shimizu, C., Agric. Biol. Chem., 53 (1989) 2377–2386.
[183] Jarori, G. K., Ray, B. D. and Nageswara Rao, B. D., Biochemistry, 28 (1989) 9343–9350.
[184] Kofron, J. L., Ash, D. E. and Reed, G. H., Biochemistry, 27 (1988) 4781–4787.
[185] Mock, W. L. and Zhuang, H., Biochem. Biophys. Res. Commun., 180 (1991) 401–406.
[186] Tanoue, A., Endo, F., Kitano, A. and Matsuda, I., J. Clin. Invest., 86 (1990) 351–355.
[187] Summerfield, A. E., Bauerle, R. and Grisham, C. M., J. Biol. Chem., 263 (1988) 18793–18801.
[188] Cammack, R., Chapman, A., Lu, W.-P., Karagouni, A. and Kelly, D. P., FEBS Lett., 253 (1989) 239–243.
[189] Kemple, M. D., Lovejoy, M. L., Ray, B. D., Prendergast, F. S. and Nageswara Rao, B. D., Eur. J. Biochem., 187 (1990) 131–135.
[190] Kemple, M. D., Ray, B. D., Jarori, G. K., Nageswara Rao, B. D. and Prendergast, F. G., Biochemistry, 23 (1984) 4383–4390.
[191] Hannick, L. I., Prasher, D. C., Schultz, L. W., Deschamps, J. R. and Ward, K. B., Proteins Struct. Funct. Genetics, 15 (1993) 103–105.
[192] Wedding, R. T., Plant Physiol., 90 (1989) 367–371.
[193] Drincovich, M. F., Iglesias, A. A. and Andreo, C. S., Physiol. Plant., 81 (1991) 462–466.
[194] Clancy, L. L., Rao, G. S. J., Finzel, B. C., Muchmore, S. W., Holland, D. R., Watenpaugh, K. D., Krishamurthy, H. M., Sweet, R. M., Cook, P. F., Harris, B. G. and Einspahr, H. M., J. Mol. Biol., 226 (1992) 565–569.
[195] Veser, J., J. Bacteriol., 169 (1987) 3696–3700.
[196] Vidgren J., Tilgmann, C., Lundstrom, K. and Liljas, A., Proteins Struct. Funct. Genetics, 11 (1991) 233–236.
[197] Sträter, N., Fröhlich, R., Schiemann, A., Krebs, B., Körner, M., Suerbaum, H. and Witzel, H., J. Mol. Biol., 224 (1992) 511–513.
[198] Green, S. M., Knutson, J. R. and Hensley, P., Biochemistry, 29 (1990) 9159–9168.
[199] Green, S. M., Ginsburg, A., Lewis, M. S. and Hensley, P., J. Biol. Chem., 266 (1991) 21474–21481.
[200] Maggini, S., Stoecklin-Tschan, F. B., Mörikofer-Zwec, S. and Walter, P., Biochem. J., 283 (1992) 653–660.
[201] Kanyo, Z. F., Chen, C. Y., Daghigh, F., Ash, D. E. and Christianson, D. W., J. Mol. Biol., 224 (1992) 1175–1177.
[202] Reczkowski, R. S. and Ash, D. E., J. Am. Chem. Soc., 114 (1992) 10992–10994.
[203] Penefsky, H. S. and Cross, R. L., Adv. Enzymol., 64 (1991) 173–214.
[204] Hiller, R. and Carmeli, C., Biochemistry, 29 (1990) 6186–6192.
[205] Mollinedo, F. and Cannistraro, S., Biochim. Biophys. Acta, 848 (1986) 224–229.
[206] Duncan, T. M. and Cross, R. L., J. Bioenerg. Biomembr., 29 (1992) 453–461.
[207] Allison, W. S., Jault, J. M., Zhuo, S. and Paik, S. R., J. Bioenerg. Biomembr., 29 (1992) 469–477.
[208] Grisham, C. M., Dev. Bioenerg. Biomembr., 6 (Struct. Funct. Membr. Proteins) (1983) 63–70.
[209] Haddy, A. E., Frasch, W. D. and Sharp, R. R., Biochemistry, 28 (1989) 3664–3669.
[210] Haddy, A. E. and Sharp, R. R., Biochemistry, 28 (1989) 3656–3664.
[211] Devlin, C. and Grisham, C. M., Biochemistry, 29 (1990) 6192–6203.
[212] Carmeli, C., Huang, J. Y., Mills, D. M., Jagendorf, A. T. and Lewis, A., J. Biol. Chem., 261 (1986) 16969–16975.

[213] Codd, R., Cox, G. B., Guss, J. M., Solomon, R. G. and Webb, D., J. Mol. Biol., 228 (1992) 306–309.
[214] Malebrán, L. P. and Cardemil, E., Biochim. Biophys. Acta, 915 (1987) 385–392.
[215] Cheng, K.-C. and Nowak, T., J. Biol. Chem., 264 (1989) 3317–3324.
[216] Burnell, J. N., Aust. J. Plant Physiol., 13 (1986) 577–587.
[217] Abeysinghe, S. I. B., Baker, P. J., Rice, D. W., Rodgers, H. F., Stillman, T. J., Ko, Y. H., McFadden, B. A. and Nimmo, H. G., J. Mol. Biol., 220 (1991) 13–16.
[218] Giachetti, E., Vanni, P., Biochem. J., 276 (1991) 223–230.
[219] Tahama, H., Shinoyama, H., Ando, A. and Fujii, T., Agric. Biol. Chem., 54 (1990) 3177–3183.
[220] Hones, I., Simon, M. and Weber, H., J. Basic Microb., 31 (1991) 251–258.
[221] Hoyt, J. C., Johnson, K. E. and Reeves, H. C., J. Bacteriol., 173 (1991) 6844–6848.
[222] Rúa, J., Robertson, A. G. S. and Nimmo, H. G., Biochim. Biophys. Acta, 1122 (1992) 212–218.
[223] Kelly, C. T., Nash, A. M. and Fogarty, W. M., Appl. Microb. Biotechnol., 19 (1984) 61–66.
[224] Tulloch, A. G. and Pato, M. D., J. Biol. Chem., 266 (1991) 20168–20174.
[225] Kinnett, D. G. and Wilcox, F. H., Int. J. Biochem., 14 (1982) 977–981.
[226] Brissette, R. E., Swislocki, N. L. and Cunningham, E. B., Am. J. Hematol., 38 (1991) 166–173.
[227] Vincent, J. B. and Averill, B. A., FEBS Lett., 263 (1990) 265–268.
[228] Vincent, J. B., Crowder, M. W. and Averill, B. A., Trends in Biol. Sci., (1992) 105–110.
[229] Hunter, T., Cell, 58 (1989) 1013–1016.
[230] Mayer, R. E., Khew-Goodall, Y. and Syone, S. R., Adv. Prot. Phosphateses, 6 (1991) 265–286.
[231] Cohen, P., Annu. Rev. Biochem., 58 (1989) 453–508.
[232] Pot, D. A. and Dixon, J. E., Biochim. Biophys. Acta, 1136 (1992) 35–44.
[233] Li, H. C., in Horecker, B. L. Stadtman, E. R. (eds.), *Curr. Topics in Cellular Regulation*, Academic Press, N.Y., 1982, pp. 129–174.
[234] Stremmer, P. C. and Klee, C. B., Curr. Opin. Neurobiol., 1 (1991) 53–56.
[235] Dietrich, M., Münstermann, D., Sauerbaum, H. and Witzel, H., Eur. J. Biochem., 199 (1991) 105–113.
[236] Doi, K., Antanaitis, B. C. and Aisen, P., Struct. Bonding, 70 (1988) 1–26.
[237] Vincent, J. B. and Averill, B. A., FASEB J., 4 (1990) 3009–3014.
[238] Beck, J. L., McArthur, M. J., De Jersey, J. and Zerner, B., Inorg. Chim. Acta, 153 (1988) 39–44.
[239] Beck, J. L., De Jersey, J., Zerner, B., Hendrich, M. P. and Debrunner, P. G., J. Am. Chem. Soc., 110 (1988) 3317–3318.
[240] Uehara, K., Fujimoto, S. and Taniguchi, T., J. Biochem. (Tokyo), 70 (1971) 183–185.
[241] Hefler, S. K. and Averill, B. A., Biochem. Biophys. Res. Commun., 146 (1987) 1173–1177.
[242] Kawabe, H., Sugiura, Y., Terauchi, M. and Tanaka, H., Biochim. Biophys. Acta, 784 (1984) 81–89.
[243] Fujimoto, S., Ohara, A. and Uehara, K., Agric. Biol. Chem., 44 (1980) 1659–1660.
[244] Fujimoto, S., Yamada, K., Kuroda, T., Tanaka, T. and Ohara, A., Nippon Nogei Kagaku Kaishi, 60 (1986) 605–608.
[245] Holz, R. C., Que Jr., L. and Ming, L. J., J. Am. Chem. Soc., 114 (1992) 4434–4436.
[246] Sugiura, Y., Kawabe, H. and Tanaka, H., J. Am. Chem. Soc., 102 (1980) 6581–6582.
[247] Fujimoto, S., Murakami, K., Ohara, A., J. Biochem. (Tokyo), 97 (1985) 1777–1784.
[248] Fujimoto, S., Takebayashi, M., Kawamoto, N. and Ohara, A., Chem. Pharm. Bull., 35 (1987) 2011–2015.
[249] Sugiura, Y., Kawabe, H., Tanaka, H., Fujimoto, S. and Ohara, A., J. Am. Chem. Soc., 103 (1981) 963–964.
[250] Sugiura, S., Kawabe, H., Tanaka, H., Fujimoto, S. and Ohara, A., J. Biol. Chem., 256 (1981) 10664–10670.
[251] Kawabe, H., Sugiura, Y. and Tanaka, H., Biochem. Biophys. Res. Commun., 103 (1981) 327–331.
[252] Uehara, K., Fujimoto, S. and Taniguchi, T., J. Biochem. (Tokyo), 75 (1974) 627–638.

[253] Fujimoto, S., Nakagawa, T., Ishimitsu, S. and Ohara, A., Chem. Pharm. Bull., 25 (1977) 1459–1462.
[254] Igaue, I., Watabe, H., Takahashi, K., Takekoshi, M. and Morota, A., Agric. Biol. Chem., 40 (1976) 823–825.
[255] Hayakawa, T., Toma, Y. and Igaue, I., Agric. Biol. Chem., 53 (1989) 1475–1483.
[256] Fujimoto, S., Nakagawa, T. and Ohara, A., Chem. Pharm. Bull., 25 (1977) 3283–3288.
[257] LeBansky, B. R., McKnight, T. D. and Griffing, L. R., Plant Physiol., 99 (1992) 391–395.
[258] David, S. S. and Que Jr., L., J. Am. Chem. Soc., 112 (1990) 6455–6463.
[259] Davis, J. C. and Averill, B. A., Proc. Natl. Acad. Sci. USA, 79 (1982) 4623–4627.
[260] Vincent, J. B., Crowder, M. W. and Averill, B. A., Biochemistry, 30 (1991) 3025–3034.
[261] Hayman, A. R., Warburton, M. J., Pringle, J. A. S. and Chambers, T. J., Biochem. Soc. Trans., 16 (1988) 895–895.
[262] Brauer, M. and Sykes, B. D., Biochemistry, 21 (1982) 5934–5939.
[263] Kabsch, W., Mannherz, H. G., Suck, D., Pai, E. F. and Holmes, K. C., Nature (London), 347 (1990) 37–44.
[264] Ohlendorf, D. H., Lipscomb, J. D. and Weber, P. C., Nature (London), 336 (1988) 403–405.
[265] True, A. E., Orville, A. M., Pearce, L. L., Lipscomb, J. D. and Que Jr., L., Biochemistry, 29 (1990) 10847–10854.
[266] Que Jr., L., Widom, J. and Crawford, R. L., J. Biol. Chem., 256 (1981) 10941–10944.
[267] Kalbitzer, H. R., Feuerstein, J., Goody, R. S. and Wittinghofer, A., Eur. J. Biochem., 188 (1990) 355–359.
[268] Eccleston, J. F., Webb, M. R., Ash, D. E. and Reed, G. H., J. Biol. Chem., 256 (1981) 10774–10777.
[269] Kalbitzer, H. R. and Wittinghofer, A., Biochim. Biophys. Acta, 1078 (1991) 133–138.
[270] Cool, R. H. and Parmeggiani, A., Biochemistry, 30 (1991) 362–366.
[271] Kjeldgaard, M. and Nyborg, J., J. Mol. Biol., 223 (1992) 721–742.
[272] Reshetnikova, L. S., Reiser, C. O. A., Schirmer, N. K., Berchtold, H., Storm, R., Hilgenfeld, R. and Sprinzl, M., J. Mol. Biol., 221 (1991) 375–377.
[273] Lippmann, C., Betzel, C., Dauter, Z., Wilson, K. and Erdmann, V. A., FEBS Lett., 240 (1988) 139–142.
[274] Mistou, M.-Y., Cool, R. H. and Parmeggiani, A., Eur. J. Biochem., 204 (1992) 179–185.
[275] Bourne, H. R., Landis, C. and Masters, S. B., Proteins Struct. Funct. Genetics, 6 (1989) 222–230.
[276] Harmark, K., Anborgh, P. H., Merola, M., Clark, B. F. C. and Parmeggiani, A., Biochemistry, 31 (1992) 7367–7372.
[277] Pai, E. F., Krengel, U., Petsko, G. A., Goody, R. S., Kabsch, W. and Wittinghofer, A., EMBO J., 9 (1990) 2351–2359.
[278] Pai, E. F., Kabsch, W., Krengel, U., Holmes, K. C., John, J. and Wittinghofer, A., Nature (London), 341 (1990) 209–214.
[279] Hurley, J. H., Dean, A. M., Thorsness, P. E., Koshland Jr., D. E. and Stroud, R. M., J. Biol. Chem., 265 (1990) 3599–3602.
[280] Ehrlich, R. S. and Colman, R. F., Biochemistry, 28 (1989) 2058–2065.
[281] Hurley, J. H., Dean, A. M., Sohl, J. L., Koshland Jr., D. E. and Stroud, R. M., Science, 249 (1990) 1012–1016.
[282] Hunter, T., Adv. Enzymol., 200 (1991) 3–37.
[283] Schulman, H., Curr. Opin. Neurobiol., 1 (1991) 43–52.
[284] Wagner, K. R., Mei, L. and Huganir, R. L., Curr. Opin. Neurobiol., 1 (1991) 65–73.
[285] Wilson, M. E. and Consigli, R. A., Virology, 143 (1985) 516–525.
[286] Schnyder, T., Gross, H., Winkler, H., Eppenberger, H. M. and Wallimann, T., J. Cell Biol., 112 (1991) 95–101.
[287] Reed, G. H. and Leyh, T. S., Biochemistry, 19 (1980) 5472–5480.

[288] Leyh, T. S., Goodhart, P. J., Nguyen, A. C., Kenyon, G. L. and Reed, G. H., Biochemistry, 24 (1985) 308–311.
[289] Jarori, G. K., Ray, B. D. and Nageswara Rao, B. D., Biochemistry, 24 (1985) 3487–3494.
[290] Berger, A., Schiltz, E. and Schulz, G. E., Eur. J. Biochem., 184 (1989) 433–443.
[291] Nageswara Rao, B. D., Cohn, M. and Noda, L., J. Biol. Chem., 253 (1978) 1149–1158.
[292] O'Sullivan, W. J. and Noda, L., J. Biol. Chem., 243 (1968) 1425–1433.
[293] Vetter, I. R., Konrad, M. and Rösch, P., Biochemistry, 30 (1991) 4137–4142.
[294] Egner, U., Tomasselli, A. G. and Schulz, G. E., J. Mol. Biol., 195 (1987) 649–658.
[295] Müller, C. W. and Schulz, G. E., J. Mol. Biol., 224 (1992) 159–177.
[296] Schultz, G. E., Elzinga, M., Marx, F. and Schirmer, R. H., Nature (London), 250 (1974) 120–123.
[297] Dreusicke, D., Karplus, P. A. and Schulz, G. E., J. Mol. Biol. 199 (1988) 359–371.
[298] Dreusicke, D. and Schulz, G. E., J. Mol. Biol., 203 (1988) 1021–1028.
[299] Müller, C. W. and Schulz, G. E., Proteins Struct. Funct. Genetics, 15 (1993) 42–49.
[300] Stehle, T. and Schulz, G. C., J. Mol. Biol., 224 (1992) 1127–1141.
[301] Stehle, T. and Schulz, G. E., Acta Cryst. B Struct. Sci., 48B (1992) 546–548.
[302] Kronman, M. J., Crit. Rev. Biochem. Mol. Biophys., 24 (1989) 565–667.
[303] Desmet, J. and Van Cauwelaert, F., Biochim. Biophys. Acta, 957 (1988) 411–419.
[304] Desmet, J., Van Dael, H., Van Cauwelaert, F., Nitta, K. and Sugai, S., J. Inorg. Biochem., 37 (1989) 185–191.
[305] Desmet, J., Haezebrouck, P. and Van Cauwelaert, F., J. Inorg. Biochem., 42 (1991) 139–145.
[306] Desmet, J., Tieghem, E., Van Dael, H. and Van Cauwelaert, F., Eur. Biophys. J., 20 (1991) 263–268.
[307] Gerken, T. A., Biochemistry, 23 (1984) 4688–4697.
[308] Murakami, K., Andree, P. J. and Berliner, L. J., Biochemistry, 21 (1982) 5488–5494.
[309] Aramini, J. M., Drakenberg, T., Hiraoki, T., Ke, Y., Nitta, K. and Vogel, H. J., Biochemistry, 31 (1992) 6761–6768.
[310] Prestrelski, S. J., Byler, D. M. and Thompson, M. P., Biochemistry, 30 (1991) 8797–8804.
[311] Acharya, K. R., Stuart, D. I., Walker, N. P. C., Lewis, M. and Phillips, D. C., J. Mol. Biol., 208 (1989) 99–127.
[312] Acharya, K. R., Ren, J., Stuart, D. I., Phillips, D. C. and Fenna, R. E., J. Mol. Biol., 221 (1991) 571–581.
[313] Harata, K. and Muraki, M., J. Biol. Chem., 267 (1992) 1419–1421.
[314] Lönnerdal, B., Keen, C. L. and Hurley, L. S., Am. J. Clin. Nutr., 41 (1985) 550–559.
[315] Ainscough, E. W., Brodie, A. M. and Plowman, J. E., Inorg. Chim. Acta, 33 (1979) 149–153.
[316] Anderson, B. F., Baker, H. M., Norris, G. E., Rice, D. W. and Baker, E. N., J. Mol. Biol., 209 (1989) 711–734.
[317] Norris, G. E. and Anderson, B. F., Baker, E. N., Acta Cryst. Sect. B B47 (1991) 998–1004.
[318] Smith, C. A., Baker, H. M. and Baker, E. N., J. Mol. Biol., 219 (1991) 155–159.
[319] Baker, E. N., personal communication (1992).
[320] Gerlt, J. A., Kenyon, G. L., Kozarich, J. W., Neidhart, D. J., Petsko, G. A. and Powers, V. M., Curr. Opin. Struct. Biol., 2 (1992) 736–742.
[321] Fee, J. A., Hegeman, G. D. and Kenyon, G. L., Biochemistry, 13 (1974) 2528–2532.
[322] Neidhart, D. J., Howell, P. L., Petsko, G. A., Powers, V. M., Li, R., Kenyon, G. L. and Gerlt, J. A., Biochemistry, 30 (1991) 9264–9273.
[323] Powers, V. M., Koo, C. W., Kenyon, G. L., Gerlt, J. A., Kozarich, J. W., Biochemistry, 30 (1991) 9255–9263.
[324] Landro, J. A., Kallarakal, A. T., Ransom, S. C., Gerlt, J. A., Kozarich, J. W., Niedhart, D. J. and Kenyon, G. L., Biochemistry, 30 (1991) 9274–9281.
[325] Neidhart, D. J., Kenyon, G. L., Gerlt, J. A. and Petsko, G. A., Nature (London), 347 (1990) 692–694.
[326] Lewis, N. G. and Yamamoto, E., Ann. Rev. Plant Physiol. Plant Mol. Biol., 41 (1990) 455–496.

[327] Higuchi, T., Wood Sci. Technol., 24 (1990) 23–63.
[328] Haertig, C., Meyer, D. and Fischer, K., Z. Chem., 30 (1990) 233–239.
[329] Hammel, K. E. in Sigel H., Sigel, A. (eds.), *Metal Ions in Biological Systems*, Vol 28, Marcel Dekker, New York, 1992, pp. 41–60.
[330] Michel Jr., F. C., Dass, S. B., Grulke, E. A. and Reddy, C. A., Appl. Environ. Microb., 57 (1991) 2368–2375.
[331] Linko, S., Biotechnol. Adv., 10 (1992) 191–236.
[332] Galliano, H., Gas, G., Seris, J. L. and Boudet, A. M., Enzyme Microb. Technol., 13 (1991) 478–482.
[333] Leatham, G. F., Appl. Microbiol. Biotechnol., 24 (1986) 51–58.
[334] Galliano, H., Gas, G. and Boudet, A. M., Plant Physiol. Biochem., 26 (1988) 619–627.
[335] Sarkanen, S., Razal, R. A., Piccariello, T., Yamamoto, E. and Lewis, N. G., J. Biol. Chem., 266 (1991) 3636–3643.
[336] Wariishi, H., Valli, K. and Gold, M. H., Biochem. Biophys. Res. Commun., 176 (1991) 269–275.
[337] Leisola, M. S. A., Kozulic, B., Meussdoeffer, F. and Fiechter, A., J. Biol. Chem., 262 (1987) 429–424.
[338] Gold, M. H., Kuwahara, J., Chiu, A. A. and Glenn, J. K., Arch. Biochem. Biophys., 234 (1984) 353–362.
[339] Tien, M. and Kirk, T. K., Proc. Natl. Acad. Sci. USA, 81 (1984) 2280–2284.
[340] Glenn, J. K., Akileswaran, L. and Gold, M. H., Arch. Biochem. Biophys., 251 (1986) 688–696.
[341] Harris, R. Z., Wariishi, H., Gold, M. H. and Ortiz de Demontellano, P. R., J. Biol. Chem., 266 (1991) 8751–8758.
[342] Daniel, G., Jellison, J., Goodell, B., Paszczynski, A. and Crawford, R., Appl. Microb. Biotech., 35 (1991) 674–680.
[343] Daniel, G., Pettersson, B. and Nilsson, T., Volc, J., Canad. J. Botany, 68 (1990) 920–933.
[344] Wariishi, H., Valli, K. and Gold, M. H., J. Biol. Chem., 267 (1992) 23688–23695.
[345] Kenten, R. H. and Mann, P. J. G., Biochem. J., 46 (1950) 67–73.
[346] Bono, J.-J., Golulas, P., Boe, J.-F., Portet, N. and Seris, J.-L., Eur. J. Biochem., 192 (1990) 189–193.
[347] Archibald, F. and Roy, B., Appl. Environ. Microbiol., 58 (1992) 1496–1499.
[348] Popp, J. L., Kalyanaraman, B. and Kirk, T. K., Biochemistry, 29 (1990) 10475–10480.
[349] Tuor, U., Wariishi, H., Schoemaker, H. E. and Gold, M. H., Biochemistry, 31 (1992) 4986–4995.
[350] Wariishi, H., Valli, K. and Gold, M. H., Biochemistry, 28 (1989) 6017–6023.
[351] Wariishi, H., Akileswaran, L. and Gold, M. H., Biochemistry, 27 (1988) 5365–5370.
[352] Wariishi, H., Dunford, H. B., MacDonald, I. D. and Gold, M. H., J. Biol. Chem., 264 (1989) 3335–3340.
[353] Paszczynski, A. and Huynh, V.-B., Crawford, R., Arch. Biochem. Biophys., 244 (1986) 750–765.
[354] Omohara, I. M., Matsumoto, Y. and Itsushi, A., Mokuzai Gakkaishi, 36 (1990) 588–590.
[355] Valli, K., Wariishi, H. and Gold, M. H., Biochemistry, 29 (1990) 8535–8539.
[356] Kersten, P. J., Kalyanaraman, B., Hammel, K. E., Reinhammar, B. and Kirk, T. K., Biochem. J., 268 (1990) 475–480.
[357] Popp, J. L. and Kirk, T. K., Arch. Biochem. Biophys., 288 (1991) 145–148.
[358] Valli, K., Brock, B. J., Joshi, D. K. and Gold, M. H., Appl. Environ. Microb., 58 (1992) 221–228.
[359] Valli, K. and Gold, M. H., J. Bacteriol., 173 (1991) 345–352.
[360] Valli, K., Wariishi, H. and Gold, M. H., J. Bacteriol., 174 (1992) 2131–2137.
[361] Lackner, R., Srebotnik, E. and Messner, K., Biochem. Biophys. Res. Commun., 178 (1991) 1092–1098.
[362] Millis, C. D., Cai, D., Stankovich, M. T. and Tien, M., Biochemistry, 28 (1989) 8484–8489.
[363] Banci, L., Bertini, I., Pease, E. A., Tien, M. and Turano, P., Biochemistry, 31 (1992) 10009–10017.

[364] de Ropp, J. S., La Mar, G. N., Wariishi, H. and Gold, M. H., J. Biol. Chem., 266 (1991) 15001–15008.
[365] Sinclair, R., Yamazaki, I., Bumpus, J., Brock, B., Chang, C.-S., Albo, A. and Powers, L., Biochemistry, 31 (1992) 4892–4900.
[366] Piontek, K., Glumoff, T. and Winterhalter, K., FEBS Lett., 315 (1993) 119–124.
[367] Ngai, K.-L. and Kallen, R. G., Biochemistry, 22 (1983) 5231–5236.
[368] Ngai, K.-L., Ornston, N. and Kallen, R. G., Biochemistry, 22 (1983) 5223–5230.
[369] Goldman, A., Ollis, D. L. and Steitz, T. A., J. Mol. Biol., 194 (1987) 143–153.
[370] Goux, W. J. and Venkatasubramanian, P. N., Biochemistry, 25 (1986) 84–94.
[371] Stein, P. E., Leslie, A. G. W., Finch, J. T. and Carrell, R. W., J. Mol. Biol., 221 (1991) 941–959.
[372] Grand, R. J. A. and Owen, D., Biochem J., 279 (1991) 609–631.
[373] Bourne, H. R., Sanders, D. A. and McCormick, F. M., Nature (London), 349 (1991) 117–127.
[374] Brünger, X. T., Milburn, M. V., Tong, L., deVos, A. M., Jancarik, J., Yamaizumi, Z., Nishimura, S., Ohtsuka, E. and Kim, S.-H., Proc. Natl. Acad. Sci. USA, 87 (1990) 4849–4853.
[375] Privié, G. G., Milburn, M. V., Tong, L., deVos, A. M., Yamaizumi, Z., Nishimura, S. and Kim, S.-H., Proc. Natl. Acad. Sci. USA, 89 (1992) 3649–3653.
[376] Foley, C. K., Pedersen, L. G., Charifson, P. S., Darden, T. A., Wittinghofer, A., Pai, E. F. and Anderson, M. W., Biochemistry, 31 (1992) 4951–4959.
[377] Schlichting, I., Almo, S. C., Rapp, G., Wilson, K., Petratos, K., Lentfer, A., Wittinghofer, A., Kabsch, W., Pai, E. F., Petsko, G. A. and Goody, R. S., Nature (London), 345 (1991) 309–315.
[378] Milburn, M. V., Tong, L., deVos, A. M., Brünger, A., Yamaizumi, Z., Nishimura, S. and Kim, S.-H., Science, 247 (1990) 939–945.
[379] Tong, L., deVos, A. M., Milburn, M. V. and Kim, S. H., J. Mol. Biol., 217 (1991) 503–516.
[380] Feuerstein, J., Kalbitzer, H. R., John, J., Goody, R. S. and Wittinghofer, A., Eur. J. Biochem., 162 (1987) 49–55.
[381] Latwesen, D. G., Poe, M., Leigh, I. G. and Reed, G. A., Biochemistry, 31 (1992) 4946–4950.
[382] Smithers, G. W., Poe, M., Latwesen, D. G. and Reed, G. H., Arch. Biochem. Biophys., 280 (1990) 416–420.
[383] Larsen, R. G., Halkides, C. J., Redfield, A. G. and Singel, D. J., J. Am. Chem. Soc., 114 (1992) 9608–9611.
[384] Mogel, S. N. and McFadden, B. A., Biochemistry, 29 (1990) 8333–8337.
[385] Houtz, R. L., Nable, R. O. and Cheniae, G. M., Plant Physiol., 86 (1988) 1143–1149.
[386] Lee, G. J. and McFadden, B. A., Biochemistry, 31 (1992) 2304–2308.
[387] Berger, S. A. and Evans, P. R., Biochemistry, 31 (1992) 9237–9242.
[388] Harpel, M. R., Larimer, F. W. and Hartman, F. C., J. Biol. Chem., 266 (1991) 24734–24740.
[389] Miziorko, H. M. and Sealy, R. C., Biochemistry, 19 (1980) 1167–1171.
[390] Pierce Jr., J. W., Plant Physiol., 81 (1984) 943–945.
[391] Pierce Jr., J. W. and Prestegard, J. H., Arch. Biochem., 245 (1986) 483–493.
[392] Gutteridge, S., Sigel, I., Thomas, B., Arentzen, R., Cordova, A. and Lorimer, G., EMBO J., 3 (1984) 2737–2743.
[393] Schneider, G., Lindqvist, Y. and Lundqvist, T., J. Mol. Biol., 211 (1990) 989–1008.
[394] Lundqvist, T. and Schneider, G., J. Biol. Chem., 266 (1991) 12604–12611.
[395] Andersson, T., Knight, S., Schneider, G., Lindqvist, Y., Lindqvist, T., Bränden, C.-I. and Lorimer, G. H., Nature 337 (1989) 229–234.
[396] Lundqvist, T. and Schneider, G., Biochemistry, 30 (1991) 904–908.
[397] Söderlind, E., Schneider, G. and Gutteridge, S., Eur. J. Biochem., 206 (1992) 729–735.
[398] Knight, S., Andersson, I. and Bränden, C.-I., J. Mol. Biol., 215 (1990) 113–160.
[399] Chapman, M. S., Suh, S. W., Curmi, P. M. G., Cascio, D., Smith, W. W. and Eisenberg, D. S., Science, 241 (1988) 71–74.
[400] Curmi, P. M. G., Cascio, D., Sweet, R. M., Eisenberg, D. and Schreuder, H., J. Biol. Chem., 267 (1992) 16980–16994.

[401] Newman, J. and Gutteridge, S., J. Biol. Chem., 265 (1990) 15154–15159.
[402] Jack, A., Ladner, J. E., Rhodes, D., Brown, R. S. and Klug, A., J. Mol. Biol., 111 (1977) 315–328.
[403] Schweizer, M. P., De, N., Pulsipher, M., Brown, M., Reddy, P. R., Petrie III, C. R. and Chheda, G. B., Biochim. Biophys. Acta, 802 (1984) 352–361.
[404] Miller, M. D., Benedik, M. J., Sullivan, M. C., Shipley, N. S. and Krause, K. L., J. Mol. Biol., 222 (1992) 27–30.
[405] Åqvist, J. and Warshel, A., J. Am. Chem. Soc., 112 (1990) 2860–2868.
[406] Weber, D. J., Mullen, G. P. and Mildvan, A. S., Biochemistry, 30 (1991) 7425–7431.
[407] Mildvan, A. S. and Serpersu, E. H., in Sigel, H. Sigel, A. (eds.), *Metal Ions in Biological Systems*, Vol. 25, Marcel Dekker, New York, 1989, pp. 309–334.
[408] Serpersu, E. H., Hibler, D. W., Gerlt, J. A. and Mildvan, A. S., Biochemistry, 28 (1989) 1539–1548.
[409] Wang, J., Hinck, A. P., Loh, S. N. and Markley, J. L., Biochemistry, 29 (1990) 4242–4253.
[410] Stanczyk, S. M. and Bolton, P. H., Biochemistry, 31 (1992) 6396–6401.
[411] Cotton, F. A., Hazen Jr., E. E. and Legg, M. J., Proc. Natl. Acad. Sci. USA, 76 (1979) 2551–2555.
[412] Loll, P. J. and Lattman, E. E., Proteins Struct. Funct. Genetics, 5 (1989) 183–201.
[413] Hynes, T. R. and Fox, R. O., Proteins Struct. Funct. Genetics, 10 (1991) 92–105.
[414] Stites, W. E., Gittis, A. G., Lattman, E. E. and Shortle, D., J. Mol. Biol., 221 (1991) 7–14.
[415] Loll, P. and Lattman, E. E., Biochemistry, 29 (1990) 6866–6873.
[416] Bowler, C., Van Montagu, M. and Inze, D., Annu. Rev. Plant Physiol. Plant Mol. Biol., 43 (1992) 83–116.
[417] Touati, D., Curr. Commun. Cell Mol. Biol., 5 (Mol. Biol. Free Radical Scavenging Syst.) (1992) 231–261.
[418] Eliasson R., Jörnvall, H. and Reichard, P., Proc. Natl. Acad. Sci. USA, (1986) 83, 2373–2377.
[419] Fontecave, M. F., Gräslund, A. and Reichard, P., J. Biol. Chem., 262 (1987) 12332–12336.
[420] Fontecave, M. F. and Reichard, P., Adv. Exp. Med. Biol., 264 (1990) 29–35.
[421] Visner, G. A., Dougall, W. C., Wilson, J. M., Burr, I. A. and Nick, H. S., J. Biol. Chem., 265 (1990) 2856–2864.
[422] Kawaguchi, T., Takeyasu, A., Matsunobu, K., Uda, T., Ishizawa, M., Suzuki, K., Nishiura, T., Ishikawa, M. and Taniguchi, N., Biochem. Biophys. Res. Commun., 171 (1990) 1378–1386.
[423] Ono, M., Kohda, H., Kawaguchi, T., Ohhiro, M., Sekiya, C., Namiki, M., Takeyasu, A. and Taniguchi, N., Biochem. Biophys. Res. Commun., 182 (1992) 1100–1107.
[424] St. Clair, D. K. and Holland, J. C., Cancer Res., 51 (1991) 939–943.
[425] Beyer, W., Imlay, J. and Fridovich, I., Prog. Nucleic Acid Res. Mol. Biol., 40 (1991) 221–233.
[426] Fridovich, I., J. Biol. Chem., 264 (1989) 7761–7764.
[427] Fridovich, I., Adv. Enzymology, 58 (1986) 61–97.
[428] Steinman, H. M., Basic Life Sci., 49 (1988) 341–346.
[429] Ludwig, M. L., Pattridge, K. A. and Stallings, W. C., in Schramm, V. L. Wedler, F. C. (eds.), *Manganese in Metabolism and Enzyme Function*, Academic Press, Orlando, Fla., 1986, pp. 405–430.
[430] Bannister, J. V., Bannister, W. H. and Rotilio, G., CRC Crit. Rev. Biochem., 22 (1987) 111–180.
[431] Kitagawa, Y., Tanaka, N., Hata, Y., Kusunoki, M., Lee, G. P., Katsube, Y., Asada, K., Aibara, S. and Morita, Y., J. Biochem. (Tokyo), 109 (1991) 477–485.
[432] Parge, H. E., Hallewell, R. A. and Tainer, J. A., Proc. Natl. Acad. Sci. USA, 89 (1992) 6109–6113.
[433] Djinovic, K., Gatti, G., Coda, A., Antolini, L., Pelosi, G., Desideri, A., Falconi, M., Marmocchi, F., Rotilio, G. and Bolognesi, M., J. Mol. Biol., 225 (1992) 791–809.
[434] Roberts, V. A., Fisher, C. L., Redford, S. M., McRee, D. E., Parge, H. E., Getzoff, E. D. and Tainer, J. A., Free Radical Res. Commun., 12–13 (1991) 269–278.
[435] Archibald, F. S. and Fridovich, I., J. Bacteriol., 146 (1981) 928–936.
[436] Archibald, F. S. and Fridovich, I., Arch. Biochem. Biophys., 215 (1982) 589–596.
[437] Archibald, F. S. and Fridovich, I., Arch. Biochem. Biophys., 214 (1982) 452–463.

[438] Archibald, F. S. and Fridovich, I., J. Bacteriol., 145 (1980) 442–451.
[439] Gregory, E. M. and Dapper, C. H., J. Bacteriol., 144 (1980) 967–974.
[440] Beyer Jr., W. F., Reynolds, J. A. and Fridovich, I., Biochemistry, 28 (1989) 4403–4409.
[441] Juan, J.-Y., Keeney, S. N. and Gregory, E. M., Arch. Biochem. Biophys., 286 (1991) 257–263.
[442] Ose, D. E. and Fridovich, I., Arch. Biochem. Biophys., 194 (1979) 360–364.
[443] Beyer Jr., W. F. and Fridovich, I., J. Biol. Chem., 266 (1991) 303–308.
[444] Privalle, C. T. and Fridovich, I., J. Biol. Chem., 267 (1992) 9140–9145.
[445] Fee, J. A., Mol. Microb., 5 (1991) 2599–2610.
[446] Pennington, C. D. and Gregory, E. M., J. Bacteriol., 166 (1986) 528–532.
[447] Gregory, E. M. and Dapper, C. H., Arch. Biochem. Biophys., 220 (1983) 293–300.
[448] Gregory, E. M., Arch. Biochem. Biophys., 238 (1985) 83–89.
[449] Meier, B., Barra, D., Bossa, F., Calabrese, L. and Rotilio, G., J. Biol. Chem., 257 (1982) 13977–13980.
[450] Meier, B., Free Radical Res. Commun., 12–13 (1991) 211–214.
[451] Martin, M. E., Byers, B. R., Olsen, M. O. J., Salin, M. L., Arceneaux, J. E. L. and Tolbert, C., J. Biol. Chem., 261 (1986) 9361–9367.
[452] Choi, J. I., Takahashi, N., Kato, T. and Kuramitsu, H. K., Infect. Immun., 59 (1991) 1564–1566.
[453] Matsumoto, T., Terauchi, K., Isobe, T., Matsuoka, K. and Yamakura, F., Biochemistry, 30 (1991) 3210–3216.
[454] Amano, A., J. Osaka Univ. Dental Soc., 35 (1990) 465–485.
[455] Amano, A., Shizukuishi, S., Tamagawa, H., Iwakura, K., Tsunasawa, S. and Tsunemitsu, A., J. Bacteriol., 172 (1990) 1457–1463.
[456] Nakayama, K., Gene, 96 (1990) 149–150.
[457] Privalle, C. T. and Fridovich, I., J. Biol. Chem., 265 (1990) 21966–21970.
[458] Sato, S. and Nakazawa, K., J. Biochem. (Tokyo), 83 (1978) 1165–1171.
[459] Borsari, M. and Azab, H. A., Bioelectrochem. Bioenerg., 27 (1992) 229–234.
[460] Azab, H. A., Banci, L., Borsari, M., Luchinat, C., Sola, M. and Viezzoli, M. S., Inorg. Chem., 31 (1992) 4649–4655.
[461] Whittaker, J. W. and Whittaker, M. M., J. Am. Chem. Soc., 113 (1991) 5528–5540.
[462] Peterson, J., Fee, J. A. and Day, E. P., Biochim. Biophys. Acta, 1079 (1991) 161–168.
[463] Lavelle, F., McAdam, M. E., Fielden, E. M. and Roberts, P. B., Biochem. J., 161 (1977) 3–11.
[464] Bull, C., Niederhoffer, E. C., Yoshida, T. and Fee, J. A., J. Am. Chem. Soc., 113 (1991) 4069–4076.
[465] Bull, C. and Fee, J. A., J. Am. Chem. Soc., 107 (1985) 3295–3304.
[466] Stallings, W. C., Bull, C., Fee, J. A., Lah, M. S. and Ludwig, M. L., Curr. Commun. Cell Mol. Biol., (Mol. Biol. Free Radical Scavenging Syst.) 5 (1992) 193–211.
[467] Gray, B. and Carmichael, A. J., Biochem. J., 281 (1992) 795–802.
[468] Brock, C. J. and Walker, J. E., Proc. Fed. Eur. Biochem. Soc. Symp. No. 62, (1980) 237–241.
[469] Chambers, S. P., Brehm, J. K., Michael, N. P., Atkinson, T. and Minton, N. P., FEMS Microb. Lett., 91 (1992) 277–284.
[470] Steinman, H. M., J. Biol. Chem., 253 (1978) 8708–8720.
[471] Takeda, Y., Avila, H., Nucl. Acids Res., 14 (1986) 4577–4589.
[472] Salin, M. L., Oesterhelt, D., Arch. Biochem. Biophys., 260 (1988) 806–810.
[473] Takao, M., Kobayashi, T., Oikawa, A. and Yasui, A., J. Bacteriol., 171 (1989) 6323–6329.
[474] Salin, M. L., Duke, M. V., Ma, D.-P. and Boyle, J. A., Free Radical Res. Commun., 12–13 (1991) 443–450.
[475] May, B. P. and Dennis, P. P., J. Biol. Chem., 264 (1989) 12253–12258.
[476] Marres, C. A. M., van Loon, A. P. G. M., Oudshoorn, P., van Steeg, H., Grivell, L. A. and Stater, E. C., Eur. J. Biochem., 147 (1985) 153–161.
[477] Haas, A. and Goebel, W., Mol. Gen. Genet., 231 (1992) 313–322.
[478] Brehm, K., Haas, A., Goebel, W. and Kreft, J., Gene, 118 (1992) 121–126.

[479] Sato, S., Nakada, Y. and Nakazawa-Tomizawa, K., Biochim. Biophys. Acta, 912 (1987) 178–184.
[480] Barra, D., Schininà, M. E., Simmaco, M., Bannister, J. V., Bannister, W. H., Rotilio, G. and Bossa, F., J. Biol. Chem., 259 (1984) 12595–12601.
[481] Heckl, K., Nucl. Acids Res., 16 (1988) 6224.
[482] Beck, Y., Oren, R., Amit, B., Levanon, A., Gorecki, M. and Hartman, J. R., Nucleic Acids Res., 15 (1987) 9076.
[483] Church, S. L., Biochim. Biophys. Acta, 1087 (1990) 250–252.
[484] Ho, Y.-S. and Crapo, J. D., FEBS Lett., 229 (1988) 256–260.
[485] Wispé, J. R., Clark, J. R., Burhans, M. S., Kropp, K. E., Korfhagen, T. R. and Whitsett, J. A., Biochim. Biophys. Acta, 994 (1989) 30–36.
[486] White, J. A., Scandalios, J. G., Biochim. Biophys. Acta, 951 (1986) 61–70.
[487] Bowler, C., Alliotte, T., De Loose, M., Van Montagu, M., Inzé, D., EMBO J., 8 (1989) 31–38.
[488] Wong-Vega, L., Burke, J. J. and Allen, R. D., Plant Mol. Biol., 17 (1991) 1271–1274.
[489] Hallewel, R. A., Mullenbach, G. T., Stempien, M. M. and Bell, G. I., Nucl. Acids Res., 14 (1986) 9539.
[490] Ho, Y.-S. and Crapo, J. D., Nucl. Acids Res., 15 (1987) 10070.
[491] Smith, M. W. and Doolittle, R. F., J. Mol. Evolut., 34 (1992) 175–184.
[492] Ditlow, C., Johansen, J. T., Martin, B. M. and Svendsen, I., Carlsberg. Res. Commun., 47 (1982) 81–91.
[493] Takao, M., Oikawa, A. and Yasui, A., Arch. Biochem. Biophys., 283 (1990) 210–216.
[494] Takao, M., Yasui, A. and Oikawa, A., J. Biol. Chem., 266 (1991) 14151–14154.
[495] Thangaraj, H. S., Lamb, F. I., Davis, E. O. and Colston, M. J., Nucl. Acids Res., 17 (1989) 8378.
[496] Thangaraj, H. S., Lamb, I. F., Davis, E. O., Jenner, P. J., Jeyakuma, L. H. and Colston, M. J., Infect. Immun., 58 (1990) 1937–1942.
[497] Zhang, Y., Lathigra, R., Garbe, T., Catty, D. and Young, D., Mol. Microb., 5 (1991) 381–391.
[498] Carlioz, A., Ludwig, M. L., Stallings, W. C., Fee, J. A., Steinman, H. M. and Touati, D., J. Biol. Chem., 263 (1988) 1555–1562.
[499] Schininà, M. E., Maffey, L., Barra, D., Bossa, F., Puget, K. and Michelson, A. M., FEBS Lett., 221 (1987) 87–90.
[500] Isobe, T., Fang, Y.-I., Muno, D., Okuyama, T., Ohmori, D. and Yamakura, F., FEBS Lett., 223 (1987) 92–96.
[501] Barra, D., Schininà, M. E., Bannister, W. H., Bannister, J. V. and Bossa, F., J. Biol. Chem., 262 (1987) 1001–1009.
[502] Laudenbach, D. E., Trick, C. G. and Strauss, N. A., Mol. Gen. Genet., 216 (1989) 455–461.
[503] Crowell, D. N. and Amasino, R. M., Plant Physiol., 96 (1991) 1393–1394.
[504] Van Camp, W., Bowler, C., Villarroel, R., Tsang, E. W. T., Van Montagu, M. and Inzé, D., Proc. Natl. Acad. Sci. USA, 87 (1990) 9903–9907.
[505] Barra, D., Schininà, M. E., Bossa, F., Puget, K., Durosay, P., Guissani, A. and Michelson, A. M., J. Biol. Chem., 265 (1990) 17680–17687.
[506] Amano, A., Shizukuishi, S., Tsunemitsu, A., Maekawa, K. and Tsunasawa, S., FEBS Lett., 272 (1990) 217–220.
[507] Heinzen, R. A., Frazier, M. E. and Mallavia, L. P., Nucl. Acids Res., 18 (1990) 6437.
[508] Heinzen, R. A., Frazier, M. E. and Mallavia, L. P., Infection and Immunity, 60 (1992) 3814–3823.
[509] Tannich, E., Bruchhaus, I., Walter, R. D. and Hoffman, R. D., Mol. Biochem. Parasitol., 49 (1991) 61–71.
[510] Lee, S. O., Kim, S. W., Uno, I. and Lee, T. H., Biosci. Biotech. Biochem., 57 (1993) 1454–1460.
[511] Parker, M. W., Schininà, M. E., Bossa, F. and Bannister, J. V., Inorg. Chim. Acta, 91 (1984) 307–317.
[512] Harris, J. I., Auffret, A. D., Northrop, F. D. and Walker, J. E., Eur. J. Biochem., 106 (1980) 297–303.

[513] Parker, M. W. and Blake, C. C. F., J. Mol. Biol., 199 (1988) 649–661.
[514] Chain, V. W. F., Bjerrum, M. J. and Borders, C. L., Arch. Biochem. Biophys., 279 (1990) 195–201.
[515] Borders Jr., C. L., Chain, V. W. F. and Bjerrum, M. J., Free Radical Res. Commun., 12–13 (1991) 279–285.
[516] Kim, S. W., Lee, S. O. and Lee, T. H., Agric. Biol. Chem., 55 (1991) 101–108.
[517] Parker, M. W. and Blake, C. C. F., FEBS Lett., 229 (1988) 377–382.
[518] Isobe, T., Fang, Y., Muno, D., Okuyama, T., Ohmori, D. and Yamakura, F., Biochem. Int., 16 (1988) 495–501.
[519] Stallings, W. C., Metzger, A. L., Pattridge, K. A., Fee, J. A. and Ludwig, M. L., Free Radical Res. Commun., 12–13 (1991) 259–268.
[520] Stallings, W. C., Pattridge, K. A., Strong, R. K. and Ludwig, M. L., J. Biol. Chem., 260 (1985) 16424–16432.
[521] Ludwig, M. L., Metzger, A. L., Pattridge, K. A. and Stallings, W. C., J. Mol. Biol., 219 (1991) 335–358.
[522] Bridgen, J., Harris, J. I. and Kolb, E., J. Mol. Biol., 105 (1976) 333–335.
[523] Wagner, U. G., Werber, M. M., Beck, Y., Hartman, J. R., Frolow, F. and Sussman, J. L., J. Mol. Biol., 206 (1989) 787–788.
[524] Borgstahl, G. E. O., Parge, H. E., Hickey, M. J., Beyer Jr., W. F., Hallewell, R. A. and Tainer, J. A., Cell, 71 (1992) 107–118.
[525] Beem, K. M., Richardson, J. S. and Richardson, D. C., J. Mol. Biol., 105 (1976) 327–332.
[526] Matsuda, Y., Higashiyama, S., Kijima, Y., Suzuki, K., Kawano, K., Akiyama, M., Kawata, S., Tarui, S., Deutsch, H. F. and Taniguchi, N., Eur. J. Biochem., 194 (1990) 713–720.
[527] Deutsch, H. F., Hoshi, S., Matsuda, Y., Suzuki, K., Kawano, K., Kitagawa, Y., Katsube, Y. and Taniguchi, N., J. Mol. Biol., 219 (1991) 103–108.
[528] Ringe, D., Petsko, G. A., Yamakura, F., Suzuki, K. and Ohmori, D., Proc. Natl. Acad. Sci. USA, 80 (1983) 3879–3883.
[529] Stoddard, B. L., Howell, P. L., Ringe, D. and Petsko, G. A., Biochemistry, 29 (1990) 8885–8893.
[530] Stoddard, B. L., Ringe, D. and Petsko, G. A., Protein Eng., 4 (1990) 113–119.
[531] Stallings, W. C., Powers, T. B., Pattridge, K. A., Fee, J. A. and Ludwig, M. L., Proc. Natl. Acad. Sci. USA, 80 (1983) 3884–3888.
[532] Morris, D. C., Searcy, D. G. and Edwards, F. F. G., J. Mol. Biol., 186 (1985) 213–214.
[533] Viglino, P., Orsega, E. F., Argese, E., Stevanto, R. and Rigo, A., Eur. Biophys. J., 15 (1987) 225–230.
[534] Heinrich, P. C., Steffen, H., Janser, P. and Wiss, O., Eur. J. Biochem., 30 (1972) 533–541.
[535] Lindqvist, Y., Schneider, G., Ermler, U. and Sundström, M., EMBO J., 11 (1992) 2373–2379.
[536] Prescott, D. J. and Vagelos, P. R., Adv. Enzymol., 36 (1972) 269–311.
[537] Holak, T. A. and Prestegard, J. H., Biochemistry, 25 (1986) 5766–5774.
[538] Holak, T. A., Kearsley, S. K., Kim, Y. and Prestegard, J. H., Biochemistry, 27 (1988) 6135–6142.
[539] Mayo, K. H. and Prestegard, J. H., Biochemistry, 24 (1985) 7834–7838.
[540] Jones, P.-J., Holak, T. A. and Prestegard, J. H., Biochemistry, 26 (1987) 3493–3500.
[541] Prestegard, J. H. and Kim, Y., in Renugoplakrishnam, V. (ed.), *Proteins*, ESCOM, Leiden, Netherlands, 1991, pp. 45–52.
[542] Kim, Y. and Prestegard, J. H., Proteins Struct. Funct. Genetics, 8 (1990) 377–385.
[543] Tener, D. M. and Mayo, K. H., Eur. J. Biochem., 189 (1990) 559–565.
[544] Frederick, A. F., Kay, L. E. and Prestegard, J. H., FEBS Lett., 238 (1988) 43–48.
[545] McRee, D. E., Richardson, J. S. and Richardson, D. C., J. Mol. Biol., 182 (1985) 467–468.
[546] Murshudov, G. N., Melik-Adamyan, W. R., Grebenko, A. I., Barynin, V. V., Vagin, A. A., Vainshtein, B. K., Dauter, Z. and Wilson, K. S., FEBS Lett., 312 (1992) 127–131.
[547] Vainshtein, B. K., Melik-Adamyan, W. R., Barynin, V. V., Vagin, A. A., Grebenko, A. I., Borisov, V. V., Bartels, K. S., Fita, I. and Rossman, M. G., J. Mol. Biol., 188 (1986) 49–61.

[548] Melik-Adamyan, W. R., Barynin, V. V., Vagin, A. A., Borisov, V. V., Vainshtein, B. K., Fita, I., Murthy, M. R. N. and Rossman, M. G., J. Mol. Biol., 188 (1986) 63–72.
[549] Kono, Y. and Fridovich, I., J. Biol. Chem., 258 (1983) 6015–6019.
[550] Kono, Y. and Fridovich, I., J. Biol. Chem., 258 (1983) 13646–13648.
[551] Allgood, G. S. and Perry, J. J., J. Bacteriol., 168 (1986) 563–567.
[552] Fronko, R. M., Penner-Hahn, J. E. and Bender, C. J., J. Am. Chem. Soc., 110 (1988) 7554–7555.
[553] Penner-Hahn, J. E., in Pecoraro, V. L. (ed.), *Manganese Redox Enzymes*, VCH Publishers Inc., New York, 1992, pp. 29–45.
[554] Beyer Jr.W. F. and Fridovich, I., in Schramm, V. L. Wedler, W. F. (eds.), *Manganese in Metabolism and Enzyme Function*, Academic Press, Orlando Fla., 1986, pp. 193–220.
[555] Beyer Jr., W. F. and Fridovich, I., Biochemistry, 24 (1985) 6460–6467.
[556] Khangulov, S. V., Barynin, V. V. and Antonyuk-Barynina, S. V., Biochim. Biophys. Acta, 1020 (1990) 25–33.
[557] Khangulov, S. V., Barynin, V. V., Voevodskaya, N. V. and Grebenko, A. I., Biochim. Biophys. Acta, 1020 (1990) 305–310.
[558] Waldo, G. S., Fronko, R. M. and Penner-Hahn, J. E., Biochemistry, 30 (1991) 10486–10490.
[559] Khangulov, S. V., Voyevodskaya, N. V., Barynin, V. V., Grebenko, A. I. and Melik-Adamyan, V. R., Biofizika, 32 (1987) 960–966.
[560] Barynin, V. V., Vagin, A. A., Melik-Adamyan, V. R., Grebenko, A. I., Khangulov, S. V., Popov, A. N., Andrianova, M. E. and Vainshtein, B. K., Soviet Phys. Dokl., 31 (1986) 457–459.
[561] Dikanov, S. A., Tsevetkov, Y. D., Khangulov, S. V. and Gol'dfeld, M. G., Dokl. Akad. Nauk. SSSR, 302 (1988) 1255–1257.
[562] Waldo, G. S., Yu, S. and Penner-Hahn, J. E., J. Am. Chem. Soc., 114 (1992) 5869–5870.
[563] Mildvan, A. S. and Loeb, L. A., Adv. Inorg. Biochem., 3 (1981) 103–123.
[564] Beese, L. and Steitz, T. A., in Eckstein, F. Lilley, D. M. J. (ed.), *Nucleic Acids in Molecular Biology*, Vol. 3, Springer-Verlag, Berlin, 1989, pp. 28–43.
[565] Freemont, P. S., Lane, A. N. and Sanderson, M. R., Biochem. J., 278 (1991) 1–23.
[566] Hübscher, U., Experientia, 39 (1983) 1–26.
[567] Sellman, E., Schroder, K. L., Knoblich, I. M. and Westerman, P., J. Bacteriol., 174 (1992) 4350–4355.
[568] Becker, J.-L., Barre-Sinoussi, F., Dormont, D., Best-Belpomme, M. and Chermann, J.-C., Cell Mol. Biol., 33 (1987) 225–235.
[569] Jouve, H., Jouve, H., Melgar, E. and Lizárraga, B., J. Biol. Chem., 250 (1975) 6631–6635.
[570] Esteban, J. A., Bernad, A., Salas, M. and Blanco, L., Biochemistry, 31 (1992) 350–359.
[571] Downey, K. M. and So, A. G., in Sigel, H. Sigel, A. (eds.), *Metal Ions in Biological Systems* Vol. 25, Marcel Dekker, New York, 1989, pp. 1–30.
[572] Beckman, R. A., Mildvan, A. S. and Loeb, L. A., Biochemistry, 24 (1985) 5810–5817.
[573] Lai, M.-D. and Beattie, K. L., Mutat. Res., 198 (1988) 27–36.
[574] Eger, B. T. and Benkovic, S. J., Biochemistry, 31 (1992) 9227–9236.
[575] Mullen, G. P., Serpersu, E. H., Ferrin, L. J., Loeb, L. A. and Mildvan, A. S., J. Biol. Chem., 265 (1990) 14327–14334.
[576] Tabor, S. and Richardson, C. C., J. Biol. Chem., 265 (1990) 8322–8328.
[577] Kristensen, T., Voss, H., Ansorge, W. and Prydz, H., Trends in Genetics, 6 (1990) 2–3.
[578] Ollis, D. L., Brick, P., Hamlin, R., Xuong, N. G. and Steitz, T. A., Nature (London), 313 (1985) 762–766.
[579] Freemont, P. S., Friedman, J. M., Beese, L. S., Sanderson, M. R. and Steitz, T. A., Proc. Natl. Acad. Sci. USA, 85 (1988) 8924–8928.
[580] Derbyshire, V., Freemont, P. S., Sanderson, M. R., Beese, L., Friedman, J. M., Joyce, C. M. and Steitz, T. A., Science, 240 (1988) 199–201.
[581] Derbyshire, V., Grindley, N. D. F. and Joyce, C. M., EMBO J., 10 (1991) 17–24.
[582] Beese, L. S. and Steitz, T. A., EMBO J., 10 (1991) 25–33.

[583] Han, H., Rifkind, J. M. and Mildvan, A. S., Biochemistry, 30 (1991) 11104–11108.
[584] Bubnov, N. V., Basnakyan, A. G. and Votrin, I. I., Bull. Exp. Biol. Med., 110 (1990) 1344–1347.
[585] Suck, D., Oefner, C. and Kabsch, W., EMBO J., 3 (1984) 2423–2430.
[586] Suck, D., Lahm, A. and Oefner, C., Nature (London), 332 (1988) 464–468.
[587] Suck, D. and Oefner, C., Nature 321 (1986) 620–625.
[588] Weston, S. A., Lahm, A. and Suck, D., J. Mol. Biol., 226 (1992) 1237–1256.
[589] Nel, W. and Terblanche, S. E., Int. J. Biochem., 24 (1992) 1267–1283.
[590] Entwistle, G. and ap Rees, T., Biochem. J., 271 (1990) 467–472.
[591] Chardot, T., Oueirozclaret, E. and Meunier, J. C., Biochemie, 73 (1991) 1205–1209.
[592] Biswas, T., Mujumder, A. and Mujumder, A. L., Ind. J. Biochem. Biophys., 22 (1985) 355–359.
[593] Mörikofer-Zwez, S., Arch. Biochem. Biophys., 223 (1983) 572–583.
[594] Vargas, A. M., Sola, M. M. and Bounias, M., J. Biol. Chem., 265 (1990) 15368–15370.
[595] Liu, F. and Fromm, H. J., J. Biol. Chem., 265 (1990) 7401–7406.
[596] Prado, F. E., Lázaro, J. L. and Gorgé, J. L., Plant Physiol., 96 (1991) 1026–1033.
[597] Liu, F., Roy, M. and Fromm, H. J., Biochem. Biophys. Res. Commun., 161 (1989) 689–695.
[598] Scheffler, J. E. and Fromm, H. J., Biochemistry, 25 (1986) 6659–6665.
[599] Liu, F. and Fromm, H. J., J. Biol. Chem., 264 (1989) 18320–18325.
[600] Liu, F. and Fromm, H. J., J. Biol. Chem., 266 (1991) 11774–11778.
[601] Ke, H., Thorpe, C. M., Seaton, B. A., Marcus, F. and Lipscomb, W. N., Proc. Natl. Acad. Sci. USA, 86 (1989) 1475–1479.
[602] Ke, H., Thorp, C. M., Seaton, B. A., Lipscomb, W. N. and Marcus, F., J. Mol. Biol., 212 (1990) 513–539.
[603] Ke, H., Zhang, Y. and Lipscomb, W. N., Proc. Natl. Acad. Sci. USA, 87 (1990) 5243–5247.
[604] Ke, H., Liang, J.-Y., Zhang, Y. and Lipscomb, W. N., Biochemistry, 30 (1991) 4412–4420.
[605] Liang, J.-Y., Huang, S., Zhang, Y., Ke, H. and Lipscomb, W. N., Proc. Natl. Acad. Sci. USA, 89 (1992) 2404–2408.
[606] Ke, H., Zhang, Y., Liang, J.-Y. and Lipscomb, W. N., Proc. Natl. Acad. Sci. USA, 88 (1991) 2989–2993.
[607] Rhee, S. G., Chock, P. B. and Stadtman, E. R., Adv. Enzymol., Relat. Areas Mol. Biol., 62 (1989) 37–92.
[608] Villafranca, J. J., Ransom, S. C. and Gibbs, E. J., Curr. Top. Cell Regul., 26 (1985) 207–219.
[609] Wedler, F. C. and Toms, R., in Schramm, V. L. Wedler, F. C. (eds.), *Manganese in Metabolism and Enzyme Function*, Academic Press, Orlando, Florida, 1986, pp. 221–238.
[610] Tholey, G., Bloch, S., Ledig, M., Mandel, P. and Wedler, F. C., Neurochem. Res., 12 (1987) 1041–1047.
[611] Valentine, R. C., Shapiro, B. M. and Stadtman, E. R., Biochemistry, 7 (1978) 2143–2152.
[612] Maurizi, M. R., Pinkofsky, H. B. and Ginsburg, A., Biochemistry, 26 (1987) 5023–5031.
[613] Shapiro, B. M., Biochemistry, 8 (1969) 659–670.
[614] Tholey, G., Ledig, M., Kopp, P., Sargentini-Maier, L., Leroy, M., Grippo, H. A. and Wedler, F. C., Neurochem. Res., 13 (1988) 1163–1167.
[615] Abell, L. M. and Villafranca, J. J., Biochemistry, 30 (1991) 1413–1418.
[616] Abell, L. M. and Villafranca, J. J., Biochemistry, 30 (1991) 6135–6141.
[617] Nakano, T. and Kimura, K., Biochem. Biophys. Res. Commun., 142 (1987) 475–482.
[618] Lin, W.-Y., Dombrosky, P., Atkins, W. M. and Villafranca, J. J., Biochemistry, 30 (1991) 3427–3431.
[619] McNemar, L. S., Lin, W.-Y., Eads, C. D., Atkins, W. M., Dombrosky, P. and Villafranca, J. J., Biochemistry, 30 (1991) 3417–3421.
[620] Balakrishnan, M. S. and Villafranca, J. J., Biochemistry, 17 (1978) 3531–3538.
[621] Hunt, J. B. and Ginsburg, A., Biochemistry, 11 (1972) 3723–3735.
[622] Villafranca, J. J., Ash, D. E. and Wedler, F. C., Biochemistry, 15 (1976) 544–553.
[623] Hunt, J. B. and Ginsburg, A., J. Biol. Chem., 255 (1980) 590–594.

[624] Hunt, J. B., Smyrniotis, P. Z., Ginsburg, A. and Stadtman, E. R., Arch. Biochem. Biophys., 166 (1975) 102–124.
[625] Maurizi, M. R. and Ginsburg, A., Biochemistry, 25 (1986) 131–140.
[626] Eads, C. D., LoBrutto, R., Kumar, A. and Villafranca, J. L., Biochemistry, 27 (1988) 165–170.
[627] Almassy, R. J., Janson, C. A., Hamlin, R., Xuong, N.-H. and Eisenberg, D., Nature (London), 323 (1986) 304–309.
[628] Yamashita, M. M., Almassy, R. J., Janson, C. A., Cascio, D. and Eisenberg, D., J. Biol. Chem., 264 (1989) 17681–17690.
[629] Lin, W.-Y., Eads, C. D. and Villafranca, J. J., Biochemistry, 30 (1991) 3421–3426.
[630] Ginsburg, A., Gorman, E. G., Neece, S. H. and Blackburn, M. B., Biochemistry, 26 (1987) 5989–5996.
[631] Shrake, A., Fisher, M. T., McFarland, P. J. and Ginsburg, A., Biochemistry, 28 (1989) 6281–6294.
[632] Ginsburg, A. and Zolkiewski, M., Biochemistry, 30 (1991) 9421–9429.
[633] Zolkiewski, M. and Ginsburg, A., Biochemistry, 31 (1992) 19991–12000.
[634] Lis, H. and Sharon, N., Annu. Rev. Biochem., 55 (1986) 35–67.
[635] Sharon, N. and Lis, H., FASEB J., 4 (1990) 3198–3208.
[636] Van Driessche, E., in Franz, H. (ed.), *Advances in Lectin Research*, VEB Verslag Volk und Gesundheit Berlin, Springer Verlag, Heidelberg, 1988, pp. 73–134.
[637] LoBrutto, R., Smithers, G. W., Reed, G. H., Orme-Johnson, W. H., Tan, S. L. and Leigh, J. S., Biochemistry, 25 (1986) 5654–5660.
[638] Antanaitis, B. C., Brown III, R. D., Chasteen, N. D., Freedman, J. H., Koenig, S. H., Lilienthal, H. R., Peisach, J. and Brewer, C. F., Biochemistry, 26 (1987) 7932–7937.
[639] Koenig, S. H. and Brown III, R. D., Biochemistry, 24 (1985) 4980–4984.
[640] McCracken, J., Peisach, J., Bhattacharyya, L. and Brewer, F., Biochemistry, 30 (1991) 4486–4491.
[641] Bhattacharyya, L., Brewer, C. F. and Brown III, R. D., Biochemistry, 24 (1985) 4985–4990.
[642] Edelman, G. M., Cunningham, B. A., Reeke Jr., G. N., Becker, J. W., Waxdal, M. J. and Wang, J. L., Proc. Natl. Acad. Sci. USA, 69 (1972) 2580–2584.
[643] Reeke Jr., G. N., Becher, J. W. and Edelman, G. M., Proc. Natl. Acad. Sci. USA, 75 (1978) 2286–2290.
[644] Derewenda, Z., Yariv, J., Helliwell, J. R., Kalb (Gilboa) A. J., Dodson, E. J., Papiz, M. Z., Wan, T. and Campbell, J., EMBO J., 8 (1989) 2189–2193.
[645] Reeke Jr., G. N., Becker, J. W. and Edelman, G. M., J. Biol. Chem., 250 (1975) 1525–1547.
[646] Hardman, K. D., Agarwal, R. C. and Freiser, M. J., J. Mol. Biol., 157 (1982) 69–86.
[647] Lobsanov, Y. D., Pletnev, V. Z. and Mokul'skii, M. A., Bioorg. Khim., 16 (1990) 1599–1606.
[648] Lobsanov, V. D., Pletnev, V. Z., Vtyurin, N. N., Lubnin, M. Y., Mokul'skii, M. A., Urzhumtsev, A. G., Lunin, V. Y. and Luzyanina, T. B., Bioorg. Khim., 16 (1990) 1589–1598.
[649] Einspahr, H., Parks, E. H., Suguna, K., Subramanian, E. and Suddath, F. L., J. Biol. Chem., 261 (1986) 16518–16527.
[650] Rini, J. M., Hardman, K. D., Einspahr, H. M., Suddath, F. L. and Carver, J. P., Trans. Am. Crystallogr. Assoc., (Publ. 1991) (Proc. Symp. Mol. Recognit. Protein Carbohydrate Interaction) 25 (1989) 51–63.
[651] Rini, J. M., Carver, J. P. and Hardman, K. D., J. Mol. Biol., 189 (1986) 259–260.
[652] Bourne, Y., Rougé, P. and Cambillau, C., J. Biol. Chem., 265 (1990) 18161–18165.
[653] Bourne, Y., Roussel, A., Frey, M., Rougé, P., Fontecilla-Camps, J.-C. and Cambillau, C., Proteins Struct. Funct. Genetics, 8 (1990) 365–376.
[654] Bourne, Y., Abergel, A., Cambillau, C., Frey, M., Rougé, P., Fontecilla-Camps, J.-C., J. Mol. Biol., 214 (1990) 571–584.
[655] Bourne, Y., Rougé, R., Cambillau, C., J. Biol. Chem., 267 (1992) 197–203.

[656] Bourne, Y., Nésa, M.-P., Rougé, P., Mazurier, J., Legrand, D., Spik, G., Montreuil, J. and Cambillau, C., J. Mol. Biol., 227 (1992) 938–941.
[657] Bourne, Y., Anguille, C., Fontecilla-Camps, J.-C., Rougé, P. and Cambillau, C., J. Mol. Biol., 213 (1990) 211–213.
[658] Wright, C. S., J. Mol. Biol., 215 (1990) 635–651.
[659] Reeke Jr., G. N. and Becker, J. W., Science, 234 (1986) 1108–1111.
[660] Delbaere, L. T., Vandonselaar, M., Prasad, L., Quail, J. W., Pearlstone, J. R., Carpenter, M. R., Smille, L. B., Nikrad, P. V., Spohr, U. and Lemieux, R. N., Canad. J. Chem., 68 (1990) 1116–1121.
[661] Shaanan, B., Lis, H. and Sharon, N., Science, 254 (1991) 862–866.
[662] Weis, W. I., Kahn, K., Fourme, R., Drickamer, K. and Hendrickson, W. A., Science, 254 (1991) 1608–1615.
[663] Weis, W. I., Drickamer, K. and Hendrickson, W. A., Nature, 360 (1992) 127–134.
[664] Loris, R., Lisgarten, J., Maes, D., Pickersgill, R., Körber, F., Reynolds, C. and Wyns, L., J. Mol. Biol., 223 (1992) 579–581.
[665] Reeke Jr., G. N. and Becker, J. W., Curr. Topics Microbiol. Immunol., 139 (1988) 35–38.
[666] Brewer, C. F., Brown III, R. D. and Koenig, S. H., J. Biomol. Struct. Dynam., 1 (1983) 961–997.
[667] Bhattacharyya, L., Brewer, C. F., Brown III, R. D. and Koenig, S. H., Biochemistry, 24 (1985) 4974–4980.
[668] Sadhu, A. and Magnuson, J. A., Biochemistry, 28 (1989) 3197–3204.
[669] Lin, S.-L., Stern, E. A., Kalb (Gilboa) A. J. and Zhang, Y., Biochemistry, 30 (1991) 2323–2332.
[670] Carpenter, F. H. and Vahl, J. M., J. Biol. Chem., 248 (1973) 294–304.
[671] Allen, M. P., Yamada, A. H. and Carpenter, F. H., Biochemistry, 22 (1983) 3778–3783.
[672] Taylor, A., Sawan, S. and James, T. L., J. Biol. Chem., 257 (1982) 11571–11576.
[673] Stirling, C. J., Colloms, S. D., Collins, J. F., Szatmari, G. and Sherrat, D. J., EMBO J., 8 (1988) 1623–1627.
[674] Federov, A. A., Barbashov, S. F., Strokopytov, B. V., Kuzin, A. P. and Fonarev, Y. D., Mol. Biol., 25 (1991) 876–887.
[675] Burley, S. K., David, P. R., Taylor, A. and Lipscomb, W. N., Proc. Natl. Acad. Sci. USA, 87 (1990) 6878–6882.
[676] Burley, S. K., David, P. R. and Lipscomb, W. S., Proc. Natl. Acad. Sci. USA, 88 (1991) 6916–6920.
[677] Burley, S. K., David, P. R., Sweet, R. M., Taylor, A. and Lipscomb, W. N., J. Mol. Biol., 224 (1992) 113–140.
[678] Cronin, C. N. and Tipton, K. F., Biochem. J., 247 (1987) 41–46.
[679] Urbina, J. A., Ysern, X. and Mildvan, A. S., Arch. Biochem. Biophys., 278 (1990) 187–194.
[680] Evans, P. R., Farrants, G. W. and Hudson, P. J., Phil. Trans. Roy. Soc. Sect. B, 293B (1981) 53–62.
[681] Hellinga, H. W. and Evans, P. R., Nature (London), 327 (1987) 437–439.
[682] Rypniewski, W. R. and Evans, P. R., J. Mol. Biol., 207 (1989) 805–821.
[683] Shirakihara, Y. and Evans, P. R., J. Mol. Biol., 204 (1988) 973–994.
[684] Schirmer, T. and Evans, P. R., Nature (London), 343 (1990) 140–145.
[685] Johnson, C. M., Cooper, A. and Brown, A. J. P., Eur. J. Biochem., 202 (1991) 1157–1164.
[686] Dryden, D. T., Varley, P. G. and Pain, R. H., Eur. J. Biochem., 208 (1992) 115–123.
[687] Laine, R., Deville-Bonne, D., Auzat, I. and Garel, J. R., Eur. J. Biochem., 207 (1992) 1109–1114.
[688] Berger, S. A. and Evans, P. R., Biochemistry, 31 (1992) 9237–9242.
[689] Blake, C. C. F. and Rice, D. W., Phil. Trans. Roy. Soc. Ser. A., 293 (1981) 93–104.
[690] Mori, N., Singer-Sam, J. and Riggs, A. D., FEBS Lett., 204 (1986) 313–317.
[691] Banks, R. D., Blake, C. C. F., Evans, P. R., Haser, R., Rice, D. W., Hardy, G. W., Merrett, M. and Phillips, A. W., Nature (London), 279 (1979) 773–777.
[692] Moore, J. M. and Reed, G. H., Biochemistry, 24 (1985) 5328–5333.

[693] Wilson, H. R., Williams, R. J. P., Littlechild, J. A. and Watson, H. C., Eur. J. Biochem., 170 (1988) 529–538.
[694] Rice, D. W. and Blake, C. C. F., J. Mol. Biol., 175 (1984) 219–223.
[695] Harlos, K., Vas, M. and Blake, C. F., Proteins Struct. Funct. Genetics, 12 (1992) 133–144.
[696] Watson, H. C., Walker, N. P. C., Shaw, P. J., Bryant, T. N., Wendell, P. L., Fothergill, L. A., Perkins, R. E., Conroy, S. C., Dobson, M. J., Tuite, M. F., Kingsman, A. J. and Kingsman, S. M., EMBO J., 1 (1982) 1635–1640.
[697] Davies, G. J., Littlechild, J. A., Watson, H. C. and Hall, L., Gene, 109 (1991) 39–46.
[698] Mas, M. T., Reslandor, Z. E. and Riggs, A. D., Biochemistry, 26 (1987) 5369–5377.
[699] Mas, M. T., Bailey, J. M. and Resplandor, Z. E., Biochemistry, 27 (1988) 1168–1172.
[700] Fairbrother, W. J., Hall, L., Littlechild, J. A., Walker, P. A., Watson, H. C. and Williams, R. J. P., FEBS Lett., 258 (1989) 247–250.
[701] Sherman, M. A., Fairbrother, W. J. and Mas, M. T., Protein Sci., 1 (1992) 752–760.
[702] Harmark, K., Anborgh, P. H., Merola, M., Clark, B. F. C. and Parmeggiani, A., Biochemistry, 31 (1992) 7367–7372.
[703] Ray, B. D. and Nageswara Rao, B. D., Biochemistry, 27 (1988) 5579–5585.
[704] Ray, B. D. and Nageswara Rao, B. D., Biochemistry, 27 (1988) 5574–5578.
[705] Chapman, B. E., O'Sullivan, W. J., Scopes, R. K. and Reed, G. H., Biochemistry, 16 (1977) 1005–1010.
[706] Kanaya, S., Miura, Y., Sekiguchi, A., Iwai, S., Inoue, H., Ohtsuka, E. and Ikehara, M., J. Biol. Chem., 265 (1990) 4615–4621.
[707] Doolittle, R. F., Feng, D.-F., Johnson, M. S. and McClure, M. A., Quart. Rev. Biol., 64 (1989) 1–30.
[708] Jou, R. W. and Cowan, J. A., J. Am. Chem. Soc., 113 (1991) 6685–6686.
[709] Kohlstaedt, L. A., Wang, J., Friedman, J. M., Rice, P. A. and Steitz, T. A., Science, 256 (1992) 1783–1790.
[710] Davies II, J. H., Hostomska, Z., Hostomsky, Z., Jordan, S. R. and Matthews, D. A., Science, 252 (1991) 88–95.
[711] Yang, W., Hendrickson, W. A., Crouch, R. J. and Satow, Y., Science, 249 (1990) 1398–1405.
[712] Katayanagi, K., Miyagawa, M., Matsushima, M., Ishikawa, M., Kanaya, S., Nakamura, H., Ikehara, M., Matsuzaki, T. and Morikawa, K., J. Mol. Biol., 223 (1992) 1029–1052.
[713] Okumura, M., Ishikawa, K., Kanaya, S., Itaya, M. and Morikawa, K., Proteins Struct. Funct. Genetics, 15 (1993) 108–111.
[714] Stubbe, J., Adv. Enzymology, 63 (1990) 349–419.
[715] Stubbe, J. A., Annu. Rev. Biochem., 58 (1989) 257–285.
[716] Fontecave, M., Nordlund, P., Eklund, H. and Reichard, P., Adv. Enzymol., 65 (1992) 147–183.
[717] Que Jr., L., Science, 253 (1991) 273–274.
[718] Auling, G., Eur. J. Microb. Biotechnol., 18 (1983) 229–235.
[719] Plönzig, J., Auling, G., Arch. Microb., 146 (1987) 396–401.
[720] Sze, I. S.-Y., McFarlan, S. C., Spormann, A., Hogenkamp, H. P. C. and Follmann, H., Biochem. Biophys. Res. Commun., 184 (1992) 1101–1107.
[721] Larsson, Å. and Sjöberg, B.-M., EMBO J., 5 (1986) 2037–2040.
[722] Nordlund, P., Sjöberg, B.-M. and Eklund, H., Nature (London), 345 (1990) 593–598.
[723] Scarrow, R. C., Maroney, M. J., Palmer, S. M., Que Jr. , L., Roe, A. L., Salowe, S. P. and Stubbe, J., J. Am. Chem. Soc., 109 (1987) 7857–7864.
[724] Atta, M., Nordlund, P., Åberg, A., Eklund, H. and Fontecave, M., J. Biol. Chem., 267 (1992) 20682–20688.
[725] Atta, M., Scheer, C., Fries, P., Fontecave, M. and Latour, J. M., Angew. Chem. Int. Ed. Engl., 31 (1992) 1513–1515.
[726] Engström, Y., Eriksson, S., Thelander, L. and Åkerman, M., Biochemistry, 18 (1979) 2941–2948.

[727] Auling, G., Thaler, M. and Diekman, H., Arch. Microb., 127 (1980) 105–114.
[728] Willing, A., Follmann, H. and Auling, G., Eur. J. Biochem., 170 (1988) 603–611.
[729] Farber, G. K., Glasfeld, A., Tiraby, G., Ringe, D. and Petsko, G. A., Biochemistry, 28 (1989) 7289–7297.
[730] Young, J. M., Schray, K. J. and Mildvan, A. S., J. Biol. Chem., 250 (1975) 9021–9027.
[731] Farber, G. K., Petsko, G. A. and Ringe, D., Protein Eng., 1 (1987) 459–466.
[732] Lehmacher, A. and Bisswanger, H., Biol. Chem. Hoppe-Seyler, 371 (1990) 527–536.
[733] Van Bastelaere, P. B. M., Callens, M., Vangrysperra, W. A. E. and Kersters-Hilderson, H. L. M., Biochem. J., 286 (1992) 724–735.
[734] Van Bastelaere, P., Vangrysperre, P. and Kersters-Hilderson, W. H., Biochem J., 278 (1991) 285–292.
[735] Batt, C. A., Jamieson, A. C. and Vandeyar, M. A., Proc. Natl. Acad. Sci. USA, 87 (1990) 618–622.
[736] Whitlow, M., Howard, A. J., Finzel, B. C., Poulos, T. P., Winborne, E. and Gilliland, G. L., Proteins Struct. Funct. Genetics, 9 (1991) 153–173.
[737] Collyer, C. A., Henrick, K. and Blow, D. M., J. Mol. Biol., 212 (1990) 211–235.
[738] Collyer, C. A., Goldberg, J. D., Viehmann, H., Blow, D. M., Ramsden, N. G., Fleet, G. W. T., Montgomery, F. J. and Grice, P., Biochemistry, 31 (1992) 12211–12218.
[739] Henrick, K., Collyer, C. A. and Blow, D. M., J. Mol. Biol., 208 (1989) 129–157.
[740] Rey, F., Jenkins, J., Janin, J., Lasters, I., Alard, P., Claessens, M., Matthyssens, G. and Wodak, S., Proteins Struct. Funct. Genetics, 4 (1988) 165–172.
[741] Mrabet, N. T., Van den Broeck, A., Van den brande, I., Stanssens, P., Laroche, Y., Lambeir, A.-M., Matthijssens, G., Jenkins, J., Chiadmi, M., van Tilbeurgh, H., Rey, F., Janin, J., Quax, W. J., Lasters, I., De Maeyer, M. and Wodak, S. J., Biochemistry, 31 (1992) 2239–2253.
[742] van Tilbeurgh, H., Jenkins, J., Chiadmi, M., Janin, J., Wodak, S. J., Mrabet, N. T. and Lambeir, A.-M., Biochemistry, 31 (1992) 5467–5471.
[743] Lambeir, A.-M., Lauwereys, M., Stanssens, P., Mrabet, N. T., Snauwaert, J., van Tilbeurgh, H., Matthyssens, G., Lasters, I., De Maeyer, M., Wodak, S. J., Jenkins, J., Chiadmi, M. and Janin, J., Biochemistry, 31 (1992) 5459–5466.
[744] Jenkins, J., Janin, J., Rey, F., Chiadmi, M., van Tilbeurgh, H., I., L., De Maeyer, M., Van Belle, D., Wodak, S. J., Lauwereys, M., Stanssens, P., Mrabet, N. T., Snauwaert, J., Matthyssens, G. and Lambeir, A.-M., Biochemistry, 31 (1992) 5449–5458.
[745] Dauter, Z., Dauter, M., Hemker, J., Witzel, H. and Wilson, K., FEBS Lett., 247 (1989) 1–8.
[746] Dauter, Z., Terry, H., Witzel, H. and Wilson, K. S., Acta Crystallog. Sect. B Struct. Sci., 46b (1990) 833–841.
[747] Carrell, H. L., Glusker, J. P., Burger, V., Manfre, F., Tritsch, D. and Biellmann, J.-F., Proc. Natl. Acad. Sci. USA, 86 (1989) 4440–4444.
[748] Callens, M., Tomme, P., Kersters-Hilderson, H., Cornelis, R., Vangrysperre, W. and De Bruyne, K., Biochem. J., 250 (1988) 285–290.
[749] Henrick, K., Blow, D. M., Carrell, H. L. and Glusker, J. P., Protein Eng., 1 (1987) 467–469.
[750] Sudfeldt, C., Schäffer, A., Kägi, J. H. R., Bogumil, R., Schulz, H. P., Wulff, S. and Witzel, H., Eur. J. Biochem., 193 (1990) 863–871.
[751] Chappelet-Tordo, D., Iwatsubo, M. and Lazdunski, M., Biochemistry, 13 (1974) 3754–3762.
[752] Weiner, R. E., Chlebowski, J. F., Haffner, P. H. and Coleman, J. E., J. Biol. Chem., 254 (1979) 9739–9746.
[753] Kim, E. E. and Wyckoff, H. W., J. Mol. Biol., 218 (1991) 449–464.
[754] Kelly, C. T., Nash, A. M. and Fogarty, W. M., Appl. Microb. Biotechnol., 19 (1984) 61–66.
[755] Kinnett, D. G. and Wilcox, F. H., Int. J. Biochem., 14 (1982) 977–981.
[756] Sowadski, J. M., Handschumacher, M. D., Krishna Murthy, H. M., Foster, B. A. and Wyckoff, H. W., J. Mol. Biol., 186 (1985) 417–433.
[757] Mangani, S., Carloni, P., Viezzoli, M. S. and Coleman, J. E., Inorg. Chim. Acta, 191 (1992) 161–166.

[758] Brewer, J. M., FEBS Lett., 182 (1985) 8–14.
[759] Lee, B. H. and Nowak, T., Biochemistry, 31 (1992) 2165–2171.
[760] Lee, M. E. and Nowak, T., Biochemistry, 31 (1992) 2172–2180.
[761] Poyner, R. R. and Reed, G. H., Biochemistry, 31 (1992) 7166–7173.
[762] Nowak, T., Mildvan, A. S. and Kenyon, G. L., Biochemistry, 12 (1973) 1690–1701.
[763] Lee, M. E. and Nowak, T., Arch. Biochem. Biophys., 293 (1992) 264–273.
[764] Lebioda, L., Stec, B. and Brewer, J. M., J. Biol. Chem., 264 (1989) 3685–3693.
[765] Lebioda, L., Stec, B., Brewer, J. M. and Tykarska, E., Biochemistry, 30 (1991) 2823–2827.
[766] Lebioda, L. and Stec, B., Biochemistry, 30 (1991) 2817–2822.
[767] Stec, B. and Lebioda, L., J. Mol. Biol., 211 (1990) 235–248.
[768] Lebioda, L. and Stec, B., J. Am. Chem. Soc., 111 (1991) 8511–8513.
[769] Nowak, T. and Maurer, P. J., Biochemistry, 20 (1981) 6901–6911.
[770] Kumar, V. D., Lee, L. and Edwards, B. F. G., FEBS Lett., 283 (1991) 311–316.
[771] Roquet, F., Declercq, J.-P., Tinant, B., Rambaud, J. and Parello, J., J. Mol. Biol., 223 (1992) 705–720.
[772] Declercq, J.-P., Tinant, B., Parello, J. and Rambaud, J., J. Mol. Biol., 220 (1991) 1017–1039.
[773] Nowak, T. and Suelter, C., Mol. Cell Biochem., 35 (1981) 65–75.
[774] Kiick, D. M. and Cleland, W. W., Arch. Biochem. Biophys., 270 (1989) 647–654.
[775] Baek, Y. H. and Nowak, T., Arch. Biochem. Biophys., 217 (1982) 491–497.
[776] Klein, D. P. and Charles, A. M., Curr. Microb., 19 (1989) 57–60.
[777] de Zwaan, A., Holwerda, D. A. and Addink, A. D. F., Comp. Biochem. Physiol. B Comp. Biochem., 52B (1975) 469–472.
[778] Carvajal, N., González, R. and Kessi, E., J. Exp. Zool., 255 (1990) 280–285.
[779] Noguchi, I. and Tanaka, T., J. Biol. Chem., 257 (1982) 1110–1113.
[780] Marie, J., Simon, M.-P., Dreyfus, J. C. and Kahn, A., Nature (London), 292 (1981) 70–72.
[781] Reed, G. H. and Cohn, M., J. Biol. Chem., 248 (1973) 6436–6442.
[782] Reed, G. H. and Morgan, S. D., Biochemistry, 13 (1974) 3537–3541.
[783] Lodato, D. T. and Reed, G. H., Biochemistry, 26 (1987) 2243–2250.
[784] Tipton, P. A., McCracken, J., Cornelius, J. B. and Peisach, J., Biochemistry, 28 (1989) 5720–5728.
[785] Buchbinder, J. L. and Reed, G. H., Biochemistry, 29 (1990) 1799–1806.
[786] Muirhead, H., Clayden, D. A., Barford, D., Lorimer, C. G., Fothergill-Gilmore, L. A., Schiltz, E. and Schmitt, W., EMBO J., 5 (1986) 475–481.
[787] Murcott, T. H. L., McNally, T., Allen, S. C., Fothergill-Gilmore, L. A. and Muirhead, H., Eur. J. Biochem., 198 (1991) 513–519.
[788] Schmidt-Bäse, K., Buchbinder, J. L., Reed, G. H. and Rayment, I., Proteins Struct. Funct. Genetics, 11 (1991) 153–157.
[789] Lundin, M., Deopujari, S. W., Lichko, L., Pereira da Silva, L. and Baltscheffsky, H., Biochim. Biophys. Acta, 1098 (1992) 217–223.
[790] Kawasaki, I., Adachi, N. and Ikeda, H., Nucl. Acids Res., 18 (1990) 5888.
[791] Stark, M. J. R. and Milner, J. S., Yeast, 5 (1989) 35–50.
[792] Lahti, R., Pitäranta, T., Valve, E., Ilta, I., Kukko-Kalske, E. and Heinonen, J., J. Bacteriol., 170 (1988) 5901–5907.
[793] Ichiba, T., Takenaka, O., Samejima, T. and Hachimori, A., J. Biochem. (Tokyo), 108 (1990) 572–578.
[794] Cohen, S. A., Sterner, R., Keim, P. S. and Heinrikson, R. L., J. Biol. Chem., 253 (1978) 889–897.
[795] Kolakowski Jr., L. F., Schloesser, M. and Cooperman, B. S., Nucl. Acids Res., 16 (1988) 10441–10452.
[796] Lundin, M., Baltscheffsky, H. and Ronne, H., J. Biol. Chem., 266 (1991) 12168–12172.
[797] Richter, O.-M. H. and Schäfer, G., Eur. J. Biochem., 209 (1992) 351–355.
[798] Kieber, J. J. and Signer, E. R., Plant Mol. Biol., 16 (1991) 345–348.

[799] Wensel, T. G. and Yang, Z., FASEB J., 6 (1992) A468.
[800] Cooperman, B. S., Baykov, A. A. and Lahti, R., Trends Biochem. Sci., 17 (1992) 262–266.
[801] Welsh, K. M. and Cooperman, B. S., Biochemistry, 23 (1984) 4947–4955.
[802] Baykov, A. A., Shestakov, A. S., Kasho, V. N., Vener, A. V. and Ivanov, A. H., Eur. J. Biochem., 194 (1991) 879–887.
[803] Baykov, A. A. and Shestakov, A. S., Eur. J. Biochem., 206 (1992) 463–470.
[804] Knight, W. B., Fitts, S. W. and Dunaway-Mariano, D., Biochemistry, 20 (1981) 4079–4086.
[805] Knight, W. B., Dunaway-Marino, D., Ransom, S. C. and Villafranca, J. J., J. Biol. Chem., 259 (1984) 2886–2895.
[806] Banerjee, A. and Cooperman, B. S., Inorg. Chim. Acta, 79 (1983) 146–148.
[807] Banerjee, A., LoBrutto, R. and Cooperman, B. S., Inorg. Chem., 25 (1986) 2417–2424.
[808] Hamm, D. J. and Cooperman, B. S., Biochemistry, 17 (1978) 4033–4040.
[809] Welsh, K. M., Jacobyansky, A., Springs, B. and Cooperman, B. S., Biochemistry, 22 (1983) 2243–2248.
[810] Lahti, R., Pohjanoksa, K., Pitäranta, T., Heikinheimo, P., Salminen, T., Meyer, P. and Heinonen, J., Biochemistry, 29 (1990) 5761–5766.
[811] Lahti, R., Salminen, T., Latonen, S., Heikinheimo, P., Pohjanoksa, K. and Heinonen, J., Eur. J Biochem., 198 (1991) 293–297.
[812] Bond, M. W., Chiu, N. Y. and Cooperman, B. S., Biochemistry, 19 (1980) 94–102.
[813] Gonzales, M. A. and Cooperman, B. S., Biochemistry, 25 (1986) 1504–1509.
[814] Kaneko, S., Ichiba, T., Hirano, N. and Hachimori, A., Biochim. Biophys. Acta, 1077 (1991) 281–284.
[815] Hirano, N., Ichiba, T. and Hachimori, A., Biochem. J., 278 (1991) 595–599.
[816] Terzyan, S. S., Vornova, A. A., Smirnova, E. A., Kuranova, I. P., Nekrasov, Y. V., Arutyunyan E. G., Vainshtein, B. K., Höhne, W. E. and Hansen, G., Bioorg. Khim., 10 (1984) 1469–1482.
[817] Kuranova, I. P., Terzyan, S. S., Voranova, A. A., Smirnova, E. A., Vainshtein, B. K., Höhne, W and Hansen, G., Bioorg. Khim., 9 (1983) 1611–1619.
[818] Chirgadze, N. Y., Kuranova, I. P., Nevskaya, N. A., Teplyakov, A. V., Wilson, K., Strokopytov, I. I., Arutyunyan, É. G. and Khene, V., Sov. Phys. Crystallogr., 36 (1991) 72–75.
[819] Chirgadze, N. Y., Strokopytov, B. V., Kuranova, I. P., Arutyunyan, É. G., Höhne, W. and Hausdorff, G., Sov. Phys. Crystallogr., 35 (1990) 451–452.
[820] Chirgadze, N. Y., Kuranova, I. P., Strokopytov, B. V., Arutyunyan, É. G. and Höhne, W. E., Sov Phys. Crystallog., 34 (1989) 867–869.
[821] Vihinen, M., Lundin, M. and Baltscheffsky, H., Biochem. Biophys. Res. Commun., 186 (1992) 122–128.
[822] Wedler, F. C., Prog. Med. Chem., 30 (1993) 89–133.
[823] *Manganese in Health and Disease*, Klimis-Tavantzis, D. J., ed., CRC: Boca Raton Fla., 1994.
[824] Kim, H. D., Sun, G. Y. and Sun, A. Y. in *Markers of Neuronal Injury and Degeneration,* J. N. Johnnessen ed., New York Acad. Sciences, New York NY 10021, 1993, pp. 376–381.
[825] Fasolato, C., Hoth, M. and Penner, R., Pflugers Arch. Eur. J. Physiol., 423 (1993) 225–231.
[826] Fasolato, C., Hoth, M., Matthews, G. and Penner, R., Proc. Natl. Acad. Sci. USA, 90 (1993) 3068–3072.
[827] Wedler, F. C., Vichnin, M. C., Ley, B. W., Tholey, G., Ledig, M. and Copin, J. C., Neurochem. Res., 19 (1994) 145–151.
[828] Rabin, O., Hegedus, L., Bourre, J. M. and Smith, Q. R. J., Neurochem., 61 (1993) 509–517.
[829] Aschner, M., Gannon, M. Brain Res. Bull., 33 (1994) 345–349.
[830] Montero, M., Garcia-Sancho, J. and Alvarez, J., Biochim. Biophys. Acta, 1177 (1993) 127–133.
[831] Bauer, P. D., Trapp, C., Drake, D., Taylor, K. G. and Doyle, R. J., J. Bacteriol., 175 (1993) 819–825.
[832] Lutz, T. A., Schroff, A. and Scharrer, E., Biol. Trace Element Res., 39 (1993) 221–227.

[833] Testolin, G., Ciappellano, S., Alberio, A., Piccinini, F., Paracchini, L. and Jotti, A., Ann. Nutr. Metab., 37 (1993) 289–294.
[834] Håkansson, K., Wehnert, A. and Liljas, A., Acta Cryst. Sect. D Biol. Cryst., D50 (1994) 93–100.
[835] Chae, M. Y., Omburo, G. A., Lindahl, P. A. and Raushel, F. M. J., Am. Chem. Soc., 115 (1993) 12173–12174.
[836] Omburo, G. A., Mullins, L. S. and Raushel, F. M., Biochemistry, 32 (1993) 9148–9155.
[837] Hayman, A. R. and Cox, T. M. J., Biol. Chem., 269 (1994) 1294–1300.
[838] Berchtold, H., Reshetnikova, L., Relser, C. O. A., Schirmer, N. K., Sprinzl, M. and Hilgenfeld, R., Nature, 365 (1993) 126–132.
[839] Kjeldgaard, M., Nissen, P., Thirup, S. and Nuborg, J., Structure, 1 (1993) 35–50.
[840] Eccleston, J. F., Molloy, D. P., Hinds, M. G., King, R. W. and Feeney, J., Eur. J. Biochem., 218 (1993) 1041–1047.
[841] Stoddard, B. L., Dean, A. and Koshland Jr., D. E., Biochemistry, 32 (1993) 9310–9316.
[842] Stoddard, B. L. and Koshland Jr., D. E., Biochemistry, 32 (1993) 9317–9322.
[843] Williams, R. L., Oren, D. A., Muñoz-Dorado, J., Inouye, S., Inouye, M. and Arnold, E. J., Mol. Biol., 234 (1993) 1230–1247.
[844] Olsen, L. R. and Reed, G. H., Arch. Biochem. Biophys., 304 (1993) 242–247.
[845] Lu, Z. C., Shorter, A. L. and Dunaway-Mariano, D., Biochemistry, 32 (1993) 2378–2385.
[846] Johnson, L. N. and Barford, D., Ann. Rev. Biophys. Biomol. Struct., 22 (1993) 199–232.
[847] Bossemeyer, D., Engh, R. A., Kinzel, V., Ponstingl, H. and Huber, R., EMBO J., 12 (1993) 849–859.
[848] Zheng, J. H., Trafny, E. A., Knighton, D. R., Xuong, N.-H., Taylor, S. S., Ten Eyck, L. F. and Sowadski, J. M., Acta Cryst. Sect. D - Biol. Cryst., 49 (1993) 362–365.
[849] Zheng, J., Knighton, D. R., Ten Eyck, L. F., Karlsson, R., Xuong, N., Taylor, S. S. and Sowadski, J. M., Biochemistry, 32 (1993) 2154–2161.
[850] Knighton, D. R., Bell, S. M., Zheng, J. H., Ten Eyck, L. F., Xuong, N. H., Taylor, S. S. and Sowadski, J. M., Acta Cryst. Sect. D - Biol. Crystallog., 49 (1993) 357–361.
[851] Madhusudan, Trafny, E. A., Xuong, N.-Y., Adams, J. A., Ten Eyck, L. F., Taylor, S. S. and Sowadski, J. M., Protein Sci., 3 (1994) 176–187.
[852] Adams, J. A. and Taylor, S. S., Protein Sci., 2 (1993) 2177–2186.
[853] Murali, N., Jarori, G. K., Landy, S. B. and Rao, B. D. N., Biochemistry, 32 (1993) 12941–12948.
[854] Müller, C. W. and Schulz, G. E., Proteins Struct. Funct. Genet., 15 (1993) 42–49.
[855] Gerstein, M., Schulz, G. and Chothia, C. J., Mol. Biol., 229 (1993) 494–501.
[856] Shi, Z. T., Byeon, I.-J. L., Jiang, R.-T. and Tsai, M.-D., Biochemistry, 32 (1993) 6450–6458.
[857] Okajima, T., Tanizawa, K. and Fukui, T., FEBS Lett., 334 (1993) 86–88.
[858] Byeon, I. J. L., Yan, H. G., Edison, A. S., Mooberry, E. S., Abildgaard, F., Markley, J. L. and Tsai, M. D., Biochemistry, 32 (1993) 12508–12521.
[859] Okajima, T., Tanizawa, K. and Fukui, T. J., Biochem., 114 (1993) 627–633.
[860] Dahnke, T. and Tsai, M. D. J., Biol. Chem., 269 (1994) 8075–8081.
[861] Ren, J. S., Stuart, D. I. and Acharya, K. R. J., Biol. Chem., 268 (1993) 19292–19298.
[862] Bunzli, J. C. G. and Pfefferle, J. M., Helv. Chim. Acta, 77 (1994) 323–333.
[863] Landro, J. A., Gerlt, J. A., Kozarich, J. W., Koo, C. W., Shah, V. J., Kenyon, G. L. and Neidhart, D. J., Biochemistry, 33 (1994) 635–643.
[864] Hammer, A., Hildenbrand, T., Hoier, H., Ngai, K.-L., Schlömann, M. and Stezowski, J. J., J. Mol. Biol., 232 (1993) 305–307.
[865] Hoier, H., Schlömann, M., Hammer, A., Glusker, J. P., Carrell, H. L., Goldman, A., Stezowski, J. J. and Heinemann, U., Acta Cryst. Sect. D - Biol. Cryst., 50 (1994) 75–84.
[866] Eriksson, K. E. L., Habu, N. and Samejima, M., Enzyme Microbiol. Technol., 15 (1993) 1002–1008.
[867] Gold, M. H. and Alic, M., Microbiol. Rev., 57 (1993) 605–622.

[868] Sterjiades, R. and Eriksson, K. E. L. In *Polyphenolic Phenomenon*; A. Scalbert, Ed.; INRA: Paris France, 1993, pp. 115–126.
[869] Ander, P., FEMS Microbiol. Rev., 13 (1994) 387–390.
[870] Orth, A. B., Royse, D. J. and Tien, M., Appl. Environ. Microbiol., 59 (1993) 4017–4023.
[871] Becker, H. G. and Sinitsyn, A. P., Biotech. Lett., 15 (1993) 289–294.
[872] Johansson, T., Welinder, K. G. and Nyman, P. O., Arch. Biochem. Biophys., 300 (1993) 57–62.
[873] Ruttimann-Johnson, C., Cullen, D. and Lamar, R. T., Appl. Environ. Microbiol., 60 (1994) 599–605.
[874] Uma, L., Kalaiselvi, R. and Subramanian, G., Biotech. Lett., 16 (1994) 303–308.
[875] Perestelo, F., Falcon, M. A., Carnicero, A., Rodriguez, A. and Delafuente, G., Biotech. Lett., 16 (1994) 299–302.
[876] Vares, T., Niemenmaa, O. and Hatakka, A., Appl. Environ. Microbiol., 60 (1994) 569–575.
[877] Schliephake, K., Lonergan, G. T., Jones, C. L. and Mainwaring, D. E., Biotech. Lett., 15 (1993) 1185–1188.
[878] Joshi, D. K. and Gold, M. H., Appl. Environ. Microbiol., 59 (1993) 1779–1785.
[879] Shah, M. M., Grover, T. A. and Aust, S. D., Biochem. Biophys. Res. Commun., 191 (1993) 887–892.
[880] Ollikka, P., Alhonmaki, K., Leppanen, V. M., Glumoff, T., Raijola, T. and Suominen, I., Appl. Environ. Microbiol., 59 (1993) 4010–4016.
[881] Goszczynski, S., Paszczynski, A., Pastigrigsby, M. B., Crawford, R. L. and Crawford, D. L. J., Bacteriol., 176 (1994) 1339–1347.
[882] Bumpus, J. A. and Tatarko, M., Curr. Microbiol., 28 (1994) 185–190.
[883] Tatarko, M. and Bumpus, J. A., Lett. Appl. Microbiol., 17 (1993) 20–24.
[884] Yadav, J. S. and Reddy, C. A., Appl. Environ. Microbiol., 59 (1993) 756–762.
[885] Yadav, J. S. and Reddy, C. A., Appl. Environ. Microbiol., 59 (1993) 2904–2908.
[886] Chang, C. W. and Bumpus, J. A., FEMS Microbiol. Lett., 107 (1993) 337–342.
[887] Paice, M. G., Reid, I. D., Bourbonnais, R., Archibald, F. S. and Jurasek, L., Appl. Environ. Microbiol., 59 (1993) 260–265.
[888] Bajpai, P., Mehna, A. and Bajpai, P. K., Process Biochem., 28 (1993) 377–384.
[889] Vazquez du Halt, R., Westlake, D. W. S. and Fedorak, P. M., Appl. Environ. Microbiol., 60 (1994) 459–466.
[890] Banci, L., Bertini, I., Kuan, I. C., Tien, M., Turano, P. and Vila, A. J., Biochemistry, 32 (1993) 13483–13489.
[891] Banci, L., Bertini, I., Bini, T. Z., Tien, M. and Turano, P., Biochemistry, 32 (1993) 5825–5831.
[892] Kuan, I.-C., Johnson, K. A. and Tien, M. J., Biol. Chem., 268 (1993) 20064–20070.
[893] Boe, J. F., Goulas, P. and Seris, J.-L., Biocatalysis, 7 (1993) 297–308.
[894] Kuan, I.-C. and Tien, M., Arch. Biochem. Biophys., 302 (1993) 447–454.
[895] Kuan, I.-C. and Tien, M., Proc. Natl. Acad. Sci. USA, 90 (1993) 1242–1246.
[896] Edwards, S. L., Raag, R., Wariishi, H., Gold, M. H. and Poulos, T. L., Proc. Natl. Acad. Sci. USA, 90 (1993) 750–754.
[897] Poulos, T. L., Edwards, S. L., Wariishi, H. and Gold, M. H., J. Biol. Chem., 268 (1993) 4429–4440.
[898] Kunishima, N., Fukuyama, K., Matsubara, H., Hatanaka, H., Shibano, Y. and Amachi, T., J. Mol. Biol., 235 (1994) 331–344.
[899] Petersen, J. F. W., Kadziola, A. and Larsen, S., FEBS Lett., 339 (1994) 291–296.
[900] Hammel, K. E., Jensen Jr., K. A., Mozuch, M. D., Landucci, L. L., Tien, M. and Pease, E. A., J. Biol. Chem., 268 (1993) 12274–12281.
[901] Franken, S. M., Scheidig, A. J., Krengel, U., Rensland, H., Lautwein, A., Geyer, M., Scheffzek K., Goody, R. S., Kalbitzer, H. R., Pai, E. F. and Wittinghofer, A., Biochemistry, 32 (1993) 8411–8420.
[902] Miller, A.-F., Halkides, C. J. and Redfield, A. G., Biochemistry, 32 (1993) 7367–7376.

[903] Kraulis, P. J., Domaille, P. J., Campbell-Burk, S. L., Van Aken, T. and Laue, E. D., Biochemistry, 33 (1994) 3515–3531.
[904] Halkides, C. J., Farrar, C. T., Larsen, R. G., Redfield, A. G. and Singel, D. J., Biochemistry, 33 (1994) 4019–4035.
[905] John, J., Rensland, H., Schlichting, I., Vetter, I., Borasio, G. D., Goody, R. S. and Wittinghofer, A., J. Biol. Chem., 268 (1993) 923–929.
[906] Hu, J.-S. and Redfield, A. G., Biochemistry, 32 (1993) 6763–6772.
[907] Chung, H.-H., Benson, D. R. and Schultz, P. G., Science, 259 (1993) 806–809.
[908] Frech, M., Darden, T. A., Pedersen, L. G., Foley, C. K., Charifson, P. S., Anderson, M. W. and Wittinghofer, A., Biochemistry, 33 (1994) 3237–3244.
[909] Hartman, F. C. and Harpel, M. R., In *Advances in Enzymology and Related Areas of Molecular Biology*; A. Meister, Ed.; John Wiley and Sons Inc., New York, 1993, Vol. 67, pp. 1–75.
[910] Spreitzer, R. J., Ann. Rev. Plant Physiol. Plant Mol. Biol., 44 (1993) 411–434.
[911] Rault, M., Giudici-Orticoni, M.-T., Gontero, B. and Ricard, J., Eur. J. Biochem., 217 (1993) 1065–1073.
[912] Gontero, B., Mulliert, G., Rault, M., Guidici-Orticoni, M.-T. and Ricard, J., Eur. J. Biochem., 217 (1993) 1075–1082.
[913] Hosur, M. V., Sainis, J. K. and Kannan, K. K., J. Mol. Biol., 234 (1993) 1274–1278.
[914] Zhang, K. Y. J., Cascio, D. and Eisenberg, D., Protein Sci., 3 (1994) 64–69.
[915] Newman, J. and Gutteridge, S. J., Biol. Chem., 268 (1993) 25876–25886.
[916] Newman, J., Bränden, C.-I. and Jones, T. A., Acta Cryst. Sect. D - Biol. Cryst., 49 (1993) 548–560.
[917] Schreuder, H. A., Knight, S., Curmi, P. M. G., Andersson, I., Cascio, D., Bränden, C.-I. and Eisenberg, D., Proc. Natl. Acad. Sci. USA, 90 (1993) 9968–9972.
[918] Schreuder, H. A., Knight, S., Curmi, P. M. G., Andersson, I., Cascio, D., Sweet, R. M., Brändén, C.-I. and Eisenberg, D., Protein Sci., 2 (1993) 1136–1146.
[919] Gutteridge, S., Rhoades, D. F. and Herrmann, C., J. Biol. Chem., 268 (1993) 7818–7824.
[920] Orozco, B. M., McClung, C. R., Werneke, J. M. and Ogren, W. L., Plant Physiol., 102 (1993) 227–232.
[921] Day, A. G., Chène, P. and Fersht, A. R., Biochemistry, 32 (1993) 1940–1944.
[922] Lorimer, G. H., Chen, Y.-R. and Hartman, F. C., Biochemistry, 32 (1993) 9018–9024.
[923] Amichay, D., Levitz, R. and Gurevitz, M., Plant Mol. Biol., 23 (1993) 465–476.
[924] Lee, G. J., McDonald, K. A. and McFadden, B. A., Protein Sci., 2 (1993) 1147–1154.
[925] Lee, E. H., Harpel, M. R., Chen, Y. R. and Hartman, F. C., J. Biol. Chem., 268 (1993) 26583–26591.
[926] Salvucci, M. E., Plant Physiol., 103 (1993) 501–508.
[927] Salvucci, M. E., Rajagopalan, K., Sievert, G., Haley, B. E. and Watt, D. S., J. Biol. Chem., 268 (1993) 14239–14244.
[928] Yokota, A. and Tokai, H., J. Biochem., 114 (1993) 746–753.
[929] Zhu, G. H. and Spreitzer, R. J., J. Biol. Chem., 269 (1994) 3952–3956.
[930] Lilley, R. M., Riesen, H. and Andrews, T. J., J. Biol. Chem., 268 (1993) 13877–13884.
[931] Hodel, A., Kautz, R. A., Jacobs, M. D. and Fox, R. O., Protein Sci., 2 (1993) 838–850.
[932] Hale, S. P., Poole, L. B. and Gerlt, J. A., Biochemistry, 32 (1993) 7479–7487.
[933] Judice, J. K., Gamble, T. R., Murphy, E. C., de Vos, A. M. and Schultz, P. G., Science, 261 (1993) 1578–1581.
[934] Weber, D. J., Serpersu, E. H., Gittis, A. G., Lattman, E. E. and Mildvan, A. S., Proteins Struct. Funct. Genet., 17 (1993) 20–35.
[935] Chuang, W.-J., Weber, D. J., Gittis, A. G. and Mildvan, A. S., Proteins Struct. Funct. Genet., 17 (1993) 36–48.
[936] Chuang, W. J., Gittis, A. G. and Mildvan, A. S., Proteins Struct. Funct. Genet., 18 (1994) 68–80.
[937] Meyerick, B. and Magnuson, M. A., Am. J. Resp. Cell Mol. Biol., 10 (1994) 113–121.

[938] Sun, Y., Hegamyer, G. and Colburn, N. H., Gene, 131 (1993) 301–302.
[939] Hassett, D. J., Woodruff, W. A., Wozniak, D. J., Vasil, M. L., Cohen, M. S. and Ohman, D. E., J. Bacteriol., 175 (1993) 7658–7665.
[940] Purdy, D. and Park, S. F., Microbiol. (Reading, UK), 140 (1994) 1203–1208.
[941] Klenk, H. P., Schleper, C., Schwass, V. and Brudler, R., Biochim. Biophys. Acta, 1174 (1993) 95–98.
[942] Kroll, J. S., Langford, P. R., Saah, J. R. and Loynds, B. M., Mol. Microbiol., 10 (1993) 839–848.
[943] Miao, Z. H. and Gaynor, J. J., Plant Mol. Biol., 23 (1993) 267–277.
[944] Sakurai, H., Kusumoto, N., Kitayama, K. and Togasaki, R. K., Plant Cell Physiol., 34 (1993) 1133–1137.
[945] DeShazar, D., Bannan, J. D., Moran, M. J. and Freidman, R. L., Gene, 142 (1994) 85–89.
[946] Duttaroy, A., Meidinger, R., Kirby, K., Carmichael, S., Hilliker, A. and Phillips, J., Gene, 143 (1994) 223–225.
[947] Streller, S., Kromer, S. and Wingsle, G., Plant Cell Physiol., 35 (1994) 859–867.
[948] Spiegelhalder, C., Gerstenecker, B., Kersten, A., Schiltz, E. and Kist, M., Infect. Immun., 61 (1993) 5315–5325.
[949] Pesci, E. C. and Pickett, C. L., Gene, 143 (1994) 111–116.
[950] Zhu, D. H. and Scandalios, J. G., Proc. Natl. Acad. Sci. USA, 90 (1993) 9310–9314.
[951] Sakamoto, A., Nosaka, Y. and Tanaka, K., Plant Physiol., 103 (1993) 1477–1478.
[952] Giglio, M.-P., Hunter, T., Bannister, J. V. Bannister, W. H. and Hunter, G. J., Biochem. Mol. Biol. Int., 33 (1994) 37–40.
[953] Ismail, S. O., Skeiky, Y. A. W., Bhatia, A., Omara-Opyene, L. A. and Gedamu, L., Infect. Immun., 62 (1994) 657–664.
[954] Church, S. L., Biochim. Biophys. Acta, 1171 (1993) 341.
[955] Wagner, U. G., Pattridge, K. A., Ludwig, M. L., Stallings, W. C., Werber, M. M., Oefner, C., Frolow, F. and Sussman, J. L., Protein Sci., 2 (1993) 814–825.
[956] Cooper, J. B., Driessen, H. P. C., Wood, S. P., Zhang, Y. and Young, D., J. Mol. Biol., 235 (1994) 1156–1158.
[957] Ming, L.-J., Lynch, J. B., Holz, R. C. and Que Jr., L., Inorg. Chem., 33 (1994) 83–87.
[958] Meier, B., Sehn, A. P., Sette, M., Paci, M., Desideri, A. and Rotillo, G., FEBS Lett., 348 (1994) 283–286.
[959] Meier, B., Sehn, A. P., Schinina, M. E. and Barra, D., Eur. J. Biochem., 219 (1994) 463–468.
[960] Meier, B., Sehn, A. P., Michel, C. and Saran, M., Arch. Biochem. Biophys., 313 (1994) 296–303.
[961] Nilsson, U., Lindqvist, Y., Kluger, R. and Schneider, G., FEBS Lett., 326 (1993) 145–148.
[962] Nikkola, M., Lindqvist, Y. and Schneider, G., J. Mol. Biol., 238 (1994) 387–404.
[963] Konig, S., Schellenberger, A., Neef, H. and Schneider, G., J. Biol. Chem., 269 (1994) 10879–10882.
[964] Ernstfonberg, M. L., Worsham, L. M. S. and Williams, S. G., Biochim. Biophys. Acta, 1164 (1993) 273–282.
[965] Khangulov, S., Sivaraja, M., Barynin, V. V. and Dismukes, G. C., Biochemistry, 32 (1993) 4912–2924.
[966] Zheng, M., Khangulov, S. V., Dismukes, G. C. and Barynin, V. V., Inorg. Chem., 33 (1994) 382–387.
[967] Beese, L. S., Derbyshire, V. and Steitz, T. A., Science, 260 (1993) 352–355.
[968] Pandey, V. N., Kaushik, N., Sanzgiri, R. P., Patil, M. S., Modak, M. J. and Barik, S., Eur. J. Biochem., 214 (1993) 59–65.
[969] Beese, L. S., Friedman, J. M. and Steitz, T. A., Biochemistry, 32 (1993) 14095–14101.
[970] Weston, S. and Suck, D., Protein Eng., 6 (1993) 349–357.
[971] Zhang, Y., Liang, J.-Y., Huang, S., Ke, H. and Lipscomb, W. N., Biochemistry, 32 (1993) 1844–1857.

[972] Liang, J. Y., Zhang, Y. P., Huang, S. H. and Lipscomb, W. N., Proc. Natl. Acad. Sci. USA, 90 (1993) 2132–2136.
[973] Lee, Y.-H., Lin, K., Okar, D., Alfano, N. L., Sarma, R., Pflugrath, J. W. and Pilkis, S. J., J. Mol. Biol., 235 (1994) 1147–1151.
[974] Wedler, F. C. and Ley, B. W., Neurochem. Res., 19 (1994) 139–144.
[975] Liaw, S.-H., Villafranca, J. J. and Eisenberg, D., Biochemistry, 32 (1993) 7999–8003.
[976] Liaw, S.-H. and Eisenberg, D., Biochemistry, 33 (1994) 675–681.
[977] Liaw, S. H., Pan, C. and Eisenberg, D., Proc. Natl. Acad. Sci. USA, 90 (1993) 4996–5000.
[978] Sharon, N., Trends Biochem. Sci., 18 (1993) 221–226.
[979] Bourne, Y., Van Tilbeurgh, H. and Cambillau, C., Curr. Opin. Struct. Biol., 3 (1993) 681–686.
[980] Delbaere, L. T. J., Vandonselaar, M., Prasad, L., Quail, J. W., Wilson, K. S. and Dauter, Z., J. Mol. Biol., 230 (1993) 950–965.
[981] Mills, A., FEBS Lett., 319 (1993) 5–11.
[982] Bourne, Y., Bolgiano, B., Nésa, M.-P., Penfold, P., Johnson, D., Feizi, T. and Cambillau, C., J. Mol. Biol., 235 (1994) 787–789.
[983] Loris, R., Thi, M. H. D., Lisgarten, J. and Wyns, L., Proteins Struct. Funct. Genet., 15 (1993) 205–208.
[984] Eiselé, J. L., Tello, D., Osinaga, E., Roseto, A. and Alzari, P. M., J. Mol. Biol., 230 (1993) 670–672.
[985] Sankaranarayanan, R., Puri, K. D., Ganesh, V., Banerjee, R., Surolia, A. and Vijayan, M., J. Mol. Biol., 229 (1993) 558–560.
[986] Wood, S. D., Reynolds, C. D., Lambert, S., McMichael, P. A. D., Allen, A. K. and Rizkallah, P. J., Acta Cryst. Sect. D - Biol. Cryst., 50 (1994) 110–111.
[987] Naismith, J. H., Habash, J., Harrop, S., Helliwell, J. R., Hunter, W. N., Wan, T. C. M., Weisgerber, S., Kalb, A. J. and Yariv, J., Acta Cryst. Sect. D Biol. Cryst., 49 (1993) 561–571.
[988] Lobsanov, Y. D. and Pletnev, V. Z., Bioorg. Khim., 19 (1993) 122–125.
[989] Lobsanov, Y. D., Gitt, M. A., Leffler, H., Barondes, S. H. and Rini, J. M., J. Biol. Chem., 268 (1993) 27034–27038.
[990] Rini, J. M., Hardman, K. D., Einspahr, H., Suddath, F. L. and Carver, J. P., J. Biol. Chem., 268 (1993) 10126–10132.
[981] Banerjee, R., Mande, S. C., Ganesh, V., Das, K., Dhanaraj, V., Mahanta, S. K., Suguna, K., Surolia, A. and Vijayan, M., Proc. Natl. Acad. Sci. USA, 91 (1994) 227–231.
[992] Bourne, Y., Mazurier, J., Legrand, D., Rouge, P., Montreuil, J., Spik, G. and Cambillau, C., Structure, 2 (1994) 209–219.
[993] Loris, R., Steyaert, J., Maes, D., Lisgarten, J., Pickersgill, R. and Wyns, L., Biochemistry, 32 (1993) 8772–8781.
[994] Schwarz, F. P., J. Inorg. Biochem., 52 (1993) 1–16.
[995] Schwarz, F. P., Puri, K. D., Bhat, R. G. and Surolia, A., J. Biol. Chem., 268 (1993) 7668–7677.
[996] Ben-Meir, D., Spungin, A., Ashkenazi, R. and Blumberg, S., Eur. J. Biochem., 212 (1993) 107–112.
[997] Kim, H. D., Burley, S. K. and Lipscomb, W. N., J. Mol. Biol., 230 (1993) 722–724.
[998] Kim, H. and Lipscomb, W. N., Biochemistry, 32 (1993) 8465–8478.
[999] Kim, H. D. and Lipscomb, W. N., Proc. Natl. Acad. Sci. USA, 90 (1993) 5006–5010.
[1000] Kim, H. and Lipscomb, W. N., Adv. Enzymol., 68 (1994) 153–213.
[1001] Davies, G. J., Gamblin, S. J., Littlechild, J. A., Dauter, Z., Wilson, K. S. and Watson, H. C., Acta Cryst. Sect. D Biol. Cryst., 50 (1994) 202–209.
[1002] Davies, G. J., Gamblin, S. J., Littlechild, J. A. and Watson, H. C., Proteins Struct. Funct. Genet., 15 (1993) 283–289.
[1003] Molnar, M. and Vas, M., Biochem. J., 293 (1993) 595–599.
[1004] Gregory, J. D. and Serpersu, E. H., J. Biol. Chem., 268 (1993) 3880–3888.

[1005] Katayanagi, K., Okumura, M. and Morikawa, K., Proteins Struct. Funct. Genet., 17 (1993) 337–346.
[1006] Katayanagi, K., Ishikawa, M., Okumura, M., Ariyoshi, M., Kanaya, S., Kawano, Y., Suzuki, M., Tanaka, I. and Morikawa, K., J. Biol. Chem., 268 (1993) 22092–22099.
[1007] Okumura, M., Ishikawa, K., Kanaya, S., Itaya, M. and Morikawa, K., Proteins Struct. Funct. Genet., 15 (1993) 108–111.
[1008] Ishikawa, K., Okumura, M., Katayanagi, K., Kimura, S., Kanaya, S., Nakamura, H. and Morikawa, K., J. Mol. Biol., 230 (1993) 529–542.
[1009] Huang, H. W., Cowan, J. A., Eur. J. Biochem., 219 (1994) 253–260.
[1010] Nordlund, P., Aberg, A., Uhlin, U. and Eklund, H., Biochem. Soc. Trans., 21 (1993) 735–738.
[1011] Rosenzweig, A. C., Frederick, C. A., Lippard, S. J. and Nordlund, P., Nature, 366 (1993) 537–543.
[1012] Atta, M., Fontecave, M. Wilkins, P. C. and Dalton, H., Eur. J. Biochem., 217 (1993) 217–223.
[1013] Hamman, S., Atta, M., Ehrenberg, A., Wilkins, P., Dalton, H., Béguin, C. and Fontecave, M., Biochem. Biophys. Res. Commun., 195 (1993) 594–599.
[1014] Meng, M. S., Bagdasarian, M. and Zeikus, J. G., Proc. Natl. Acad. Sci. USA, 90 (1993) 8459–8463.
[1015] Zheng, Y. J., Merz, K. M. and Farber, G. K., Protein Eng., 6 (1993) 479–484.
[1016] Allen, K. N., Lavie, A., Farber, G. K., Glasfeld, A. G., Petsko, G. A. and Ringe, D., Biochemistry, 33 (1994) 1481–1487.
[1017] Cha, J. H., Cho, Y. J., Whitaker, R. D., Carrell, H. L., Glusker, J. P., Karplus, P. A. and Batt, C. A., J. Biol. Chem., 269 (1994) 2687–2694.
[1018] Allen, K. N., Lavie, A., Glasfeld, A. G., Tanada, T. N., Gerrity, D. P., Carbon, S. C., Farber, G. K., Petsko, G. A. and Ringe, D., Biochemistry, 33 (1994) 1488–1494.
[1019] Bogumil, R., Kappl, R., Huttermann, J., Sudfeldt, C. and Witzel, H., Eur. J. Biochem., 213 (1993) 1185–1192.
[1020] Varsani, L., Cui, T., Rangarajan, M., Hartley, B. S., Goldberg, J., Collyer, C. and Blow, D. M., Biochem. J., 291 (1993) 575–583.
[1021] Carrell, H. L., Hoier, H. and Glusker, J. P., Acta Cryst. Sect. D Biol. Cryst., 50 (1994) 113–123.
[1022] Rasmussen, H., la Cour, T., Nyborg, J. and Schülein, M., Acta Cryst. Sect. D Biol. Cryst., 50 (1994) 124–131.
[1023] Rasmussen, H., la Cour, T., Nyborg, J. and Schülein, M., Acta Cryst. Sect. D Biol. Cryst., 50 (1994) 231–233.
[1024] LLoyd, L. F., Gallay, O. S., Akins, J. and Zeikus, J. G., J. Mol. Biol., 240 (1994) 504–506.
[1025] Bonet, M. L., Llorca, F. I. and Cadenas, E., Int. J. Biochem., 25 (1993) 7–12.
[1026] Janeway, C. M. L., Xu, X., Murphy, J. E., Chaidaroglou, A. and Kantrowitz, E. R., Biochemistry, 32 (1993) 1601–1609.
[1027] Murphy, J. E., Xu, X. and Kantrowitz, E. R., J. Biol. Chem., 268 (1993) 21497–21500.
[1028] Xu, X., Qin, X.-Q. and Kantrowitz, E. R., Biochemistry, 33 (1994) 2279–2284.
[1029] Lebioda, L., Zhang, E., Lewinski, K. and Brewer, J. M., Proteins Struct. Funct. Genet., 16 (1993) 219–225.
[1030] Zhang, E. L., Hatada, M., Brewer, J. M. and Lebioda, L., Biochemistry, 33 (1994) 6295–6300.
[1031] Brewer, J. M., Robson, R. L., Glover, C. V. C., Holland, M. J. and Lebioda, L., Proteins Struct. Funct. Genet., 17 (1993) 426–434.
[1032] McPhalen, C. A., Sielecki, A. R., Santarsiero, B. D. and James, M. N. G., J. Mol. Biol., 235 (1994) 718–732.
[1033] Pauls, T. L., Durussel, I., Cox, J. A., Clark, I. D., Szabo, A. G., Gagne, S. M., Sykes, B. D. and Berchtold, M. W., J. Biol. Chem., 268 (1993) 20897–20903.
[1034] Blancuzzi, Y., Padilla, A., Parello, J. and Cavé, A., Biochemistry, 32 (1993) 1302–1309.
[1035] Buchbinder, J. L., Baraniak, J., Frey, P. A. and Reed, G. H., Biochemistry, 32 (1993) 14111–14116.

[1036] Tan, X. L., Poyner, R., Reed, G. H. and Scholes, C. P., Biochemistry, 32 (1993) 7799–7810.
[1037] Tan, X., Poyner, R., Reed, C. H. and Scholes, C. P., J. Inorg. Biochem., 51 (1993) 125.
[1038] Obmolova, G., Kuranova, I. and Teplyakov, A., J. Mol. Biol., 232 (1993) 312–313; Teplyakov, A., Obmolova, G., Wilson, K. S., Ishii, K., Kaji, H., Samejima, T. and Kuranova, I., Protein Sci., 3 (1994) 1098–1107.
[1039] Lamzin, V. S. and Wilson, K. S., Acta Cryst. Sect. D Biol. Cryst., 49 (1993) 129–147.
[1040] Oganessyan, V. Y., Kurilova, S. A., Vorobyeve, N. N., Nazarova, T. I., Popov, A. N., Lebedev, A. A., Avaeva, S. M. and Harutyunyan, E. H., FEBS Lett., 348 (1994) 301–304.
[1041] Kanake, J., Neal, G. S., Salminen, T., Glumhoff, T., Cooperman, B. S., Lahti, R. and Goldman, A., Protein Eng., 7 (1994) 823–830.

REPERTORIES OF METAL IONS AS LEWIS ACID CATALYSTS IN ORGANIC REACTIONS

Junghun Suh

OUTLINE

1. **Introduction** **116**
2. **Catalysis by Metal Ion Itself** **117**
   2.1 Activation of Electrophiles 117
   2.2 Activation of Leaving Groups 119
   2.3 Activation of Acids 121
   2.4 Blockade of Inhibitory Reverse Paths 122
   2.5 Recognition and Masking of Anions 123
   2.6 Provision of Productive Conformations 123
3. **Catalysis by Metal-Bound Water** **125**
   3.1 Attack as a Nucleophile 125
   3.2 Assistance as a General Acid 126
4. **Catalysis by Metal-Bound Hydroxide Ion** **127**
   4.1 Attack as a Nucleophile 127
   4.2 Assistance as a General Base 128

Perspectives on Bioinorganic Chemistry
Volume 3, pages 115–149

ISBN: 1-55938-642-8

5. **Catalysis by Binuclear Metal Ions** 129
6. **Cooperation of Metal Ions with Organic Functional Groups** 129
7. **Operation of Multiple Catalytic Repertories** 132
8. **Effects of Nature of Metal Ion and Structure of Ligand on Catalytic Efficiency** 133
9. **Catalytic Repertories of Metal Ions in Metalloenzymes: Carboxypeptidase A as an Example** 136
10. **Catalytic Repertories of Metal Ions in Artificial Metalloenzymes** 141
Acknowledgments 146
References 147

1. INTRODUCTION

Catalysis by metal ions in organic reactions may be divided into three categories: Lewis acid catalysis, catalysis via redox reactions, and catalysis via organometallic compounds [1]. Lewis acid catalysis by metal ions plays important roles also in the action of many metalloenzymes. Virtually all types of organic reactions are catalyzed by such metalloenzymes. For example, Zn(II) metalloenzymes are the most typical metalloenzymes involving metal ions as the Lewis acid catalysts. Zn(II) is now known to be an essential component of more than 200 enzymes isolated from different species [2, 3]. They participate in biological reactions encompassing synthesis and degradation of all major metabolites, i.e., carbohydrates, lipids, proteins, and nucleic acids; at least one example of a Zn(II) enzyme is found in each class of the six categories designated by the International Union of Biochemistry on the basis of the types of organic reactions [3].

Knowledge of the principles of Lewis acid catalysis by metal ions is important for the study of mechanisms of not only inorganic but also organic and enzymatic reactions. Moreover, it can be exploited in the design of biomimetic catalysts. Metal ions can participate both in the process of molecular recognition of substrate structures by artificial metalloenzymes and in the process of catalysis within the supramolecular complexes formed therein.

Although a large number of studies have been performed on the Lewis acid catalysis by metal ions during the last two decades [4–7], many mechanistic points are to be cleared in order to understand details of the catalytic action of metal ions. What kinds of catalytic roles can be played by metal ions? What kinds of catalytic features can be manifested by metal-bound water molecules and metal-bound hydroxide ions? How metal ions and metal-bound water molecules/hydroxide ions cooperate with organic catalytic groups? How is the catalytic efficiency affected by the nature of metal ions, by the structure of ligands, or by the configuration around metal ions? These are some of the questions to be answered in the study of Lewis acid catalysis by metal ions in organic reactions and in biological processes. In this article, the catalytic repertories of metal ions acting as Lewis acid catalysts in organic reactions are described. In addition, how the catalytic roles of the active-site metal ion of a metalloenzyme are elucidated is discussed using carboxypeptidase A as an example. How the principles of Lewis acid catalysis by metal ions are exploited in the design of artificial metalloenzymes is also presented.

2. CATALYSIS BY METAL ION ITSELF

Unlike the proton, another Lewis acid, metal ions can be present in relatively high concentrations even at neutral or basic pHs and can possess a multiple number of positive charges. In addition, metal ions can bind specific parts of organic molecules through chelate formation. By acting as Lewis acids, metal ions bound to organic molecules can perform several catalytic roles, even without assistance from other catalytic factors.

2.1 Activation of Electrophiles

Upon coordination to metal ions, organic electrophiles become more electrophilic. For example, negative charges developed on carbonyl oxygen atoms during the reactions of carbonyl compounds or acyl derivatives can be stabilized by interaction with metal ions.

Coordination of the carbonyl oxygen to the metal ion and attack of hydroxide ion at the metal-bound carbonyl group (**1**) are assumed to occur in the hydrolysis of a number of esters or amides. In most cases, however, this mechanism is not easily differentiated from the kinetically equivalent attack by the metal-bound hydroxide ion at the carbonyl carbon (**2**) [8]. In the case of substitutionally inert complexes of Co(III), ^{18}O-tracer experiments revealed that both of the two mechanisms occur in the hydrolysis of the bound amino acid esters and amides [9–11].

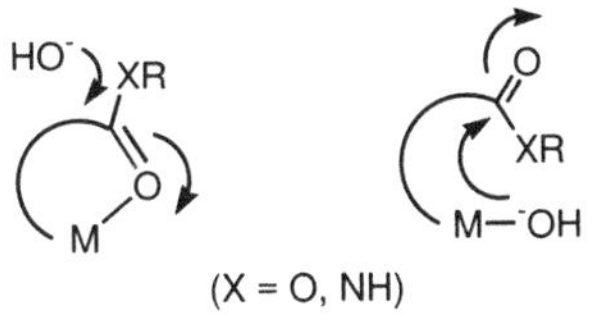

HO⁻ XR O M

O XR M—⁻OH

(X = O, NH)

1 **2**

In addition to ester (**3**) [12, 13] or amide (**4**) [9, 14] hydrolysis, reduction of aldehydes or ketones (**5**) [15, 16], nitrile hydration (**6**) [17–19], and oxaloacetate decarboxylation (**7**) [20, 21] are among the reactions catalyzed by this effect. Nucleophilic attack (**8**) by carbon nucleophiles at carbonyl groups activated by metal ions has also been reported [22]. Stability of hydrated form of aldehydes can be also enhanced through coordination of the oxy anion to metal ions (**9**) [23].

In the Ni(II) or Cu(II)-complex of 2-cyano-1,10-phenanthroline or 2-cyanopyridine, the metal ion is not geometrically allowed to interact with the nitrile group.

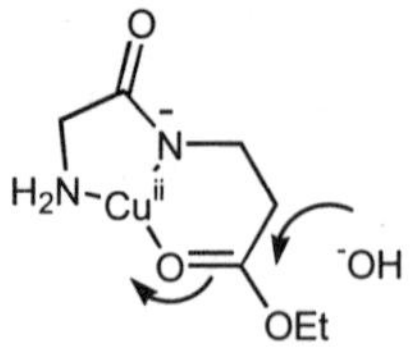

O
N
H_2N–Cu^{ii}
O
^-OH
OEt

3

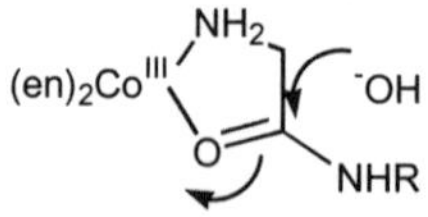

NH_2
$(en)_2Co^{III}$
^-OH
O
NHR

4

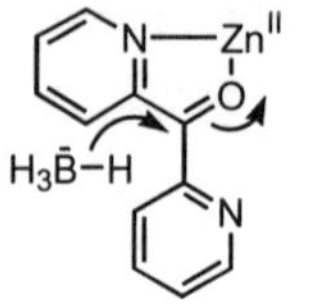

N—Zn^{II}
O
$H_3\bar{B}$–H
N

5

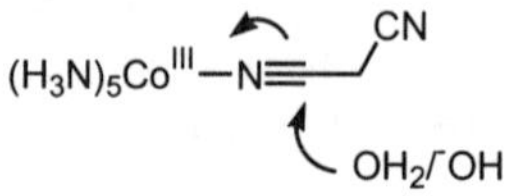

$(H_3N)_5Co^{III}$—N≡
CN
$OH_2/^-OH$

6

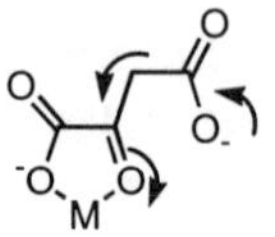

O
O
O
^-O
M
O

M: Zn^{II}, Cu^{II}

7

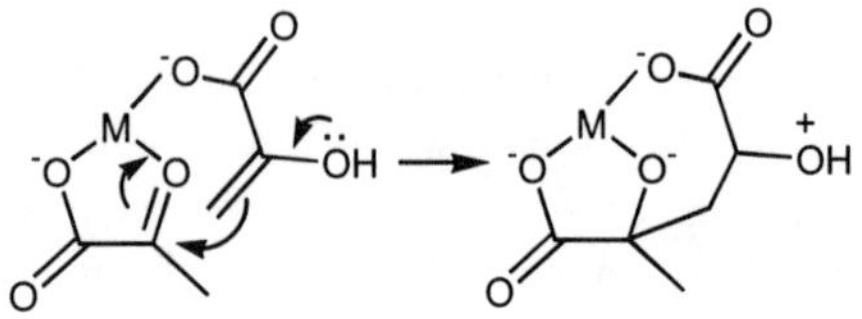

O
^-O
M
^-O
O
OH
O
→
O
^-O
M
^-O
O^-
+
OH
O

M: Zn^{II}, Mg^{II} 8

OH
N
H
Cu^{II}—O

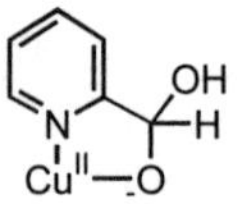

9

The metal ion, however, can bind the nitrogen atom in the transition state (**10**) for the hydration reaction, stabilizing the negative charge developed on the nitrogen atom [17, 18].

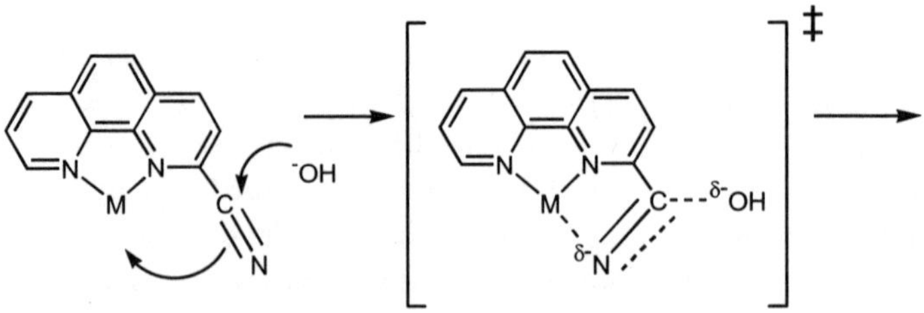

N
M
N
C
N
^-OH
→
‡
N
M
N
C
δ^-OH
δ^-N
→

M: Ni^{II}, Cu^{II}

10

Negative charges developed on the phosphoryl oxygen atoms in the nucleophilic reactions of phosphoryl derivatives are also stabilized by interaction with metal ions (**11**) [24, 25].

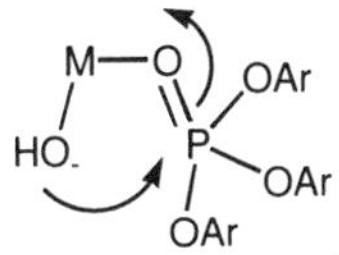

M: Zn^{II}, Cu^{II}

11

2.2 Activation of Leaving Groups

When the alkoxy oxygen atom is bound to a metal ion, its leaving ability from electrophilic carbon centers is enhanced, leading to catalysis in ester hydrolysis (**12**) [26], acetal hydrolysis (**13**) [27], or epoxide hydrolysis (14) [28].

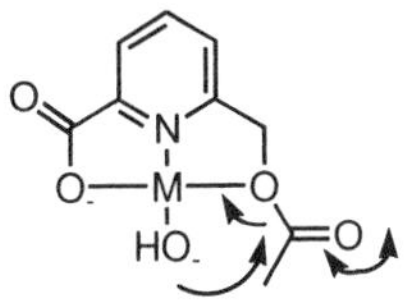

M: Ni^{II}, Co^{II}

12

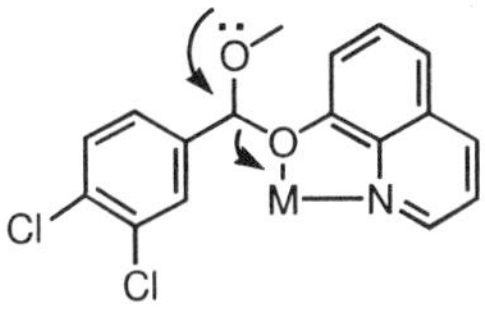

M: Cu^{II}, Co^{II}, Ni^{II}, Mn^{II}, Zn^{II}

13

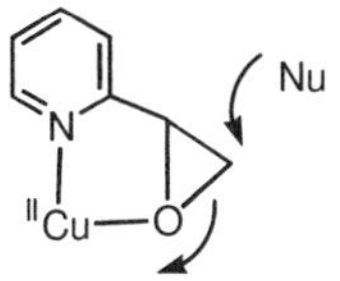

Nu: H_2O, MeOH, EtOH, Cl^-, Br^-, AcO^-

14

Even the leaving ability of hydroxide and oxide ions from acyl centers is raised upon coordination to metal ions. For example, metal-bound carboxylate ion can be hydrolyzed through the attack of hydroxide ion at the acyl carbon of the bound carboxylate ion (**15**) [8]. Oxide ion is much more basic than hydroxide ion, with the leaving group ability being extremely low. It is remarkable, therefore, to note that even oxide ion can be activated upon coordination to a metal ion leading to effective expulsion from a procarbonyl carbon.

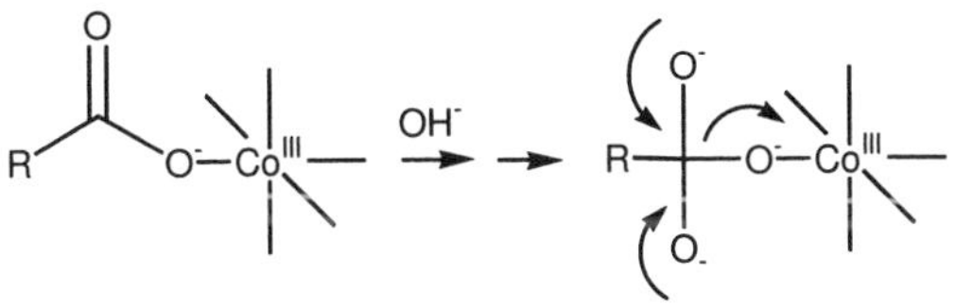

15

When an oxime is the leaving group, metal binding is much more efficient compared with alkoxy leaving groups as the oxime nitrogen atom can be the coordinating site. Enhancement in the leaving ability of the oxime oxygen from

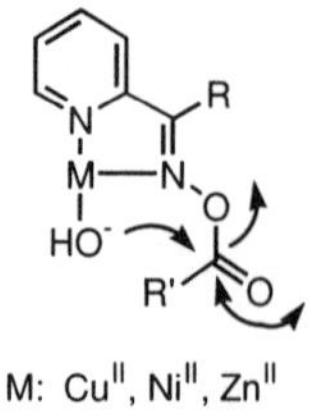

M: Cu^{II}, Ni^{II}, Zn^{II}

16 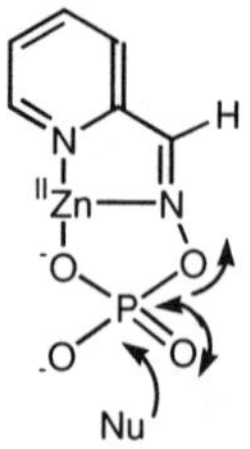 **17** 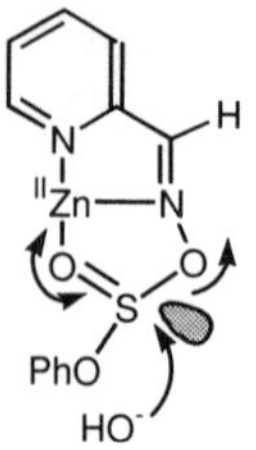 **18**

acyl (**16**) [29–31], phosphoryl (**17**) [32], or sulfinyl (**18**) [33] centers has been achieved with metal ions.

Coordination of an amide nitrogen atom to a metal ion may lead to catalysis in amide hydrolysis. The nitrogen atoms of strained β-lactams may be basic enough to coordinate to metal ions (**19**) [34, 35]. On the other hand, amide nitrogen atoms of ordinary amides might be able to bind metal ions after the acyl carbons lose the trigonal character and become tetrahedral (**20**) [36]. Coordination of an intermediate amine to a metal ion, however, can inhibit the protonation of the nitrogen atom resulting in retardation of expulsion of the amine moiety from the procarbonyl carbon (**21**) [37].

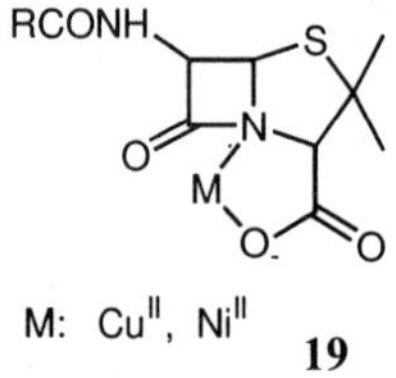

M: Cu^{II}, Ni^{II}

19 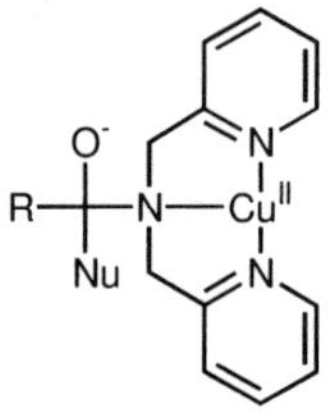 **20**

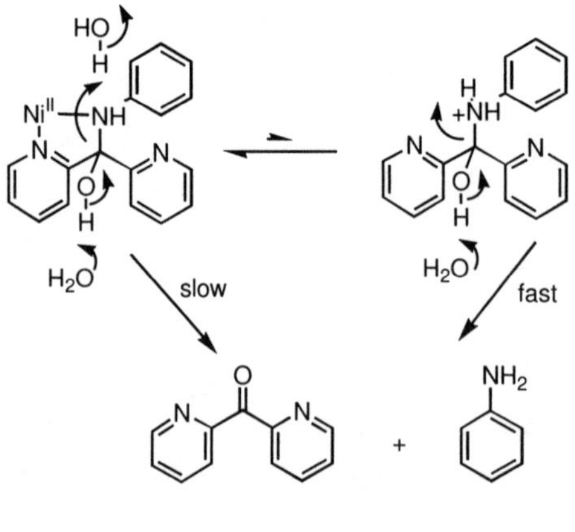

slow fast

Coordination of other leaving atoms such as sulfur (**22**) [38, 39] or halogens (**23**) [40] to metal ions can also raise the leaving ability.

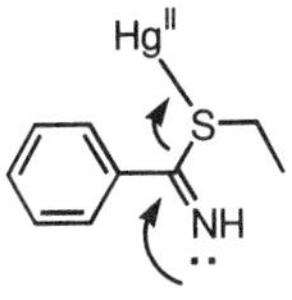

22 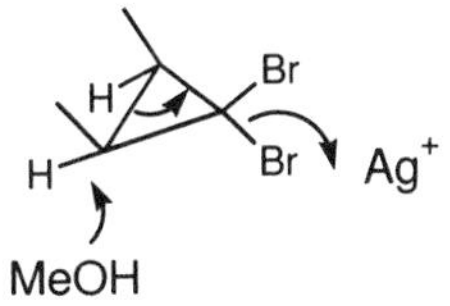 **23**

2.3 Activation of Acids

Upon coordination to metal ions, the acidity of water molecule is enhanced (**24**) [41–43]. Similarly, deprotonation of alcohols (**25**) [44], oximes (**26**) [45–47], or amines (**27**) [48–51] is facilitated by coordination of the alkoxy oxygen, imine nitrogen, or amine nitrogen atoms, respectively, to metal ions. Then, the conjugate bases are produced in considerable concentrations at neutral or even acidic pHs. The conjugate bases would become much less nucleophilic upon coordination to metal ions. This, however, is dominated by counterbalancing effects such as the generation of the basic species in greater concentrations in the presence of the metal ions and the template effect exerted by the metal ions, leading to overall rate-enhancement for nucleophilic attack at carbonyl, olefinic, or phosphoryl centers.

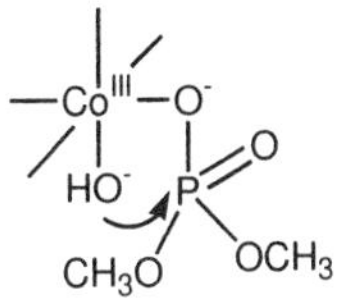

24 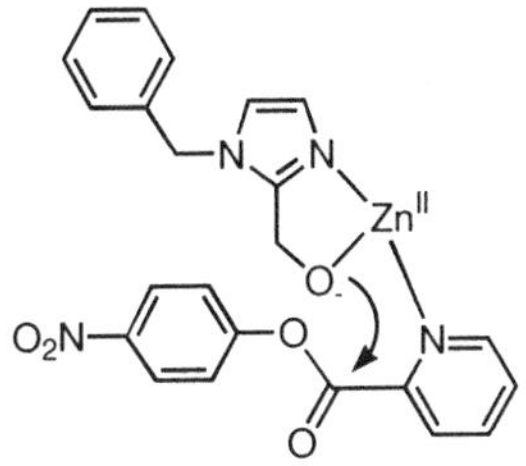 **25**

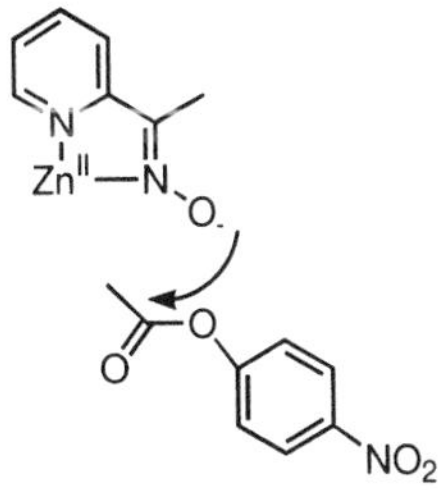

26 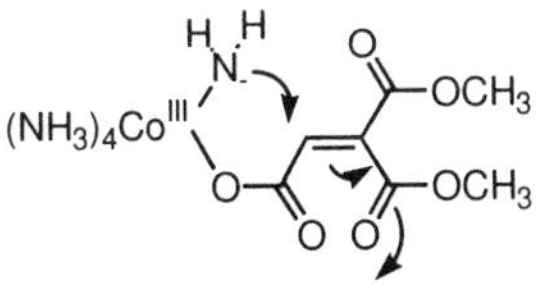 **27**

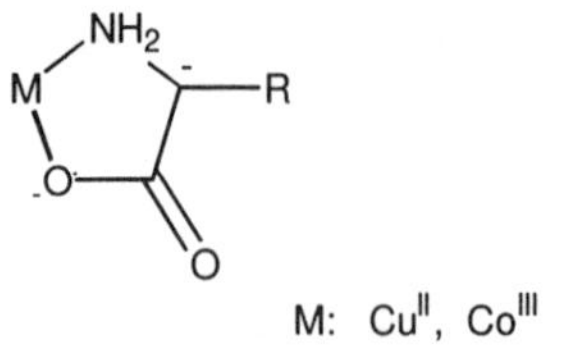

M: Cu^{II}, Co^{III}

28

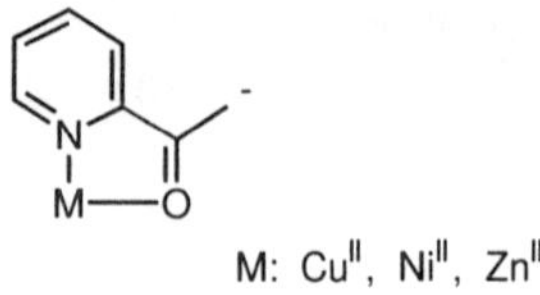

M: Cu^{II}, Ni^{II}, Zn^{II}

29

Metal ions can also activate protons attached to remote parts of the ligand. For example, the C_α–H of chelated acyl or carbonyl compounds can be activated, resulting in proton exchange, epimerization, enolization, or aldol condensation via carbanion intermediates (**28, 29**) [52–54].

2.4 Blockade of Inhibitory Reverse Paths

The hydrolysis of 3-carboxyaspirin is greatly accelerated by Fe(III) or Al(III) ion [55]. The catalysis is solely due to the blockade of an inhibitory reverse path by the metal ion. In the anhydride intermediate (**30**) formed by the nucleophilic attack (path *a*) of the carboxylate group at the ester linkage, the reverse attack (path *b*) of the phenolate anion at the anhydride linkage is very effective. In the absence of the metal ions, therefore, hydrolysis of the substrate occurs through an alternative route, via general base catalysis (path *c*) by the intramolecular carboxylate group. When Fe(III) or Al(III) ion is added, the metal ion binds the salicylate portion of the anhydride intermediate (**31**), blocking the phenolate anion. In the presence of the metal ions, the substrate is hydrolyzed through the anhydride intermediate, instead of the general base pathway.

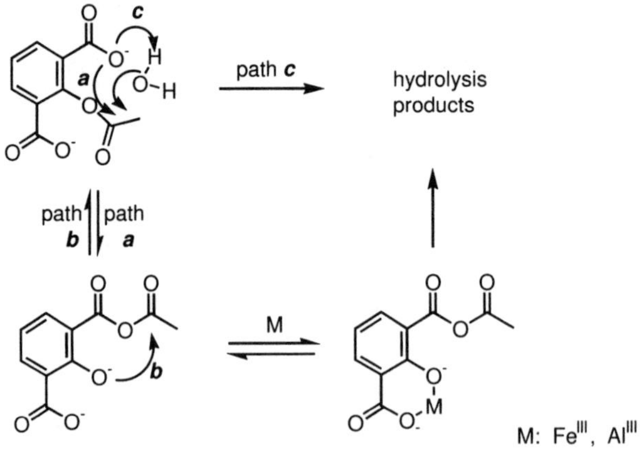

path ***c*** → hydrolysis products

path ***b*** / path ***a***

M

M: Fe^{III}, Al^{III}

30 **31**

A great number of enzymatic reactions involve covalent intermediates [56]. When the enzymatic reaction is a substitution reaction, the leaving group of the substrate remains in the vicinity of the reaction site after it is cleaved by the attack of the enzymatic group. The reverse attack of the leaving group at the resultant intermediate, however, should also be very efficient if the leaving group remains in close proximity to the reaction site. Since this retards the overall reaction, the enzyme must separate the leaving group from the reaction site or block the reactivity of the leaving group. Thus, metal ions of some metalloenzymes might participate in catalysis simply by blocking the inhibitory reverse paths through binding at the leaving groups.

2.5 Recognition and Masking of Anions

Metal ions possess strong affinity toward anions. In metalloenzymes, recognition of anions and other bases by metal ions would be utilized in folding of protein backbones to obtain catalytically active enzyme conformations and in complex formation with anionic substrates. Examples of highly effective recognition of anionic substrates by metal centers of synthetic molecules will be presented later in Section 10 with macrocyclic complexes built on a polymer.

Phosphate esters play many important biological roles [57]. Phosphate monoesters are dianionic, phosphate diesters are monoanionic, while phosphate triesters are neutral. Phosphate diesters, the chemical linkage used in DNA, are extremely stable against hydrolysis. For example, the half-life of dimethyl phosphate is estimated to be 10^{10} years at room temperature and neutral pHs. Some Co(III) complexes accelerate hydrolysis of dimethyl phosphate by 10^{10} times (**24**) [43]. Phosphate monoesters are more labile than the diesters but they are also quite stable against hydrolysis. On the other hand, phosphate triesters are readily hydrolyzed. The stability of phosphate diesters and monoesters is partly attributed to electrostatic repulsion between the anionic esters and hydroxide anion [57]. In the metal ion-catalyzed hydrolysis of anionic phosphoryl derivatives, including Mg^{2+}-catalyzed ATP hydrolysis [57], rate acceleration can be explained at least partly in terms of masking (**24**) of the anions to render the anionic substrate analogous to an uncharged species and more susceptible to nucleophilic attack by hydroxide ion [58].

2.6 Provision of Productive Conformations

In order to raise catalytic efficiency remarkably, very high "effective molarities" [59] of reactants are to be achieved. Metal ions, acting as templates, can convert intermolecular reactions into intramolecular ones, raising the effective molarities. In addition, metal ions can assemble organic molecules (**32**) [60] by recognizing anions and other Lewis bases, producing various host molecules.

In metalloenzymes, metal ions may induce conformational changes of proteins leading to very fine alignment of catalytic groups. This effect has been demonstrated with macrocyclic complexes built on poly(ethylenimine), as will be discussed later in Section 10.

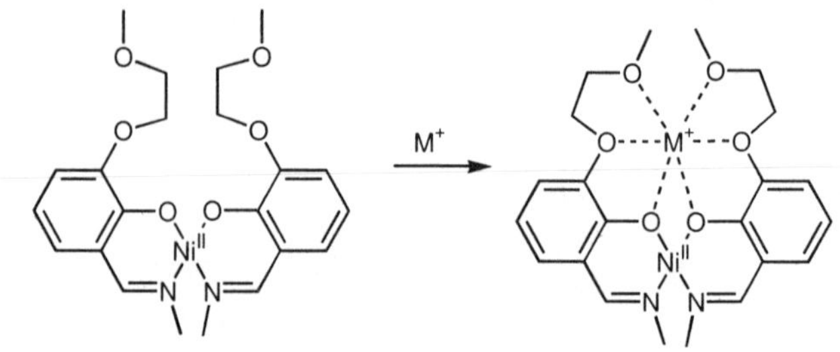

32

In enzymatic reactions, the binding forces between the substrate and the enzyme may be directly utilized to induce strain or distortion in the substrate structure, which facilitates the reactions [61]. The binding force attained through metal chelation of the 2,2′-dipyridyl moiety has been used to induce strain in the reaction centers attached to the 3 and 3′ positions (**33**), resulting in acceleration of cyclization, elimination, and racemization reactions [62]. Similarly, chelation of di-2-pyridyl ketone to Co(II) ion appears to induce strain in the carbonyl group, which is relieved by hydration of the carbonyl group (**34**) [15]. A phosphate monoaryl ester chelated to a Co(III) complex is hydrolyzed by the attack of an external hydroxide ion via the chelated five-coordinate phosphorane intermediate. The strain in the four-membered ring of the chelated ester (**35**) would be relieved in the transition state when the tetrahedral phosphorus atom is converted into a trigonal bipyramidal one [58, 63].

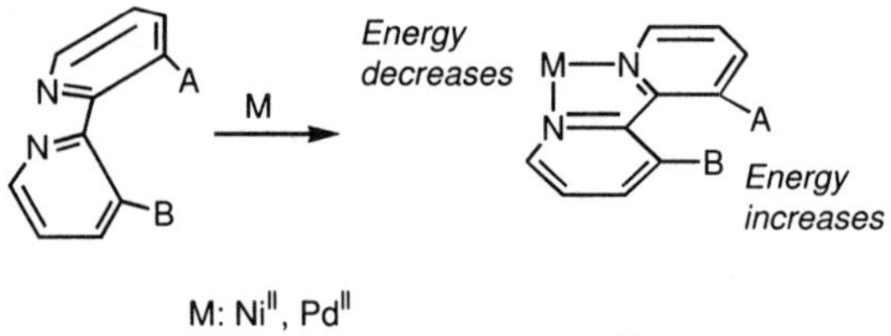

Energy decreases

Energy increases

M: Ni^{II}, Pd^{II}

33

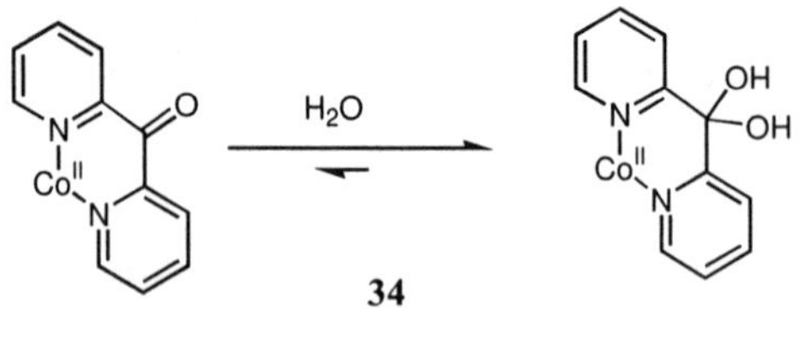

34

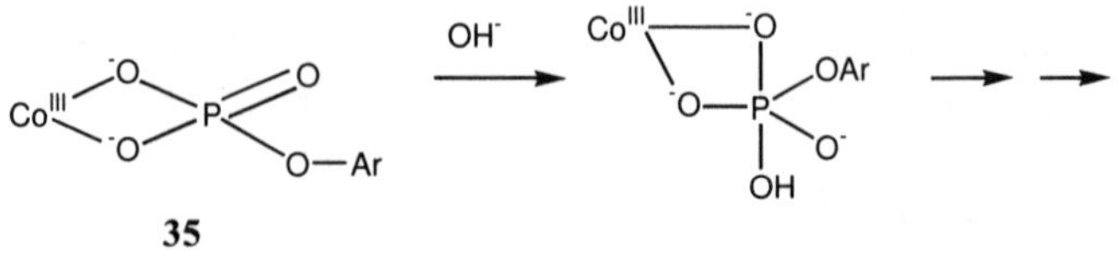

35

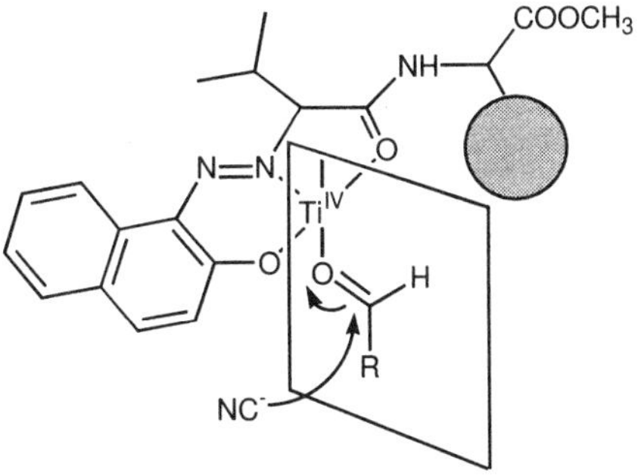

36

Metal ions are used as Lewis acid catalysts in many enantioselective synthetic reactions [64–66]. The enantioselectivity is frequently achieved by the diastereomeric transition states whose conformations are selectively provided by the catalytic metal complexes (**36**).

3. CATALYSIS BY METAL-BOUND WATER

When metal ions are dissolved in water, metal-bound water molecules and metal-bound hydroxide ions are also produced and play important catalytic roles.

3.1 Attack as a Nucleophile

In the Cu(II)-catalyzed hydrolysis of the acetyl ester of 2-pyridinecarboxaldoxime, two reaction paths were observed: rate of one path was dependent on hydroxide concentration and that of the other independent of pH [67]. Detailed analysis of the kinetic data indicated that the metal-bound water molecule makes nucleophilic attack at the carbonyl carbon of the bound ester (**37**), instead of the kinetically equivalent attack by an external water molecule at the ester linkage bound by the metal ion. The basicity and nucleophilicity of water would decrease remarkably upon coordination to metal ions. Efficient nucleophilic attack by the Cu(II)-bound water in **37** is attributable to the general base assistance from another

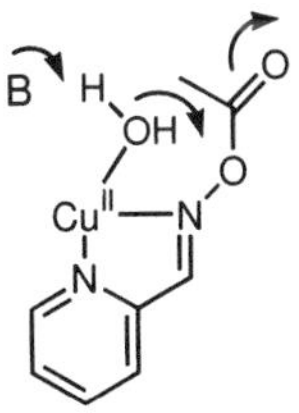

37

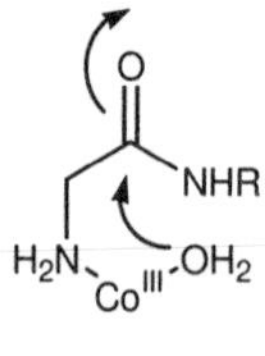

O NHR H_2N Co^{III} OH_2

38

O NHR H_2N Co^{III} OH ⇌ O^- NHR OH H_2N Co^{III} ⇌ O^- $\overset{+}{N}H_2R$ OH_2 H_2N Co^{III} →

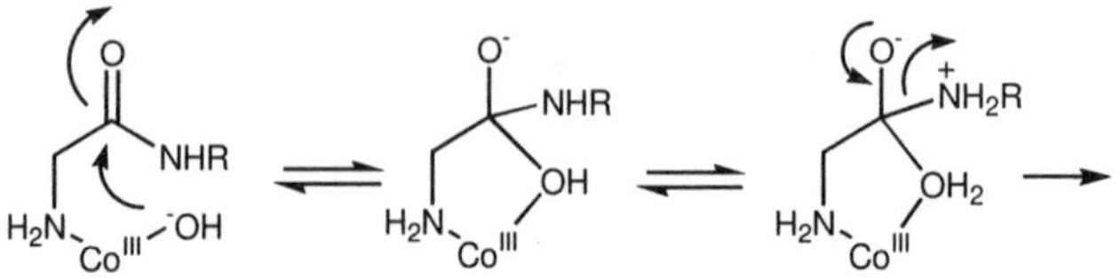

39

water molecule and to the efficient intramolecular reaction between the nucleophile and the ester.

In lactonization or amide hydrolysis reactions of metal-bound carboxylic acids or amides, rates of some reaction paths are independent of pH. This may be ascribed to the intramolecular nucleophilic attack of metal-bound water molecules at the carboxyl or amide groups (**38**) [68, 69]. This path, however, is also consistent with the attack by the metal-bound hydroxide ion and participation of the specific acid in the rate-controlling expulsion of hydroxy or amine leaving groups (**39**) [70].

3.2 Assistance as a General Acid

Upon coordination to metal ions, water becomes weak acids and may act as general acids. This catalytic role has been demonstrated in the hydrolysis of an alkyl amide cocatalyzed by a metal ion and carboxylate anion in dimethyl sulfoxide containing 5% (v/v) water (**40**) [71, 72]. Here, the alkoxide anion formed by the

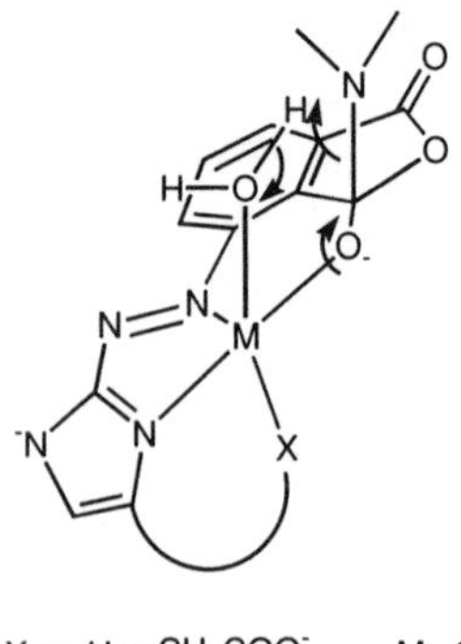

N O O H H O O^- N=N M ^-N N X

X = H or CH_2COO^-, M: Cu^{II}, Ni^{II}

40

nucleophilic attack by the carboxylate anion at the carbonyl carbon is stabilized by the metal ion. In the expulsion of the amine leaving group from the tetrahedral intermediate, the metal-bound water acts as a general acid to protonate the leaving nitrogen atom. This reaction will be discussed again later in Sections 6 and 9.

4. CATALYSIS BY METAL-BOUND HYDROXIDE ION

Depending on the nature of metal ions, metal-bound hydroxide ions can be present in considerable concentrations at neutral or even acidic pHs [41, 42].

4.1 Attack as a Nucleophile

Nucleophilic attack by metal-bound hydroxide ions is known for a variety of reactions. For example, attack by a metal-bound hydroxide at bound esters has been already illustrated with **12** and **16**. Nucleophilic attack by the metal-bound water molecule at the bound ester represented by **37** occurs together with the nucleophilic attack by the metal-bound hydroxide ion [67].

Amide hydrolysis can occur through the nucleophilic attack by metal-bound hydroxide ions as indicated by **2** [10, 68]. Very efficient amide hydrolysis has been achieved (**41**) when a metal-bound hydroxide ion is located in the vicinity of the carbonyl carbon of the scissile bond [73, 74]. The expulsion of amine nitrogens from tetrahedral intermediates requires proton donors. When a metal-bound hydroxide makes nucleophilic attack at an amide, it can also act as the proton donor needed for the expulsion of the amine leaving group (**42**).

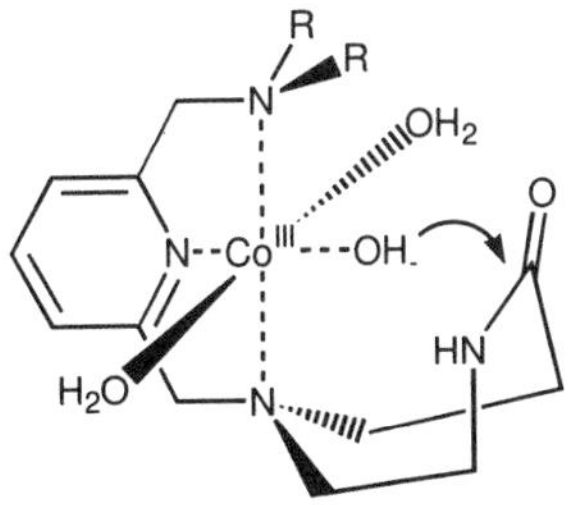

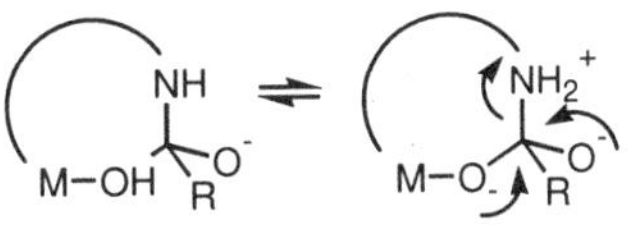

42

41

Nucleophilic attack by metal-bound hydroxide ions at other carbonyl centers such as aldehydes (**43**) [75] or anhydrides (**44**) [76] has been investigated as models of metalloenzymes such as carbonic anhydrase or carboxypeptidase A. Nucleophilic attack by metal-bound hydroxides at nitriles can lead to very efficient hydration reactions (**45**) [77].

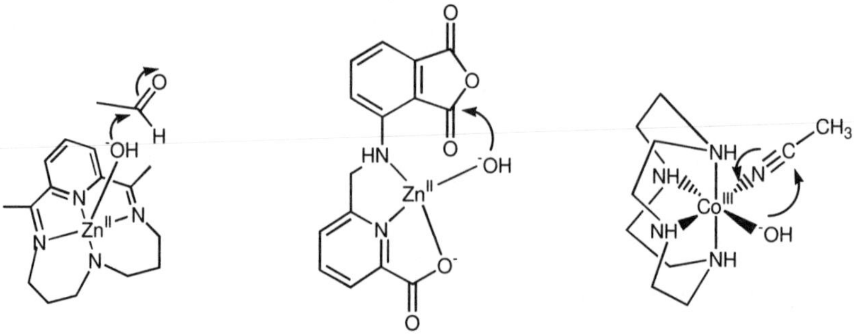

43 **44** **45**

Metal-bound hydroxide ions also act as nucleophiles toward phosphate esters. For example, phosphate diesters are subject to metal-catalyzed hydrolysis, via nucleophilic attack of metal-bound hydroxide ions to form intermediate complexes involving four-membered rings (**46**) [77–79]. Other examples are **11** and **24**.

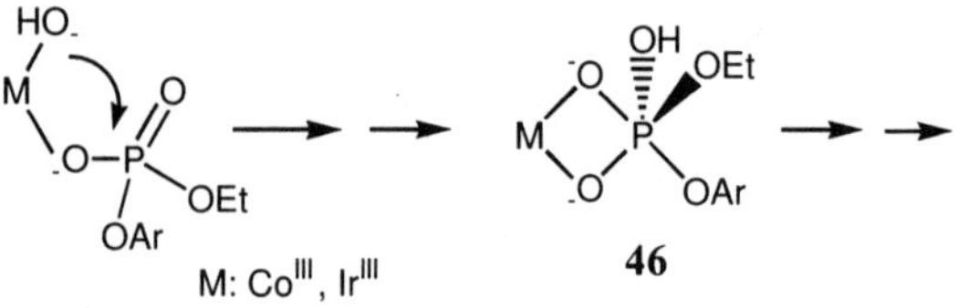

M: Co^{III}, Ir^{III}

46

Intramolecular nucleophilic attack by Co(III)-bound hydroxide ions at olefinic carbon atoms leading to alkene hydration has been observed (**47**) [80].

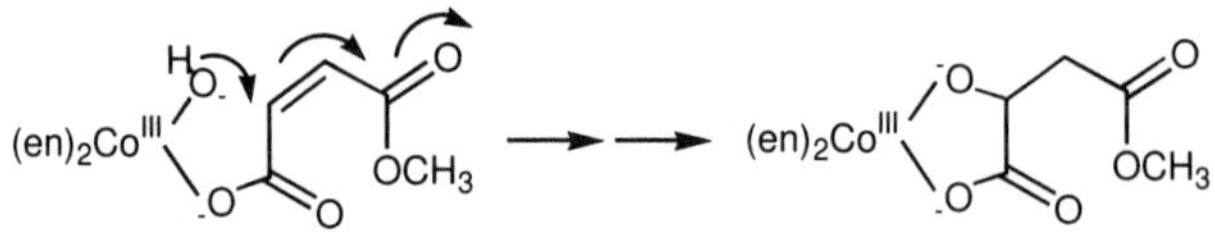

47

4.2 Assistance as a General Base

Since metal-bound hydroxide ions are weak bases, they can act as general base catalysts. This catalytic role, however, has been demonstrated only very recently in the Cu(II)-catalyzed hydrolysis of *m*-(2-imidazolylazo)phenyl *p*-toluenesulfonate [81]. In this reaction, hydroxoCu(II) ion participates as a general base in the proton transfer (**48**) between addition intermediates.

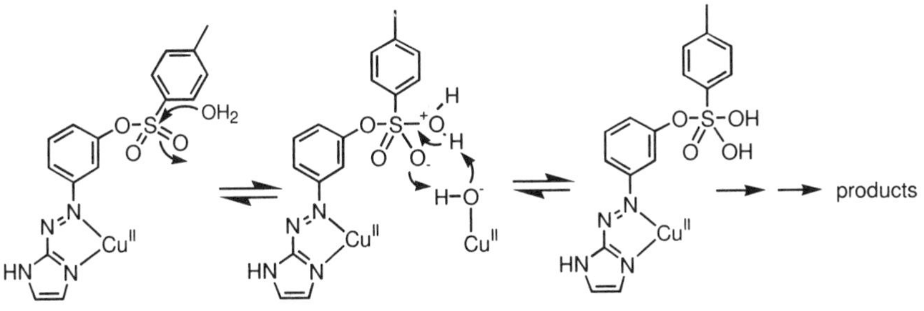

48

5. CATALYSIS BY BINUCLEAR METAL IONS

The Zn(II)-catalyzed hydrolysis of the acetyl esters of 2-acetylpyridineketoxime and 6-carboxy-2-pyridinecarboxaldoxime involves the participation of two Zn(II) ions [82]. Mechanistic analysis indicated the participation of binuclear Zn(II) ions as the catalytic unit (**49**), instead of the kinetically equivalent participation of two separate Zn(II) ions (**50**) [82]. In addition, catalysis by the mononuclear Zn(II) ion was not observed at all for the hydrolysis of the acetyl ester of 6-carboxy-2-pyridinecarboxaldoxime. The equilibrium concentration of the binuclear Zn(II) ion must be very low compared with the mononuclear species. The efficient catalysis by the binuclear Zn(II) species is apparently due to the geometry of the transition state. Some other metal-catalyzed reactions involve two metal ions in the transition state [83, 84]. In these reactions, binuclear metal ions might be involved as catalytic units, although the exact mechanisms are not known.

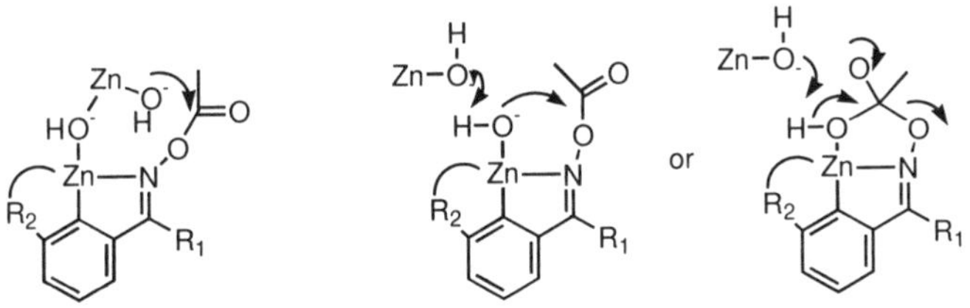

49 ($R_1 = CH_3$, $R_2 = H$; $R_1 = H$, $R_2 = COO^-$) **50**

6. COOPERATION OF METAL IONS WITH ORGANIC FUNCTIONAL GROUPS

The catalytic efficiency of metal ions would be greatly enhanced when other catalytic factors participate in cooperation with metal ions. As examples of the participation of general acids or bases in metal-catalyzed organic reactions, the

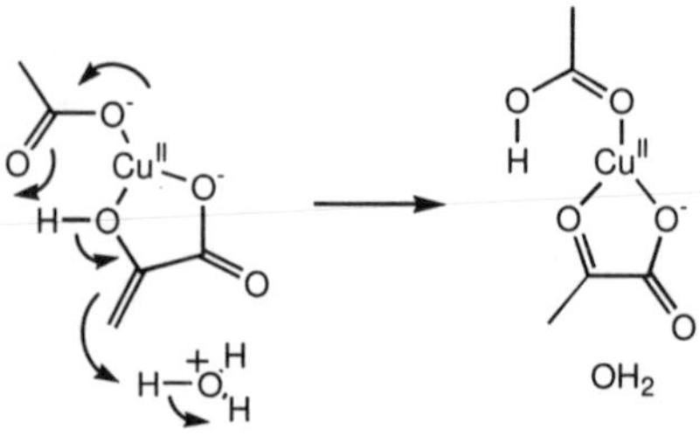

51

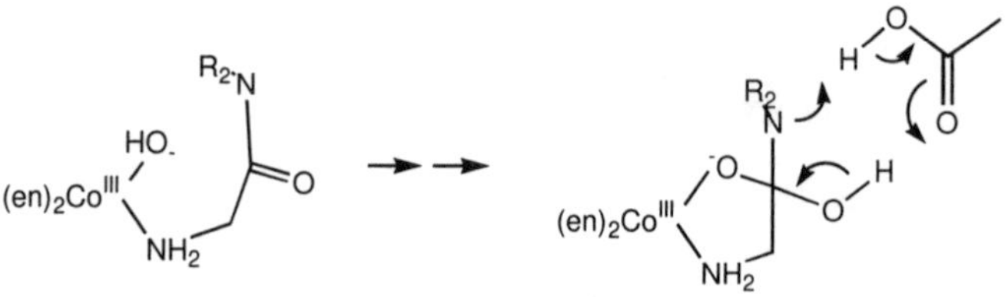

52

cocatalysis by acetic acid and Cu(II) ion in the ketonization of pyruvate enolates (**51**) [85] or by the Co(III)-bound hydroxide ion and added buffers in amide hydrolysis (**52**) [86] may be cited.

Cooperation between a metal ion and an organic functional group would become much more efficient when both the organic catalytic group and the metal-binding site are located in close proximity of the reaction center. For example, highly effective cooperation between metal ions and amide oxygen atoms has been observed in the cleavage of aryl ester bonds (**53**) [87, 88]. In the hydration of acetaldehyde catalyzed by the Zn(II)-complex of 2-pyridinecarboxaldoxime, the oximate anion of the chelated ligand acts as a general base catalyst in the attack of a water molecule at the Zn(II)-bound carbonyl group (**54**) [89]. Collaboration of metal ions with quinoline nitrogen (**55**) has been achieved in dephosphorylation of a phosphate diester [90]. Here, the metal ion activates the leaving group whereas the quinoline moiety acts as a nucleophile.

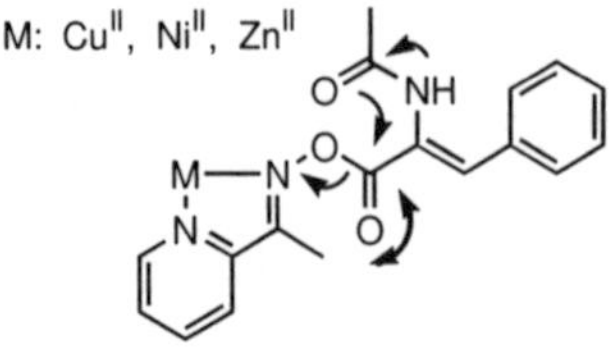

M: Co^{II}, Ni^{II}, Zn^{II}

M: Cu^{II}, Ni^{II}, Zn^{II}

53 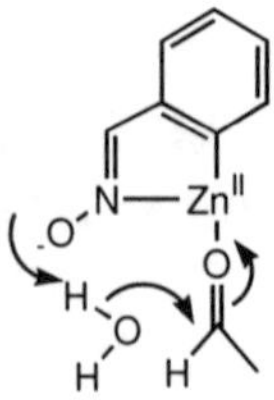 54 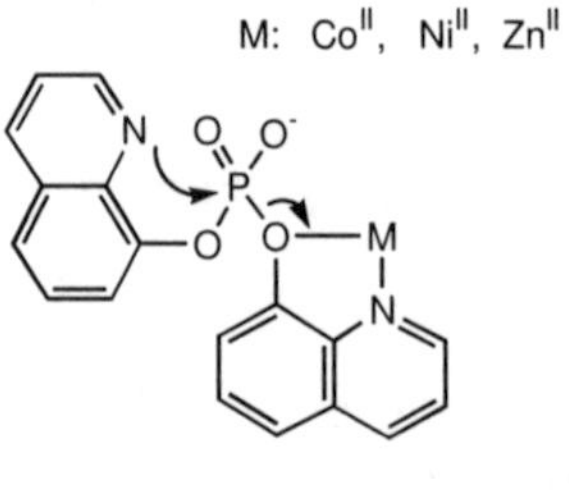 55

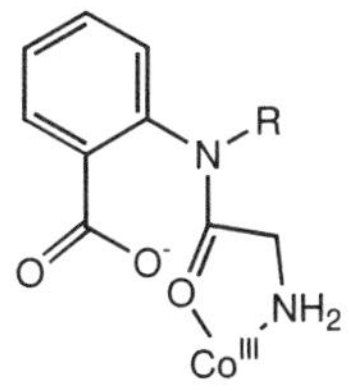

M: Cu^{II}, Ni^{II}, Co^{II}, Zn^{II}

56 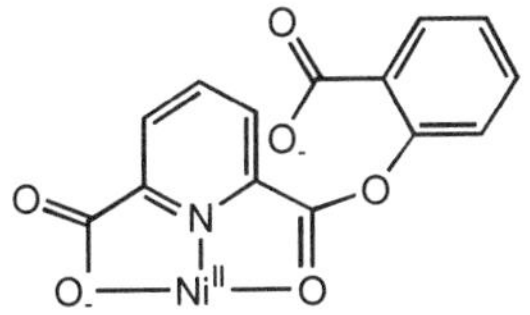 **57** 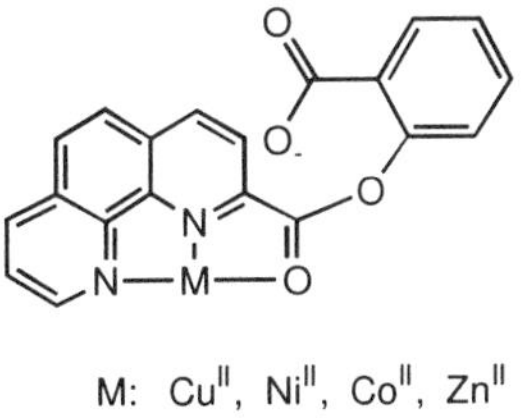 **58**

In view of the mechanism of carboxypeptidase A which involves participation of the Glu-270 carboxylate group and the active-site Zn(II) ion as will be discussed later in Section 9, several attempts have been made to achieve cooperative catalysis by metal ions and a carboxyl group in ester or amide hydrolysis. The attempts have not been successful in most of the model studies (**56**–**60**) [74, 86, 91–94]. In the absence of metal ions, the carboxyl group acts as either a nucleophile or a general base resulting in efficient catalysis in the amide or ester hydrolysis of phthalamic acids related to **56**, salicylate derivatives related to **57** and **58**, glutaryl monoesters related to **59**, or maleamic acids related to **60** [59]. It was hoped that activation of the carbonyl group or the leaving group by metal ions could lead to cooperative catalysis. In these reactions, however, cooperation by metal ions and carboxyl group was not achieved.

In order to improve the catalytic system of **41** (R = CH_3), it was attempted to introduce additional catalytic participation of carboxyl group. For this purpose, a derivative containing carboxyl groups (R = CH_2COOH) has been synthesized and kinetic data were obtained in the presence (**61**) or absence of metal ions [74]. In the absence of metal ions, the carboxyl-containing amide manifested about 15-times lower reactivity toward hydroxide ion compared with the amide without carboxyl groups. This is attributable to the unfavorable electrostatic interaction between the anionic substrate and hydroxide ion. In the presence of Co(III) ion, **61** was

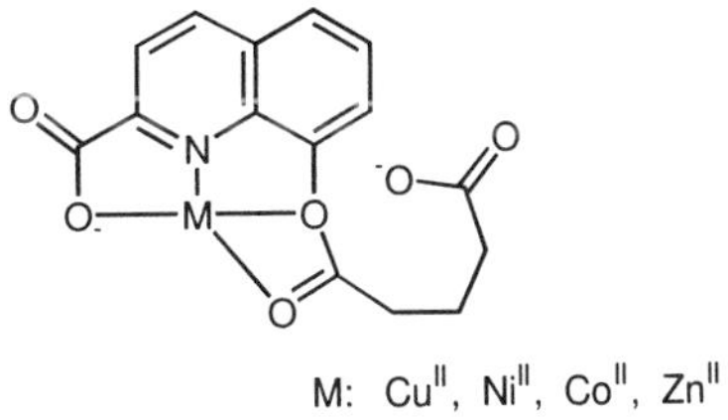

M: Cu^{II}, Ni^{II}, Co^{II}, Zn^{II}

59

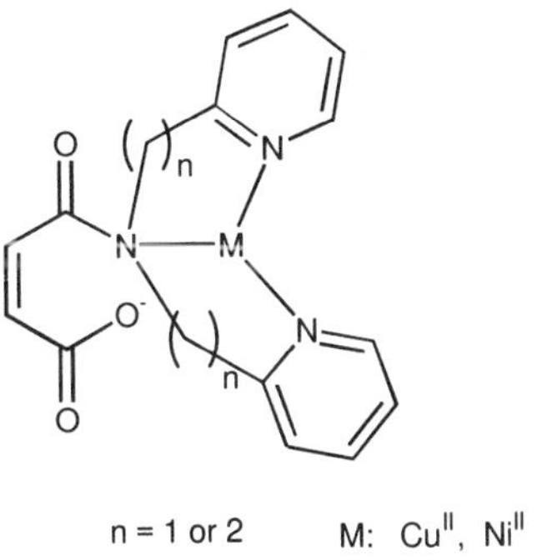

n = 1 or 2 M: Cu^{II}, Ni^{II}

60

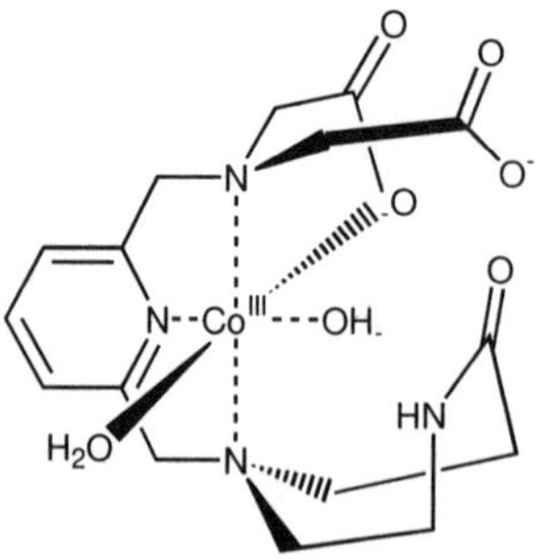

61

hydrolyzed only twice faster than **41**. Thus, no evidence is obtained for cooperation between the metal ion and the carboxyl group in amide hydrolysis of **61**, contrary to the interpretation made by the original authors [74].

Efficient collaboration of metal ions and carboxyl groups in the hydrolysis of both alkyl ester and alkyl amide linkages has been achieved by using 2-imidazolylazo moieties as the metal-chelating sites (**62**) [71, 72, 84]. Investigation with the model system containing a carboxymethyl group as an extra chelating site led to accumulation of anhydride intermediate formed by the nucleophilic attack of the catalytic carboxylate anion at the scissile ester or amide linkage [72]. This indicates that the mechanism of **62/40** is operative. Implications of this model study on the mechanism of carboxypeptidase A will be discussed again later in Section 9.

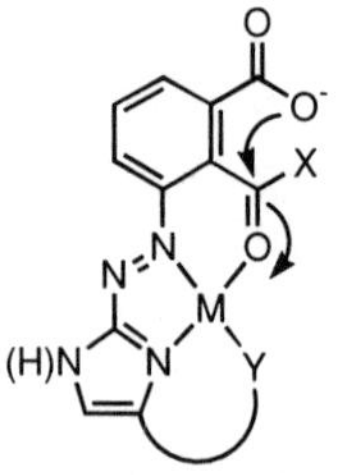

(X: OCH_3 or $N(CH_3)_2$,
Y: H or CH_2COO^-,
M: Cu^{II} or Ni^{II})

62

7. OPERATION OF MULTIPLE CATALYTIC REPERTORIES

Unlike most of organic catalytic groups, metal ions often perform more than two catalytic roles simultaneously. Repertories of metal ions played in the ester hydrolysis represented by **49** [82], for example, include template effect of the metal ion, participation of a binuclear metal ion, activation of the leaving oxime by the metal ion, enhanced ionization of water upon coordination to the metal ion, and nucleophilic attack by the metal-bound hydroxide ion.

In the amide hydrolysis represented by **62/40** [71, 72], the metal ion stabilizes the oxyanion of the tetrahedral intermediate, enhances the acidity of the metal-bound water molecule, and provides suitable geometry for the cyclic transition state. In addition, the metal-bound water molecule acts as a general acid.

In the metal-catalyzed hydrolysis of phosphinate esters or mono-, di-, and triesters of phosphoric acid, intermediates containing four-membered rings are often involved as exemplified by **24**, **35**, or **46** [63, 77–79]. Here, template effects, stabilization of the negative charges developed on the phosphoryl oxygen atoms, induction of strain in the substrates, and enhancement in the ionization of the metal-bound water molecule are achieved by the metal ions. In addition, the metal-bound hydroxide ion participates as a nucleophile.

8. EFFECTS OF NATURE OF METAL ION AND STRUCTURE OF LIGAND ON CATALYTIC EFFICIENCY

There is no doubt that the nature of metal ions as well as the nature and the exact locations of ligands play crucial roles in the catalytic action of metal ions both in biological and simple organic reactions.

Remarkable changes in catalytic efficiency have been observed in metal-catalyzed organic reactions by changing the metal ion. For example, rates for the attack by the metal-bound hydroxide ion in **63** was **b**>>**a**>>**c** when the metal ion was Cu(II) [83]. The fast rate of **63b** was attributed to the steric compression of the transition state by the methyl substituent. The large rate retardation resulting from the incorporation of the carboxy group in **63c** was interpreted in terms of steric expansion in the transition state by the carboxy group. In contrast to the results obtained with Cu(II), differences in the rates for **63a–c** were much smaller when the metal ion was Ni(II). The different behavior for the two metal ions was accounted for in terms of differences in the fit of metal ions in complex **63a–c**.

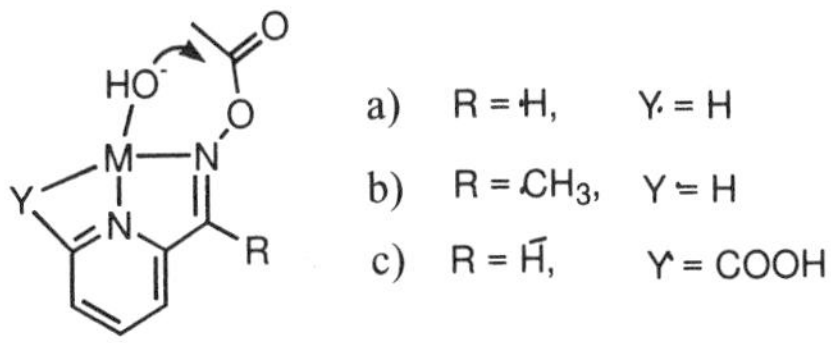

a) R = H, Y = H
b) R = CH_3, Y = H
c) R = H, Y = COOH

63

In the metal-catalyzed hydrolysis of phosphate diesters, Ir(III) was much less effective than Co(III). This was ascribed to the larger size of Ir(III) ion and, consequently, more difficult intramolecular attack (**64**) by the Ir(III)-bound hydroxide at the phosphorus center to form the four-membered ring of the intermediate [79].

Different reactivities of the Cu(II) complexes containing various ligands in ester hydrolysis have been correlated with reduction of the Lewis acidity of Cu(II) ion by strong donor ligands (**65**) [95].

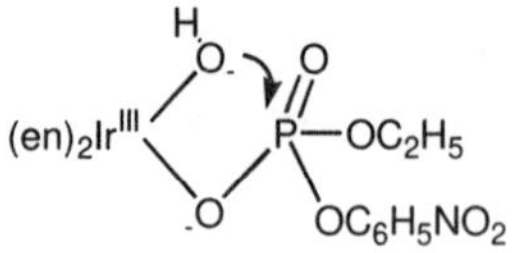

64

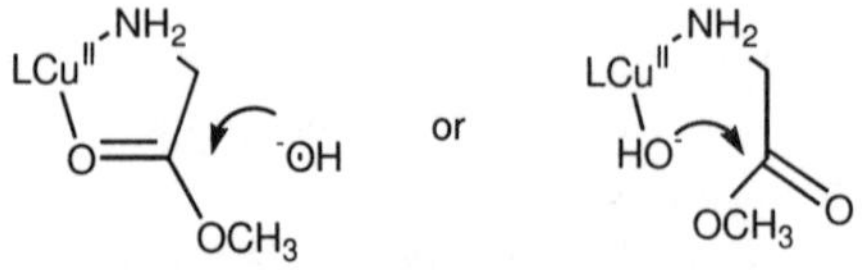

L: H_2O, $(pyCH_2)_2NH$, $HN(CH_2CH_2NH_2)_2$, $N(CH_2COO^-)_3$, $HN(CH_2COO^-)_2$

65

In the hydrolysis of phosphate diesters catalyzed by *cis*-diaquotetrazaCo(III) complexes, a large increase in the activity has been observed with a change in the tetramine ligand structure (**66–68**) [77, 78]. Relative magnitude of catalytic effects of **66–68** was 310:57:1. The catalytic reaction was proposed to proceed through the intramolecular attack by Co(III)-bound hydroxide ion at the complexed substrate. The effect of ligand structure on the catalytic efficiency was explained in terms of bond angles in the transition state containing a four-membered ring (**69**).

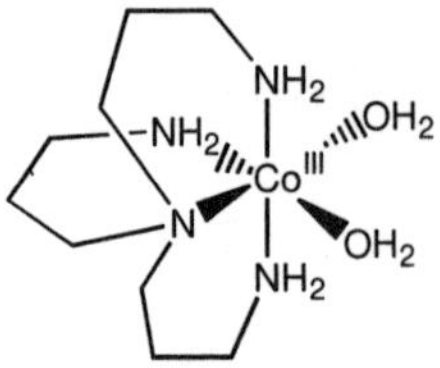

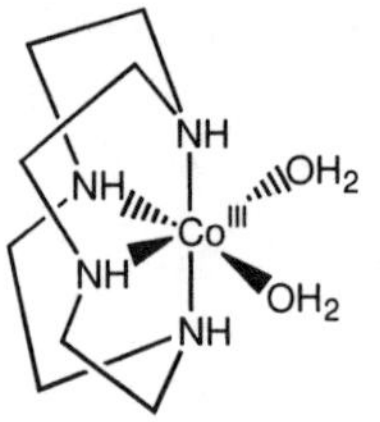

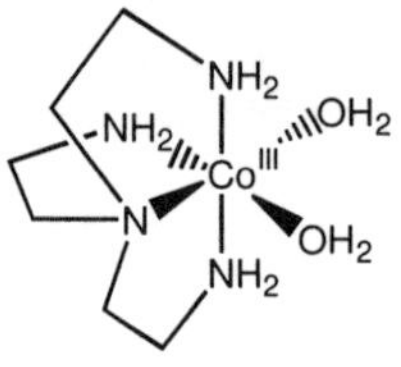

66 **67** **68**

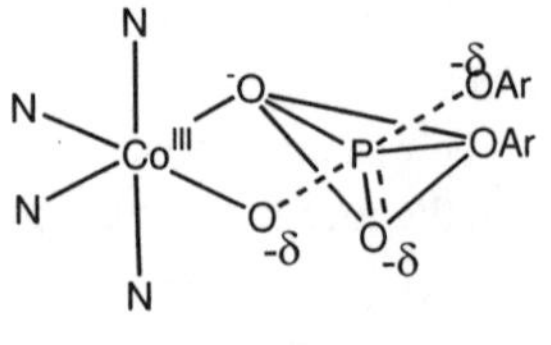

69

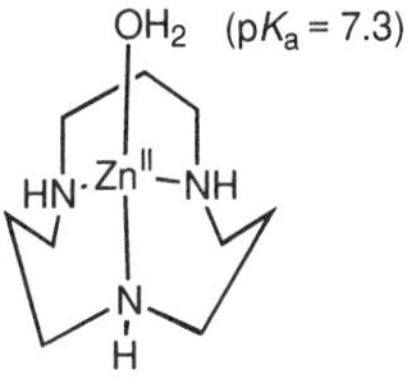

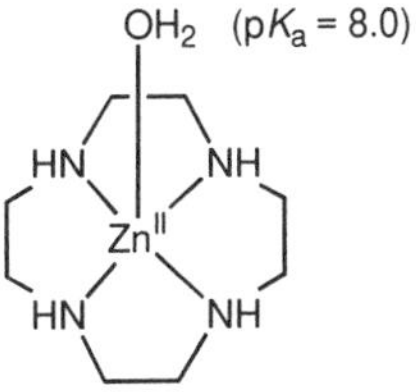

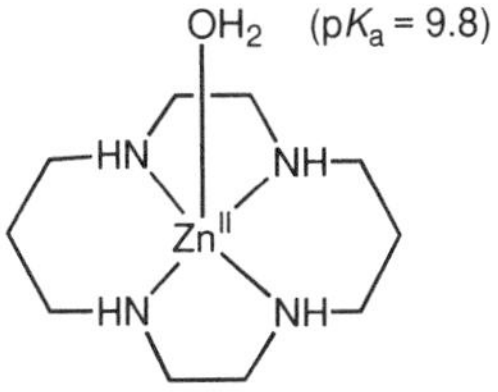

OH2 (pK_a = 7.3) OH2 (pK_a = 8.0) OH2 (pK_a = 9.8)

70 **71** **72**

The pK_a of Zn(II)-bound water molecule of polyazamacrocyclic Zn(II) complexes (**70**–**72**) shifts considerably from that (pK_a = 9.0) of aquoZn(II) ion as the ligand structure is changed [96]. In the nucleophilic attack at a phosphotriester (tris(4-nitrophenyl) phosphate), a phosphodiester (bis(4-nitrophenyl) phosphate), or a carboxy ester (4-nitrophenyl acetate), the Zn(II)-bound hydroxide ions of **70**–**72** manifested reactivity similar to that of hydroxide ion, in spite of basicity difference of up to 10^8-fold. Because a greater portion of the metal-bound water molecules ionizes at neutral pHs to produce the corresponding metal-bound hydroxide ions for **70** and **71** compared with water molecule unbound to Zn(II) ion, large rate acceleration was achieved at neutral pHs with **70** and **71**, with the degree of catalysis being sensitively dependent on the structure of the macrocyclic ligands.

Zn(II) complexes (**70**–**72**) manifested catalytic effect on the reduction of 4-nitrobenzaldehyde with 2-propyl alcohol (**73**), and, thus, can be considered as a model of alcohol dehydrogenase [97]. The catalytic efficiency was far greater for **70** compared with **71** and **72**, and this was attributed to the greatest Lewis acidity and the coordinatively least saturated structure of the Zn(II) ion of **70**. The large differences in the acidity of Zn(II)-bound water molecule and in the catalytic capability of the reduction reaction observed for **70**–**72** provide another example to demonstrate how the physicochemical and catalytic properties of a metal ion can be modified by changing the ligand structure.

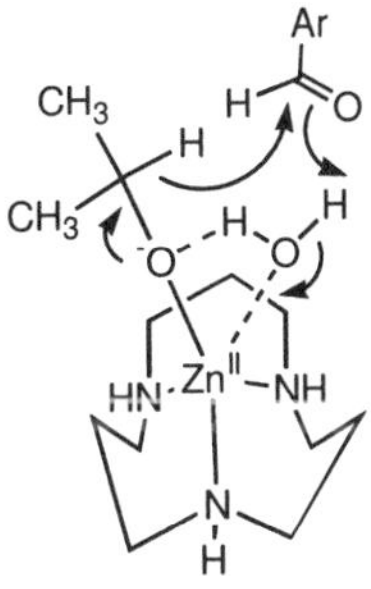

73

In many Zn(II)-metalloenzymes, the carboxylate-imidazole-Zn(II) triad exists (**74**), suggesting the modification of the properties of the Zn(II) ion through

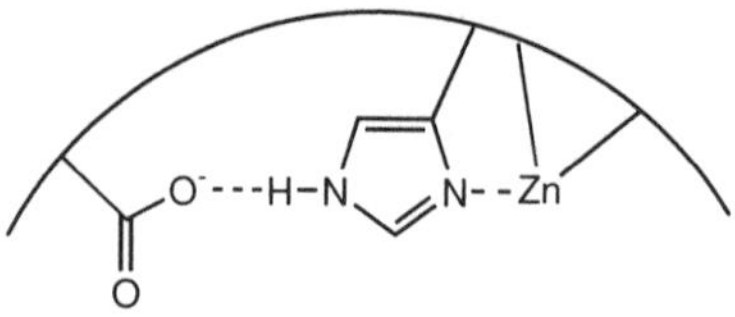

O⁻---H–N N--Zn

74

interaction of the imidazole ligand with the distant carboxylate anion [98]. In the amide hydrolysis represented by **40** and **62**, catalysis was observed only when the imidazolyl N–H was ionized [71, 72]. Apparently, the increase in the electron density on the imidazolyl ring enhances the catalytic activity of the central metal ion, in analogy with the carboxylate–imidazole–Zn(II) triad present in metalloenzymes, although the reason for the enhancement in the reactivity is not clearly understood.

9. CATALYTIC REPERTORIES OF METAL IONS IN METALLOENZYMES: CARBOXYPEPTIDASE A AS AN EXAMPLE

Lewis acid catalysis by metal ions in metalloenzymes would involve catalytic repertories discussed above. At present, no physical tools are available to elucidate the catalytic roles of such metal ions directly by using the metalloenzymes. Instead, various chemical methods are exploited to obtain circumstantial evidence for the catalytic roles. Here, attempts to determine the catalytic roles of metal ions acting as Lewis acid catalysts in metalloenzymes are described by using carboxypeptidase A as an example.

Carboxypeptidase A, a Zn(II)-exopeptidase, is the most intensively investigated among metalloenzymes. X-ray crystallographic studies on carboxypeptidase A revealed that the active-site Zn(II) ion, Glu-270, Tyr-248, and Arg-145 are among the catalytic groups located in the vicinity of the active site [99], and mechanistic studies have been focused on elucidation of the roles of these primary catalytic groups. Kinetic studies with chemically or mutagenetically modified forms of carboxypeptidase A indicated that the phenolic group of Tyr-248 does not play crucial catalytic role at physiologically important pHs [100, 101], contrary to the earlier proposal of general acid catalysis in the cleavage of amide substrates. Few mechanistic data have been obtained for Arg-145. The mechanistic analysis, therefore, has been primarily concerned with the Zn(II) ion and the Glu-270.

The mechanism of carboxypeptidase A has been the center of controversy in spite of the large number of literature. In one of the most often proposed mechanisms, the Glu-270 carboxylate makes nucleophilic attack at the carbonyl carbon of the substrate. In support of this anhydride mechanism, we have reported accumulation of intermediates during the carboxypeptidase A-catalyzed ester hydrolysis [102, 103]. Several lines of evidence were consistent with the nucleophilic attack of

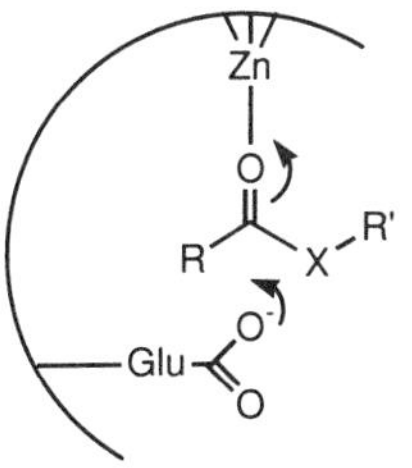

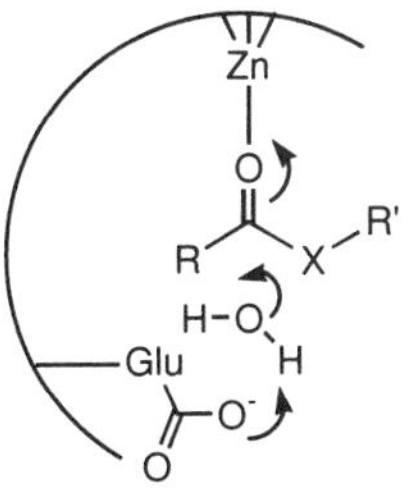

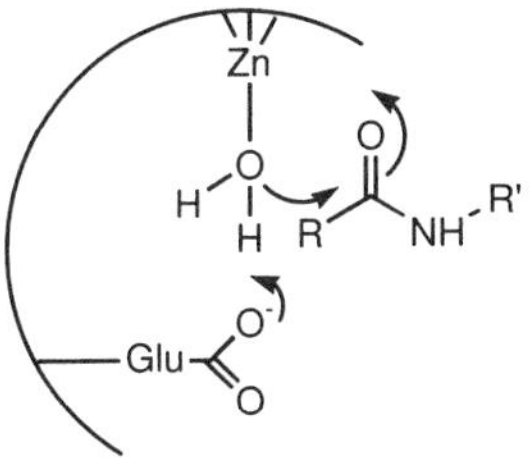

(X = O, NH)

75 **76** **77**

Glu-270 (**75**) [104, 105], although the structure of the accumulating intermediate has not been positively identified [106]. Some other research groups have also reported evidence in support of the anhydride mechanism such as resonance Raman spectroscopic measurements at subzero temperatures or trapping of the anhydride intermediate with a radioactive borohydride derivative [107, 108], although validity of these measurements has been questioned [86]. Recently, detection of anhydride intermediate by Raman spectroscopy has been reported for the esterase action of carboxypeptidase A crystals [109].

In other widely proposed mechanisms, the Glu-270 carboxylate acts as a general base to assist the attack of water at the substrate (**76**). Recently, the nucleophilic attack by the Zn(II)-bound water molecule at the substrate with the general-base assistance from Glu-270 has been proposed (**77**). This mechanism is based on the results of the X-ray crystallographic studies on complexes of carboxypeptidase A formed with pseudo-substrates or inhibitors [106].

Ketonic inhibitors are converted into a hydrated species upon complexation with the active site of carboxypeptidase A, with the *gem*-diol hydroxy groups coordinated to the Zn(II) ion (**78**). Stabilization of the hydrated form of a carbonyl compound upon coordination to the active-site metal ion is not surprising in view of the nonenzymatic example of **9**. However, the original investigators regarded this unique to carboxypeptidase A and considered it as evidence supporting the mechanism of **77**.

Some phosphonate inhibitors are bound to carboxypeptidase A with the two phosphoryl oxygens coordinated to the Zn(II) ion (**79**) [110]. Furthermore, linear correlation was seen between log K_i for the phosphonate inhibitors and log k_{cat}/K_m for analogous amide substrates [111]. These were taken to support the mechanism of **77** with the two oxygens of the hydrated form of the amide bond coordinated to the Zn(II) ion in the transition state (**80**) [110]. The effective nucleophilic attack by metal-bound hydroxide ions or metal-bound water molecules at carboxy derivatives as illustrated by **2**, **37**, or **41** [10, 67, 73, 74] may be considered as models supporting the nucleophilic attack by the Zn(II)-bound water in the carboxypeptidase A action.

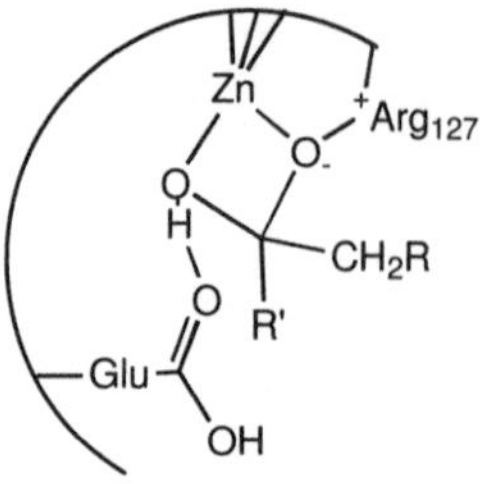

78 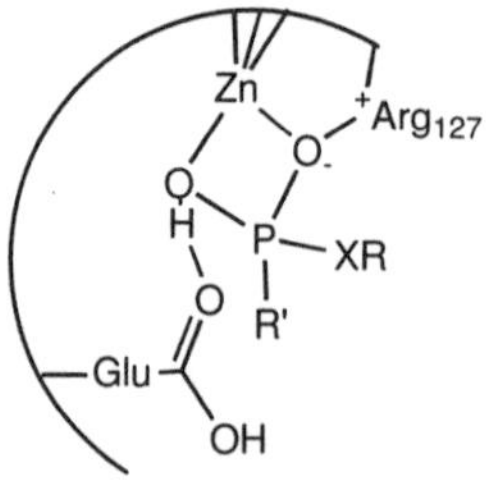 **79** (X = NH or O) 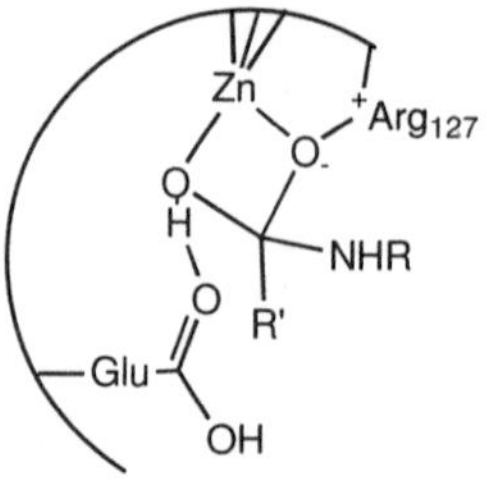 **80**

The crystallographic data which have been taken to support the mechanism of **77** are ambiguous. The crystallographic structures represent static or unproductive species instead of dynamic and productive transition state. In addition, thermodynamically stable structures are disclosed by X-ray crystallography. According to the Curtin–Hammett principle [112], a thermodynamically stable species is not necessarily a kinetically productive species. For example, although the thermodynamically most stable conformation of cyclohexyl halides contain the halogen atoms in the equatorial positions (**81**), the axial form is the reactive species for the E2 elimination reaction (**82**). Thus, the conformation revealed by the crystallographic analysis of carboxypeptidase A complexed with inhibitors or pseudosubstrates is not necessarily related to the structure of the transition state for the productive reaction path.

81 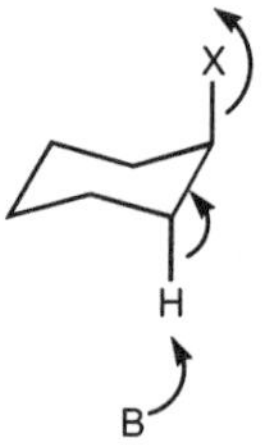 **82**

The linear correlation observed between log K_i for the phosphonate inhibitors and log k_{cat}/K_m for the analogous amide substrates does not indicate that the binding mode of the phosphonate inhibitors is similar to the structure of the transition state of amide hydrolysis. Thousands of linear correlation data (e.g., Hammett plots, Brønsted plots, etc.) between thermodynamic and kinetic free energies are available in the literature for various types of organic reactions [113]. The correlation, however, does not mean that the transition state corresponding to the activation free energy resembles the species corresponding to the thermodynamic free energy. For example, although log k for nucleophilic reactions on benzoate esters is linearly

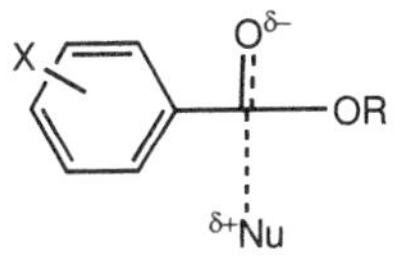

83 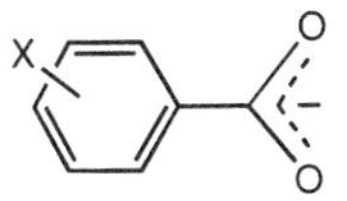 **84**

correlated with log K_a for ionization of benzoic acids, the structure of the acyl carbon in the transition state for the nucleophilic reaction (**83**) is not directly related to that in the acid anion (**84**). The linear correlation observed for the phosphonate inhibitors of carboxypeptidase A, therefore, does not provide reliable information on the structure of transition state.

Plot of log K_i for the phosphonate inhibitors of carboxypeptidase A against log K_m for the carboxypeptidase A-catalyzed hydrolysis of analogous amide substrates shows better linear correlation than that against log k_{cat}/K_m. If the linear correlation between log K_i against log k_{cat}/K_m is to be taken to indicate similar binding mode for the inhibitor and the transition state, that between log K_i against log K_m would be taken to suggest similar binding mode for the inhibitor and the substrate. This contradicts the original interpretation which supported the mechanism of **77**.

As shown by the foregoing evaluation of mechanistic data obtained directly with carboxypeptidase A, conclusive evidence is not available for the catalytic roles of the active-site Zn(II) ion. Instead, more convincing information has been obtained from model studies.

Efficient cooperation between metal ions and carboxyl group in the hydrolysis of both alkyl ester and alkyl amide linkages has been achieved with model compounds using 2-imidazolylazo moiety as a metal-chelating site (**85**–**88**) [71, 72, 114]. When DMSO containing 5% (v/v) water was employed as the reaction medium, the Cu(II) or Ni(II) ion-catalyzed deacylation of **85**–**88** manifested the catalytic features of carboxypeptidase A such as efficient cleavage of alkyl ester and alkyl amide bonds; cooperation among metal ion, carboxyl group, and medium in catalysis; and optimum reactivity attained when the catalytic carboxyl group is in the anionic state [7, 71, 72].

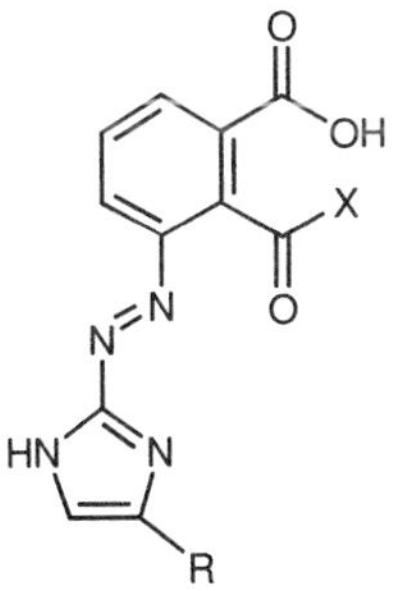

85: X = OCH_3, R = H

86: X = $N(CH_3)_2$, R = H

87: X = OCH_3, R = CH_2COOH

88: X = $N(CH_3)_2$ R = CH_2COOH

The Cu(II)- or Ni(II)-catalyzed deacylation of model compounds **87** and **88** that contain carboxymethyl group as an extra chelating site led to accumulation of anhydride intermediate **90** formed by the nucleophilic attack (**89**) of the catalytic carboxylate anion at the scissile ester or amide linkage [7, 72]. Structure of the anhydride intermediate was further confirmed by trapping of the intermediate with methanol leading to the formation of **87**. Furthermore, the metal-catalyzed deacylation of **87** and **88** involves the general-acid catalysis by the metal-bound water molecule in the breakdown of the tetrahedral intermediate (**40**).

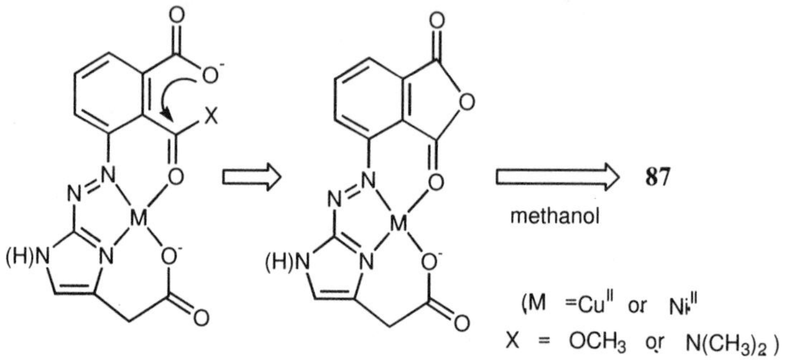

methanol → **87**

(M = Cu^{II} or Ni^{II}
X = OCH_3 or $N(CH_3)_2$)

89 **90**

The rate constant measured at 50 °C for amide **88** complexed to Cu(II) (1×10^{-3} s^{-1}) or Ni(II) ion (1×10^{-4} s^{-1}) is greater than or comparable to that for ester **87** complexed to Cu(II) (8×10^{-4} s^{-1}) or Ni(II) ion (3×10^{-4} s^{-1}). The leveling of reactivity toward the amide and the ester is noteworthy in view of the much greater stability of amide bonds compared with ester bonds. This is another important feature [115] of carboxypeptidase A reproduced by the model.

The carboxylate anion of **85–88** acts as a nucleophile in the metal ion-catalyzed deacylation as proposed by mechanism **75** for the carboxypeptidase A action. Cooperation of carboxylate anion with metal ions is required for both the model and carboxypeptidase A. In addition, 95% (v/v) DMSO resembles the microenvironment of the active site of carboxypeptidase A [116].

Results of the model study suggest that, if ester substrates are hydrolyzed through anhydride mechanism **75** in the carboxypeptidase A action, peptide substrates might be equally well hydrolyzed by the same mechanism. Quite convincing evidence for the anhydride mechanism in the ester hydrolysis by carboxypeptidase A has been recently obtained by detection of the anhydride intermediate in the esterase action of carboxypeptidase A crystals [109]. Then, it is likely that the peptide substrates are hydrolyzed by the anhydride mechanism, too. The leveling of activity towards ester and amide linkages in the model study of **89** has been attributed to the cyclic mechanism of **40** which involves the general acid participation of the metal-bound water molecule in the expulsion of the leaving amine. The same leveling of activity

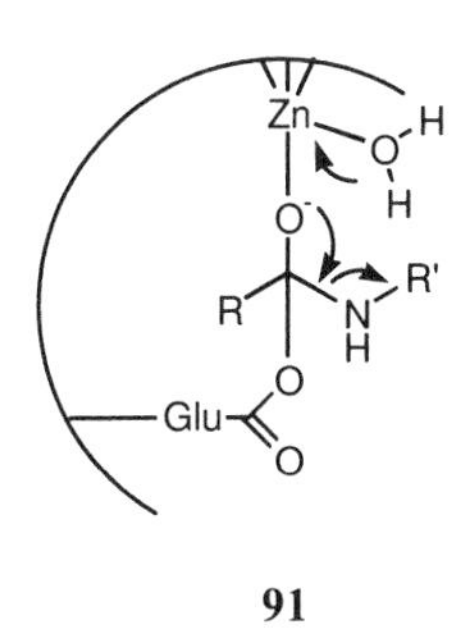

91

manifested by carboxypeptidase A may involve a similar mechanism (91). In conclusion, catalytic roles of the active-site Zn(II) ion of carboxypeptidase A in the most updated mechanism are polarization of the carbonyl groups and the stabilization of negative charge developed on the carbonyl oxygen atom in the transition state (**75**) as well as the general acid catalysis by the Zn(II)-bound water molecule in the expulsion of the leaving amine (**91**).

10. CATALYTIC REPERTORIES OF METAL IONS IN ARTIFICIAL METALLOENZYMES

Host molecules capable of tight binding of metal ions, complexation with organic substrates by molecular recognition, and acceleration of transformations within the resultant supramolecular complexes would lead to efficient artificial metalloenzymes. In the biomimetic catalytic systems involving metal ions as Lewis acid catalysts, the metal ions would play one or more catalytic repertories mentioned above.

In order to design artificial metalloenzymes, it is desirable to develop molecular architecture with very tightly bound metal ions. Tight binding of a metal ion may be achieved by employing a multiaza macrocyclic complex as a part of the catalytic unit. In addition, multiaza macrocyclic metal complexes [117–119] manifest ability of catalysis in aldehyde hydration [75, 120], aldehyde reduction [97], ester hydrolysis [120, 121], or phosphate hydrolysis [122] and molecular recognition of small organic molecules [123].

Many multiaza macrocyclic metal complexes are prepared by the condensation of carbonyl compounds with multiamines in the presence of metal ions [117–119]. In this regard, poly(ethylenimine) (**92**) can be used as a synthon of macrocyclic complexes as well as the backbone of polymeric macrocycles. In addition, several derivatives of poly(ethylenimine) have been investigated as synthetic enzymes, demonstrating their ability to form complexes with organic compounds and to catalyze several types of organic reactions [124].

The structures of typical macrocyclic complexes built on poly(ethylenimine) are **93–96** [125]. Each of **93–96** shows the nature of the metal ion and the dicarbonyl

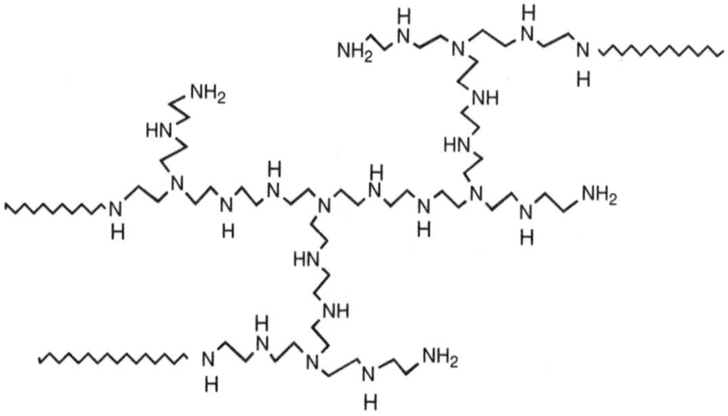

92

compound used in the condensation, instead of the exact structure of the complex. For example, **96** is prepared by the Ni(II)-template condensation of poly(ethylenimine) with butanedione.

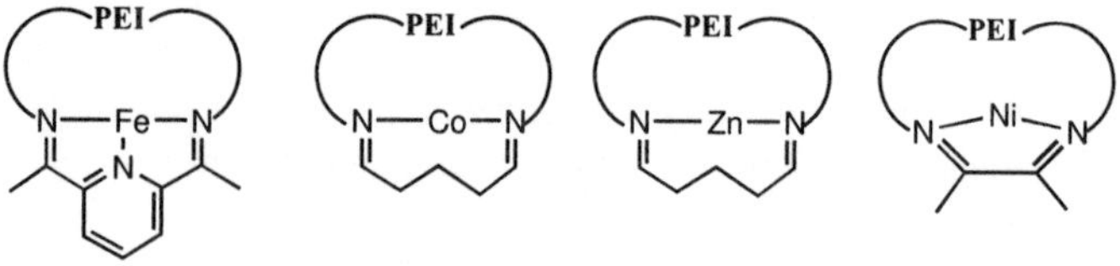

93 94 95 96

The macrocycle-containing poly(ethylenimine)s possess fixed metal centers which are not removed by repetitive dialysis. The polymeric macrocycles exhibit greater affinity toward anion **97** compared with unmodified poly(ethylenimine). Recognition of **98** by **94–96** and the consequent complex formation are reflected in the saturation behavior for the dependence on polymer concentration of pseudo-first-order rate constants for the deacylation of **98**. Deacylation of anionic ester **98** is about 100 times faster than that of neutral ester **99** in the presence of **93–96**. Deacylation of **98** inactivates the reaction centers, indicating that the amine nitrogen

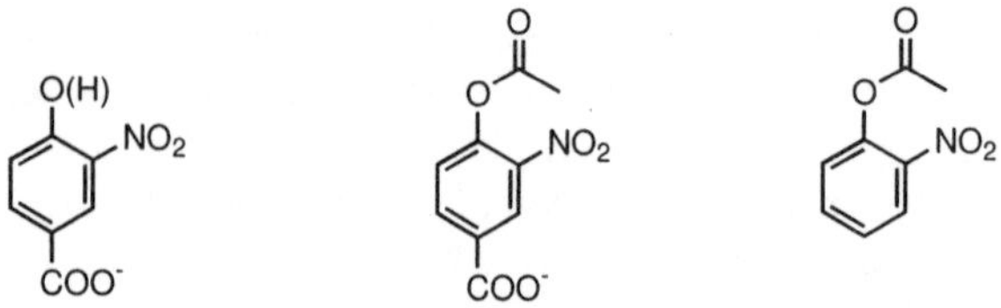

97 98 99

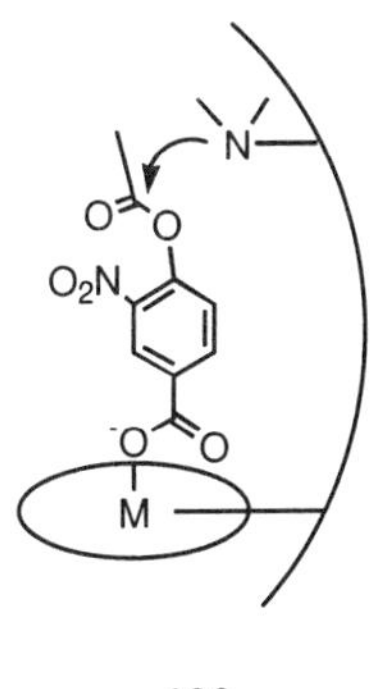

100

atom located close to the metal center attacks the bound ester on the acyl carbon (**100**). Thus, creation of tight metal centers, recognition of anions, and acceleration of deacylation of the bound anionic ester are achieved by the macrocycle-containing poly(ethylenimine) derivatives.

Macrocyclic complex **101** was built by condensation of poly(ethylenimine) with glyoxal, the simplest dicarbonyl compound, in the presence of Ni(II) ion and the reduction of **101** with sodium borohydride produced the saturated analogue **102**. Carboxyphenols **97** and **103** and esters **99** and **104** were bound by **101** much better than by **102**, whereas the transition states for the deacylation of **98** or **104** were stabilized much better by **102** than **101** [126]. Thus, **101** recognizes the substrate and the products selectively, while **102** recognizes the transition state selectively. Selective recognition of transition state is the major aim in the design of artificial enzymes.

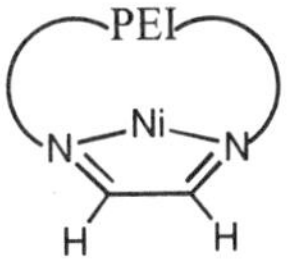

101 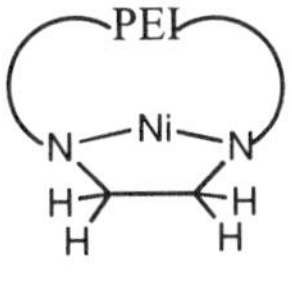 **102**

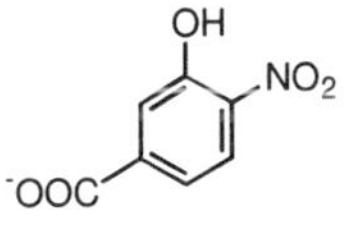

103 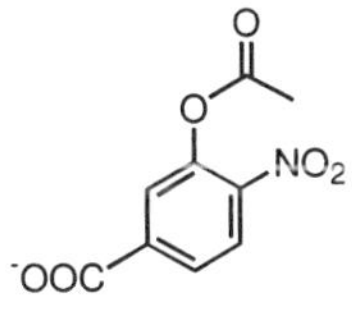 **104**

Selective recognition of the substrates and the products by **101** was explained in terms of the greater Lewis acidity of Ni(II) ion in **101** compared with that in **102** due to the lower basicity of the imine nitrogen compared with the amine nitrogen

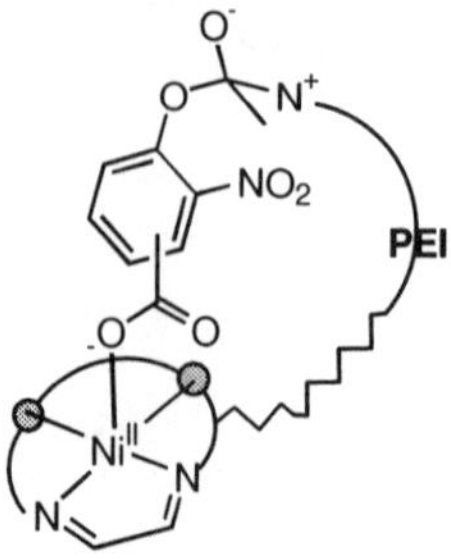

105 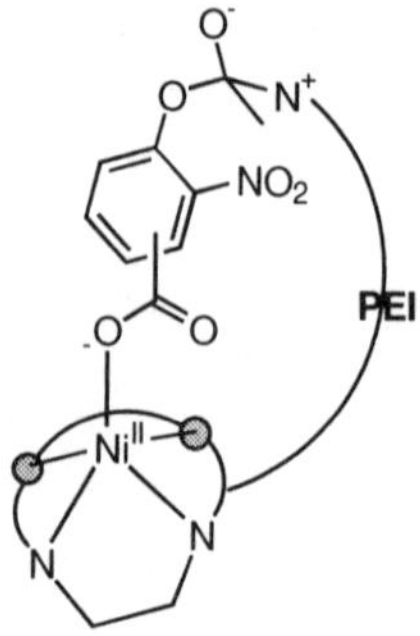 106

and the dπ-pπ back-bonding between the metal ion and the diimine bond in **101**. Selective recognition of the transition state by **102** was attributed to the smaller strain involved in the cyclic transition state due to the differences in the molecular geometry. As schematically illustrated by **105** and **106**, the C–N bond in **102** is longer than the C=N bond in **101** and the dπ-pπ back-bonding in **101** can shorten the Ni–N bonds, leading to different shapes of the cyclic transition states.

In the design of effective catalysts, minute structural adjustment of transition states is important. As mentioned in Section 2.6, metal ions of metalloenzymes might be exploited to induce conformational changes of the proteins leading to very fine alignment of the catalytic groups. Selective stabilization of the cyclic transition states of ester deacylation by **102** is a rare example in which such a catalytic role is played by metal ions in synthetic systems.

In deacylation of esters **98** and **104**, the catalytic role played by the metal centers is only recognition of anions and the consequent anchoring of the anionic esters to provide suitable conformation for the ester aminolysis. Furthermore, the deacylation leads to acylation of the polymer backbone without regeneration of the polymer entity, and, therefore, true catalysis was not achieved. Much better artificial metalloenzymes would be obtained if the metal ions are exploited both in the complexation of the substrate and in the catalytic conversion of the complexed substrate and if the catalyst is regenerated after transformation of the substrate to the product. This was achieved in the hydrolysis of phosphomonoester **107** and phosphodiester **108** by Co(II)-macrocyclic complex **109** built on poly(ethylenimine) [127].

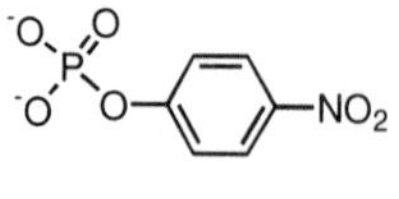

107 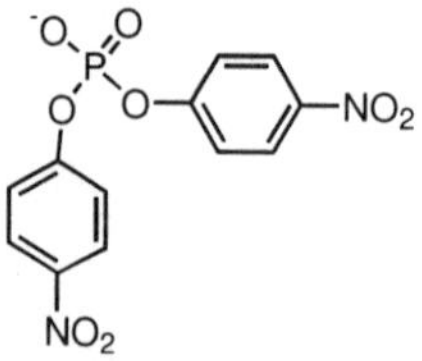 108 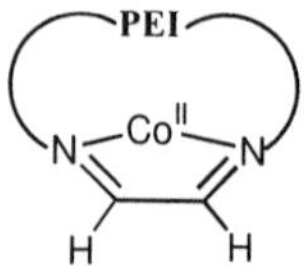 109

The hydrolysis of **107** in the presence of **109** revealed both complexation with the catalyst and regeneration of catalyst. Complexation of **107** to **109** was much stronger than that of **108**, apparently due to stronger electrostatic interaction by the dianionic ester with the metal center of **109**. The degree of rate acceleration achieved by **109** for the hydrolysis of **107** and **108** was much greater than that ever achieved by complexes of divalent metal ions and was comparable to some of the best known catalysts involving trivalent metal ions. In analogy with known catalytic systems for phosphoester hydrolysis, the mechanism of **109**-catalyzed hydrolysis of **107** and **108** was proposed as **110** and **111**. Here, the metal ions serve both in the substrate binding process and the catalytic step. Catalytic repertories of metal ions exploited in this reaction include recognition of anions, template effect, induction of strain, masking of anions, and nucleophilic attack by metal-bound hydroxide ion, as discussed in previous sections.

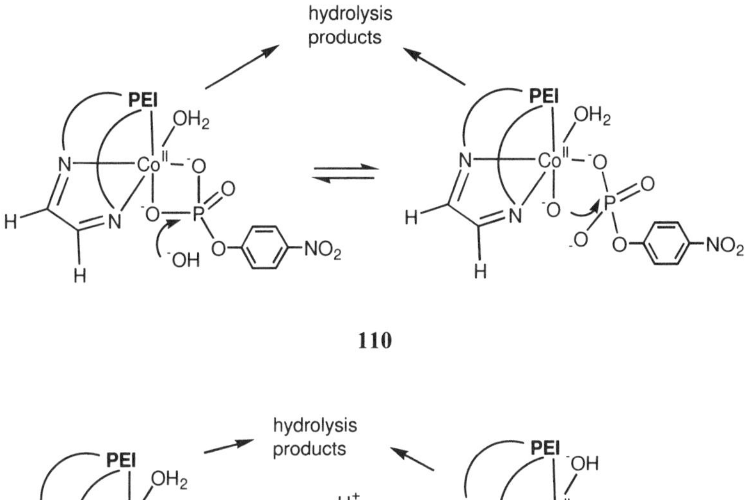

hydrolysis products

PEI OH_2 Co^{II} N H ^{-}OH P O NO_2

110

hydrolysis products

$-H^+$

PEI OH_2 Co^{II} HO^- P NO_2

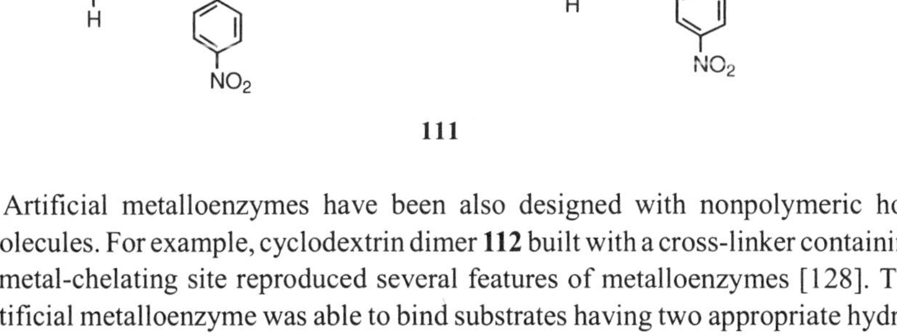

111

Artificial metalloenzymes have been also designed with nonpolymeric host molecules. For example, cyclodextrin dimer **112** built with a cross-linker containing a metal-chelating site reproduced several features of metalloenzymes [128]. The artificial metalloenzyme was able to bind substrates having two appropriate hydro-

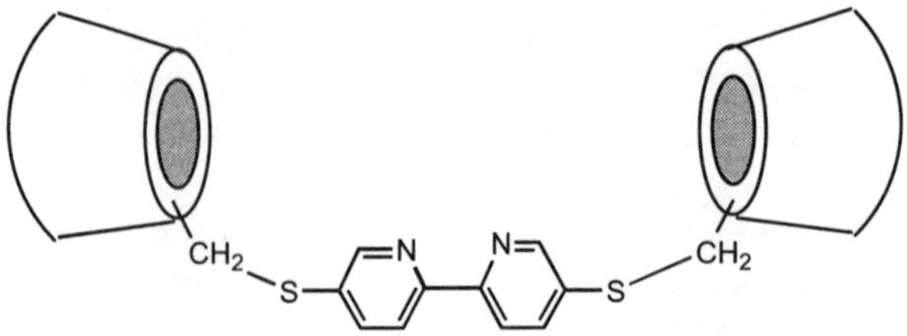

112

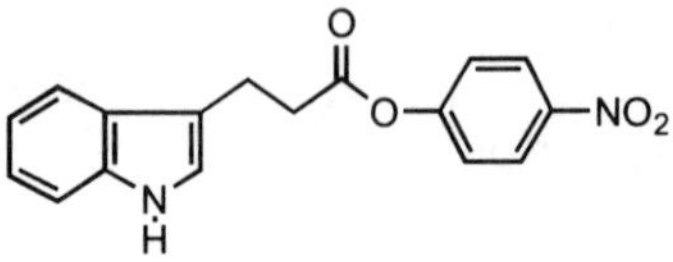

113

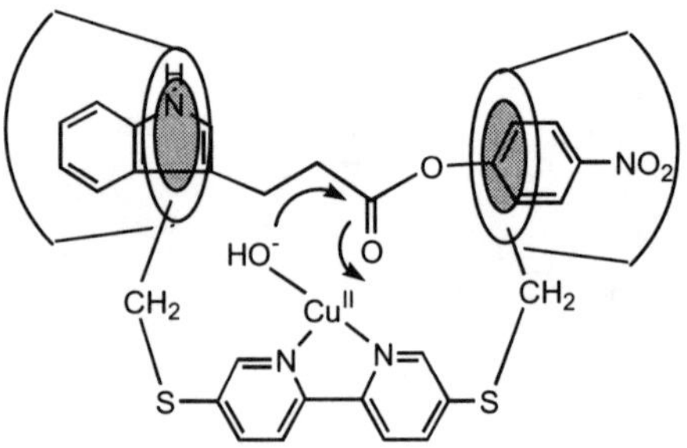

114

phobic groups (e.g. **113**) that can be accommodated into the two cyclodextrin cavities. In addition, hydrolysis of **113** was accelerated by 10^4-fold in the presence of Cu(II) ion. In the proposed mechanism (**114**) for the hydrolysis of **113** catalyzed by **112**, the metal-bound hydroxide ion acts as a nucleophile and the metal ion activates the carbonyl group by stabilizing the negative charge developed on the carbonyl oxygen in the transition state.

ACKNOWLEDGMENTS

This work was supported by the Basic Science Research Institute Program, R.O.K. Ministry of Education, and Center for Molecular Catalysis.

REFERENCES

[1] Bender, M. L., *Mechanisms of Homogeneous Catalysis from Protons to Proteins*, Wiley-Interscience, New York, 1971, p. 212.
[2] Vallee, B. L. and Wacker, W. E. C., in Fasman, G. D. (ed.), *Handbook of Biochemistry and Molecular Biology,* Vol. 2, 3rd edn., CRC Press, Cleveland, 1976, pp. 276–292.
[3] Vallee, B. L., in Bertini, I., Luchinat, C., Maret, W. and Zeppezauer, M. (eds.), *Zinc Enzymes*, Vol. 1, Birkhäuser, Boston, 1986, Ch. 1.
[4] Satchell, D. P. N. and Satchell, R. S., Annu. Rep. Prog. Chem. Sect. A, 75 (1979) 25.
[5] Hay, R. W., in Wilkinson, G. (ed.), *Comprehensive Coordination Chemistry*, Vol. 6, Pergamon, Oxford, 1987, pp. 412–485.
[6] Suh, J., Bioorg. Chem., 18 (1990) 345.
[7] Suh, J., Acc. Chem. Res., 25 (1992) 273.
[8] Wells, M. A. and Bruice, T. C., J. Am. Chem. Soc., 99 (1977) 5341.
[9] Buckingham, D. A., Sargeson, A. M. and Keene, F. R., J. Am. Chem. Soc., 96 (1974) 4981.
[10] Buckingham, D. A., Foster, D. M. and Sargeson, A. M., J. Am. Chem. Soc., 96 (1974) 1726.
[11] Sutton, P. A. and Buckingham, D. A., Acc. Chem. Res., 20 (1987) 357.
[12] Buckingham, D. A., Dekkers, J., Sargeson, A. M. and Wein, M., J. Am. Chem. Soc., 94 (1972) 4032.
[13] Hay, R. W. and Nolan, K. B., J. Chem. Soc., Dalton Trans., (1974) 2542.
[14] Zhu, L. and Kostic, N. M., Inorg. Chem., 31 (1992) 3994.
[15] Suh, M. P., Kwak, C.-H. and Suh, J., Inorg. Chem., 28 (1989) 50.
[16] Creighton, D. J., Hajdu, J. and Sigman, D. S., J. Am. Chem. Soc., 98 (1976) 4619.
[17] Breslow, R., Fairweather, R. and Keana, J., J. Am. Chem. Soc., 89 (1967) 2135.
[18] Breslow, R. and Schmir, M., J. Am. Chem. Soc., 93 (1971) 4960.
[19] Creaser, I. I., Harrowfield, J. MacB., Keene, F. R. and Sargeson, A. M., J. Am. Chem. Soc., 103 (1981) 3559.
[20] Covey, W. D. and Leussing, D. L., J. Am. Chem. Soc., 96 (1974) 3860.
[21] Kubala, G. and Martell, A. E., J. Am. Chem. Soc., 104 (1982) 6602.
[22] Cheong, M. and Leussing, D. L., J. Am. Chem. Soc., 111 (1989) 2541.
[23] El-Hilaly, A. E. and El-Ezaby, M. S., J. Inorg. Nucl. Chem., 38 (1976) 1533.
[24] Gellman, S. H., Petter, R. and Breslow, R., J. Am. Chem. Soc., 108 (1986) 2388.
[25] Menger, F. M. and Tsuno, T., J. Am. Chem. Soc., 111 (1989) 4903.
[26] Fife, T. H. and Przystas, T. J., J. Am. Chem. Soc., 104 (1982) 2251.
[27] Przystas, T. J. and Fife, T. H., J. Am. Chem. Soc., 102 (1980) 4391.
[28] Hanzlik, R. P. and Hamburg, A., J. Am. Chem. Soc., 100 (1978) 1745.
[29] Suh, J., Lee, E. and Jang, E. S., Inorg. Chem., 20 (1981) 1932.
[30] Suh, J., Suh, M. P. and Lee, J. D., Inorg. Chem., 24 (1985) 3088.
[31] Suh, J., Kwon, B. N., Lee, W. Y. and Chang, S. H., Inorg. Chem., 26 (1987) 805.
[32] Hsu, C.-M. and Cooperman, B. S., J. Am. Chem. Soc., 98 (1976) 5652.
[33] Suh, J. and Koh, D. J., Org. Chem., 52 (1987) 3446.
[34] Gensmantel, N. P., Gowling, E. W. and Page, M. I., J. Chem. Soc., Perkin Trans. II (1978) 375.
[35] Schwartz, M. A., Bioorg. Chem., 11 (1982) 4.
[36] Houghton, R. P. and Puttner, R. R., Chem. Commun., (1970) 1270.
[37] Suh, J. and Min, D. W., J. Org. Chem., 56 (1991) 5710.
[38] Satchell, D. P. N. and Secemski, I. I., J. Chem. Soc. (B) (1970) 1306.
[39] Satchell, D. P. N., Chem. Soc. Rev., 6 (1977) 345.
[40] Blackburn, G. M. and Ward, C. R. M., J. Chem. Soc., Chem. Commun., (1976) 79.
[41] Basolo, F. and Pearson, R. G., *Mechanisms of Inorganic Reactions*, 2nd edn., Wiley, New York, 1968, p. 32.
[42] Barnum, D. W., Inorg. Chem., 22 (1983) 2297.

[43] Kim, J. H. and Chin, J., J. Am. Chem. Soc., 114 (1992) 9792.
[44] Eiki, T., Kawada, S., Matsushima, K., Mori, M. and Tagaki, W., Chem. Lett., (1980) 997.
[45] Breslow, R. and Chapman, D., J. Am. Chem. Soc., 87 (1965) 4195.
[46] Lau, H.-p. and Gutsche, C. D., J. Am. Chem. Soc., 100 (1978) 1857.
[47] Suh, J., Cheong, M. and Han, H., Bioorg. Chem., 12 (1984) 188.
[48] Harrowfield, J. M., Jones, D. R., Lindoy, L. F. and Sargeson, A. M., J. Am. Chem. Soc., 102 (1980) 7733.
[49] Gainsford, A. R., Pizer, R. D., Sargeson, A. M. and Whimp, P. D., J. Am. Chem. Soc., 103 (1981) 792.
[50] Lawson, P. J., Mccarthy, N. G. and Sargeson, A. M., J. Am. Chem. Soc., 104 (1982) 6710.
[51] Dixon, N. E. and Sargeson, A. M., J. Am. Chem. Soc., 104 (1982) 6716.
[52] Akaobri, S., Otani, T. T., Marshall, R., Winitz, M. and Greenstein, J. P., Arch. Biochem. Biophys., 83 (1959) 1.
[53] Cox, B. G., J. Am. Chem. Soc., 96 (1974) 6823.
[54] Buckingham, D. A., Stewart, I. and Sutton P. A., J. Am. Chem. Soc., 112 (1990) 845.
[55] Suh, J. and Chun, K. H., J. Am. Chem. Soc., 108 (1986) 3057.
[56] Spector, L. B., *Covalent Catalysis by Enzymes*, Springer-Verlag, New York, 1982.
[57] Dugas, H., *Bioorganic Chemistry: A Chemical Approach to Enzyme Action*, 2nd edn., Springer-Verlag, New York, 1989, Ch. 3.
[58] Loran, J. S., Naylor, R. A. and Williams, A., J. Chem. Soc., Perkin Trans. II (1977) 418.
[59] Kirby, A. J., Adv. Phys. Org. Chem., 17 (1980) 183.
[60] Schepartz, A. and McDevitt, J. P., J. Am. Chem. Soc., 111 (1989) 5976.
[61] Jencks, W. P., *Catalysis in Chemistry and Enzymology*, McGraw-Hill, New York, 1969, p. 294.
[62] Rebek, J., Jr., Costello, T. and Wattley, R., J. Am. Chem. Soc., 107 (1985) 7487.
[63] Anderson, B., Milburn, R. M., Harrowfield, J. M., Robertson, G. B. and Sargeson, A. M., J. Am. Chem. Soc., 99 (1977) 2652.
[64] Finn, M. G. and Sharpless, B., in Morrison, J. D. (ed.), *Asymmetric Synthesis*, Vol. 5, Academic Press, Orlando, 1985, pp. 247–308.
[65] Noyori, R. and Takaya, H., Acc. Chem. Res., 23 (1990) 345.
[66] Nitta, H., Yu, D., Kudo, M., Mori, A. and Inoue, S., J. Am. Chem. Soc., 114 (1992) 7969.
[67] Suh, J., Cheong, M. and Suh, M. P., J. Am. Chem. Soc., 104 (1982) 1654.
[68] Boreham, C. J., Buckingham, D. A., Francis, D. J. and Sargeson, A. M., J. Am. Chem. Soc., 103 (1981) 1975.
[69] Boreham, C. J., Buckingham, D. A. and Keene, F. R., J. Am. Chem. Soc., 101 (1979) 1409.
[70] Suh, J. and Han, H., Bioorg. Chem., 12 (1984) 177.
[71] Suh, J., Hwang, B. K. and Koh, Y. H., Bioorg. Chem., 18 (1990) 207.
[72] Suh, J., Park, T. H. and Hwang, B. K., J. Am. Chem. Soc., 114 (1992) 5141.
[73] Groves, J. T. and Chambers, R. R., J. Am. Chem. Soc., 106 (1984) 630.
[74] Groves, J. T. and Baron, L. A., J. Am. Chem. Soc., 111 (1989) 5442.
[75] Woolley, P., Nature, 258 (1975) 677.
[76] Breslow, R., McClure, D. E., Brown, R. S. and Eisenach, J., J. Am. Chem. Soc., 97 (1975) 194.
[77] Chin, J., Acc. Chem. Res., 24 (1991) 145.
[78] Chin, J., Banaszezyk, M., Jubian, V. and Zou, X., J. Am. Chem. Soc., 111 (1989) 186.
[79] Hendry, P. and Sargeson, A. M., J. Am. Chem. Soc., 111 (1989) 2521.
[80] Gahan, L. R., Harrowfield, J. M., Herit, A. J., Lindoy, L. F., Whimp, P. O. and Sargeson, A. M., J. Am. Chem. Soc., 107 (1985) 6231.
[81] Suh, J., Kim, J. and Lee, C. S., J. Org. Chem., 56 (1991) 4364.
[82] Suh, J., Han, O. and Chang, B., J. Am. Chem. Soc., 108 (1986) 1839.
[83] Suh, J. and Chang, B., Bioorg. Chem., 15 (1987) 167.
[84] Suh, J., Chung, S. and Lee, S. H., Bioorg. Chem., 15 (1987) 383.
[85] Miller, B. A. and Leussing, D. L., J. Am. Chem. Soc., 107 (1985) 7146.

[86] Schepartz, A. and Breslow, R., J. Am. Chem. Soc., 109 (1987) 1814.
[87] Suh, J. and Kim, S. M., Bioorg. Chem., 17 (1989) 169.
[88] Fife, T. H., Przystas, T. J. and Pujari, M. P., J. Am. Chem. Soc., 110 (1988) 8157.
[89] Woolley, P. R., J. Chem. Soc., Chem. Commun. (1975) 579.
[90] Browne, K. A. and Bruice, T. C., J. Am. Chem. Soc., 114 (1992) 4951.
[91] Breslow, R. and McAllister, C., J. Am. Chem. Soc., 93 (1971) 7076.
[92] Fife, T. H., Przystas, T. J. and Squillacote, V. L., J. Am. Chem. Soc., 101 (1979) 3017.
[93] Suh, J. and Baek, D.-J., Bioorg. Chem., 10 (1981) 266.
[94] Suh, J., Kim, M. J. and Seong, N. J., J. Org. Chem., 46 (1981) 4354.
[95] Nakon, R., Rechani, P. R. and Angelici, R. J., J. Am. Chem. Soc., 96 (1974) 2117.
[96] Koike, T. and Kimura, E., J. Am. Chem. Soc., 113 (1991) 8935.
[97] Kimura, E., Shionoya, M., Hoshino, A., Ikeda, T. and Yamada, Y., J. Am. Chem. Soc., 114 (1992) 10134.
[98] Christianson, D. W. and Alexander, R. S., J. Am. Chem. Soc., 111 (1989) 6412.
[99] Lipscomb, W. N., Acc. Chem. Res., 3 (1970) 81.
[100] Suh, J. and Kaiser, E. T., J. Am. Chem. Soc., 98 (1976) 7060.
[101] Hilvert, D., Gardell, S. J., Rutter, W. J. and Kaiser, E. T., J. Am. Chem. Soc., 108 (1986) 5298.
[102] Suh, J., Cho, C. and Chung, S., J. Am. Chem. Soc., 107 (1985) 4530.
[103] Suh, J., Hwang, B. K., Jang, I. and Oh, E., J. Biochem. Biophys. Meth., 22 (1991) 167.
[104] Suh, J., Hong, S. B. and Chung, S., J. Biol. Chem., 108 (1986) 7112.
[105] Suh, J., Chung, S. and Choi, G. B., Bioorg. Chem., 17 (1989) 64.
[106] Christianson, D. W. and Lipscomb, W. N., Acc. Chem. Res., 22 (1989) 62.
[107] Makinen, M. W., Yamamura, K. and Kaiser, E. T., Proc. Natl. Acad. Sci. USA, 73 (1976) 3882.
[108] Sander, M. E. and Witzel, H., Biochem. Biophys. Res. Comm., 132 (1985) 681.
[109] Britt, B. M. and Peticolas, W. L., J. Am. Chem. Soc., 114 (1992) 5295.
[110] Kim, H. and Lipscomb, W. N., Biochemistry, 30 (1991) 8171.
[111] Hanson, J. E., Kaplan, A. P. and Bartlett, P. N., Biochemistry, 28 (1989) 6294.
[112] Eliel, E. L., *Stereochemistry of Carbon Compounds*, McGraw-Hill, New York, 1962, Chs. 6 and 8.
[113] Jones, R. A. Y., *Physical and Mechanistic Organic Chemistry*, 2nd edn., Cambridge University Press, Cambridge (1983) Ch. 3.
[114] Suh, J., Chung, S. and Lee, S. H., Bioorg. Chem., 15 (1987) 383.
[115] Auld, D. S., in Page, M. I. and Williams, A. (eds.), *Enzyme Mechanisms*, Royal Society of Chemistry, London, 1987, Ch. 14.
[116] Suh, J., Kim, Y., Lee, E. and Chang, S. H., Bioorg. Chem., 14 (1986) 33.
[117] Mashiko, T. and Dolphin, D., in Wilkinson, G. (ed.), *Comprehensive Coordination Chemistry*, Vol. 2, Pergamon, Oxford, 1987, Ch. 21.1.
[118] Curtis, N. F., in Wilkinson, G. (ed.), *Comprehensive Coordination Chemistry*, Vol. 2, Pergamon, Oxford, 1987, Ch. 21.2.
[119] Mertes, K. B. and Lehn, J.-M., in Wilkinson, G. (ed.), *Comprehensive Coordination Chemistry*, Vol. 2, Pergamon, Oxford (1987) Ch. 21.3.
[120] Kimura, E., Shiota, T., Koike, T., Shiro, M. and Kodama, M., J. Am. Chem. Soc., 112 (1990) 5805.
[121] Chin, J. and Zou, X., J. Am. Chem. Soc., 106 (1984) 3687.
[122] Breslow, R., Peter, R. and Gellman, S. H., J. Am. Chem. Soc., 108 (1986) 2388.
[123] Fujita, M., Yazaki, J. and Ogura, K., J. Am. Chem. Soc., 112 (1990) 5645.
[124] Klotz, I. M., in Page, M. I. and Williams, A. (eds.), *Enzyme Mechanisms*, Royal Society of Chemistry, London (1987) Ch. 2.
[125] Suh, J., Cho, Y. and Lee, K. J., J. Am. Chem. Soc., 113 (1991) 4198.
[126] Suh, J. and Kim, N., J. Org. Chem., 58 (1993) 1284.
[127] Kim, N. and Suh, J., J. Org. Chem., 59 (1994) 1561.
[128] Breslow, R. and Zhang, B., J. Am. Chem. Soc., 114 (1992) 5883.

THE MULTICOPPER–ENZYME ASCORBATE OXIDASE

Albrecht Messerschmidt

OUTLINE

	Abstract	**152**
1.	**Introduction**	**152**
2.	**Occurrence, Sequences, and Biological Function**	**154**
3.	**Molecular and Spectroscopic Properties**	**155**
4.	**X-Ray Structure of Ascorbate Oxidase**	**157**
	4.1 Crystallization	157
	4.2 Overall Description of the Structure	158
	4.3 Secondary Structure and Tetramer Contact Surface Areas	160
	4.4 Copper Site Geometries	163
5.	**Ascorbate Oxidases, Fungal Laccases, and Related Proteins**	**167**
6.	**Oxidation–Reduction Potentials**	**170**
7.	**Kinetic Properties of Laccase and Ascorbate Oxidase**	**171**
8.	**Functional Derivatives**	**176**
	8.1 General Remarks	176
	8.2 X-Ray Structure of the Type-2 Depleted (T2D) Form of Ascorbate Oxidase	177

Perspectives on Bioinorganic Chemistry
Volume 3, pages 151–197

ISBN: 1-55938-642-8

8.3 X-ray Structure of the Reduced Form of Ascorbate Oxidase 179
8.4 X-ray Structure of the Peroxide Form of Ascorbate Oxidase 180
8.5 X-ray Structure of the Azide Form of Ascorbate Oxidase 183
9. The Catalytic Mechanism **185**
10. Electron Transfer Processes **187**
10.1 Electron Transfer to the Type-1 Copper Redox Center 187
10.2 Intramolecular Electron Transfer from the Type-1 Copper Center to the Trinuclear Copper Center 190
10.3 Electron Transfer within the Trinuclear Copper Site 192
Acknowledgments **192**
References **192**

ABSTRACT

It is the aim of this review to demonstrate progress in the field of blue-copper-containing oxidases during the last 10 years. This progress is mainly due to the determination of the amino acid sequences for all members of this group and the X-ray crystal structure of ascorbate oxidase. The three-dimensional structure of ascorbate oxidase shows the nature and spatial arrangement of the copper centers and the three-domain structure. However, modern spectroscopic techniques like, e.g., low-temperature MCD and ENDOR, made invaluable contributions as well. A structurally based amino acid sequence alignment strongly suggests a three-domain structure for laccase closely related to ascorbate oxidase and a six-domain structure for ceruloplasmin. These domains demonstrate homology with the small blue copper proteins. The relationship suggests that laccase, like ascorbate oxidase, has a mononuclear blue copper in domain 3 and a trinuclear copper between domain 1 and 3, and ceruloplasmin has mononuclear copper ions in domains 2, 4, and 6 and a trinuclear copper between domains 1 and 6. X-ray structures of functional derivatives of ascorbate oxidase provide pictures of intermediate states which will be very probably passed during the catalytic cycle. A catalytic mechanism has been proposed which is based on the available mechanistic data and these new results.

1. INTRODUCTION

The activation of dioxygen in biological systems has been the focus of interest of biochemists, bioinorganic chemists, and physiologists for many years (see e.g. King et al. [1]). Enzymes involved in direct oxygen activation are oxidases and oxygenases. Oxygenases introduce either one atom of dioxygen into substrate and reduce the other atom to water (monooxygenases), or transfer two oxygen atoms into substrate (dioxygenases). Oxidases can be divided into two-electron and four-electron transferring enzymes. The first group reduces dioxygen to hydrogen peroxide, and the second one dioxygen to water. Most of the oxygenases as well as oxidases contain as prosthetic groups either flavin, iron (heme or nonheme), or copper.

Oxygen is used for respiration upon reduction to water by cytochrome-*c* oxidase, the terminal oxidase of the respiratory chain. Cytochrome-*c* oxidase is a complex metalloprotein containing three copper ions and two heme groups (a and a_3) that provides a critical function in cellular respiration in both prokaryotes and eukaryotes. The enzyme catalyzes the four-electron reduction of dioxygen with concomitant one-electron oxidation of cytochrome-*c*. Energy released in this exergonic reaction is conserved as a pH gradient and a trans-membrane potential generated in part by H^+ consumption and also by proton translocation through the protein complex (see e.g. the review on cytochrome-*c* oxidase by Capaldi [2]).

Ascorbate oxidase, laccase, and ceruloplasmin form the group of blue oxidases. These are multicopper enzymes catalyzing the four-electron reduction of dioxygen to water with concomitant one-electron oxidation of the substrate [3] which is very similar to the reaction performed by cytochrome-*c* oxidase. All three enzymes have been known for many years and an overwhelming number of papers have appeared since their discovery dealing with the many different aspects of these enzymes.

Laccase was discovered in 1883 by Yoshida [4] who found that the latex of the Chinese or Japanese lacquer tree was rapidly hardened to a plastic in the presence of oxygen, which he addressed this to the presence of a "diastase" in the lacquer. A few years later Bertrand [5] further purified this enzyme and named it laccase. He suggested that laccase is a metalloprotein containing manganese and introduced the term oxidase. About 50 years later Keilin and Mann [6] demonstrated that laccase is a copper enzyme and showed that its blue color disappears reversibly on addition of substrate. Laccase has been extensively reviewed by relevant researchers in this field over the last 20 years and a representative selection of papers is listed [3, 7–12].

Ascorbate oxidase was discovered in 1928 by Szent-Györgyi [13] who found that certain plant tissues could catalyze the aerobic oxidation of ascorbic acid. In 1930 [14] he postulated the existence of a specific enzyme in cabbage leaves. In 1940 Lovett-Janison and Nelson [15] and Stotz [16] could confirm the existence of ascorbate oxidase and its nature as copper-containing enzyme. Since that time this enzyme has also been investigated in detail and the results are summarized in several reviews [17–20].

Ceruloplasmin is a plasma protein and was first isolated by Holmberg in 1944 [21]. Between 1947 and 1951 Holmberg and Laurell [22] demonstrated that this α_2-globulin was responsible for high-affinity binding of copper in plasma. This enzyme was named ceruloplasmin. Several comprehensive reviews [3, 7–8, 23–29] have been published during the last three decades and include detailed biological, structural, spectroscopic, kinetic, physiological, and genetic information.

This review will mainly concentrate on structural and functional aspects of ascorbate oxidase. Great progress has been achieved during the last 10 years by the determination of amino acid sequences of ascorbate oxidase and related proteins from several sources and the X-ray structure of ascorbate oxidase. This new information forms the basis of a much deeper understanding of the function of ascorbate oxidase and of the related enzymes, and it is the aim of this review to demonstrate this.

2. OCCURRENCE, SEQUENCES, AND BIOLOGICAL FUNCTION

Ascorbate oxidase is found in several plants, but cucumber (*Cucumis sativus*) and green zucchini squash (*Cucurbita pepo medullosa*) are the most common sources [9]. The immunohistochemical localization of ascorbate oxidase in green zucchini reveals that ascorbate oxidase is distributed in all specimens examined ubiquitously over vegetative and reproductive organs [30]. Primary structures of ascorbate oxidase from cucumber [31], pumpkin (*Cucurbita* sp Ebisu Nankin) [32], and zucchini [33] have recently been reported.

The *in vivo* role in plants of ascorbate and ascorbate oxidase is still under debate. Since catechols and polyphenols are also substrates *in vitro* [34], ascorbate oxidase might be involved in processes like fruit ripening. A role in a redox system, as an alternative to the mitochondrial chain in growth promotion or in susceptibility to disease has also been postulated [35]. Recently, it was suggested that ascorbate oxidase might be involved in reorganization of the cell wall to allow for expansion [36]. This was deduced from the findings that ascorbate oxidase is highly expressed at a stage when rapid growth is occurring (in both fruits and leaves) and is localized in the fruit epidermis, where cells are under greatest tension during rapid growth in girth. Hayashi and Morohashi [37] studied the development of ascorbate oxidase activity in mustard cotyledons. There results are in good agreement with the hypothesis that phytochrome-mediated increase in ascorbate oxidase activity is accompanied by the synthesis of the enzyme.

The high catalytic activity of ascorbate oxidase towards an important biological molecule like L-ascorbate has been used for practical purposes. The main application has been the removal of ascorbate and/or oxygen from biological samples [38] and feedstuff. It has also been used for the automated detection of ascorbic acid in biological fluids for industrial and clinical applications.

Ascorbate oxidase is highly specific for the reducing substrate, ascorbate, and other compounds with an enediol group adjacent to a carbonyl group. A novel chromophoric subtrate, 2-amino-ascorbic acid, was tested for ascorbate oxidase [39]. It is oxidized to the chromophoric 2,2′-nitrilodi-2(2′)-deoxy-L-ascorbic acid (red pigment) by the enzyme. The kinetics show that it is a well-behaved alternative subtrate for the enzyme. It is suggested that this novel reactivity of the enzyme could be used to design a sensitive, convenient, and continuous spectrophotometric assay for ascorbate oxidase.

3. MOLECULAR AND SPECTROSCOPIC PROPERTIES

Molecular as well as spectroscopic properties of ascorbate oxidase and related blue multicopper oxidases are summarized in Tables 1 and 2, respectively. Data from other spectroscopic techniques used on blue oxidases such as fluorescence, CD, or resonance Raman spectroscopy have not been included because the implications of most of these studies are not so striking. The interested reader is referred to the existing review articles [10, 18, 19, 26].

Laccases are monomeric glycoenzymes. The molecular masses for the fungal laccases range between 59 and 65 kDa. The reported carbohydrate content is about 10%. The lengths of the published amino acid sequences are between 500 and 600 for the mature enzymes. The isoelectric point usually lies in the acidic pH region. But the preparations of *Polyporus versicolor* and *Neurospora crassa* laccase consist of several isoenzymes with isoelectric points up to neutral pHs. The molecular properties of tree laccases show some differences. The molecular weights are from 110 to 140 kDa with varying carbohydrate contents. The maximum carbohydrate content has been determined to 45%. The protein part of the enzyme was assessed to be about 65 kDa. The isoelectric point of tree laccase is in the basic pH region. Ascorbate oxidase, also a glycoprotein, is present in solution as a homodimer with molecular masses between 130 and 140 kDa. One monomer is built up by a single

Table 1. Molecular Properties of Representative Members of Blue Copper Oxidases

Blue Copper Oxidase	*Amino Acids*	*Molecular Mass of Protein (kDa)*	*Molecular Mass of Whole Enzyme (kDa)*	*Carbohydrate Content (%)*	*IP*	*Ref.*
Laccase						
Fungal: *Polyporus versicolor*						
A	n.d.[a]	n.d.	64.4	10–14	3.1	[40, 41]
B			64.7	10	4.6–6.8	[40–42]
Neurospora crassa	570	63.161	64	11–12	5.0–7.2	[43–45]
Phlebia radiata	527	56.433	64	11–12	4.9[b]	[46, 47]
Armillaria mellea	n.d.	n.d.	59	n.d.	4.1	[48]
Tree: *Rhus venicifera*	n.d.	n.d.	110–141	45	8.6	[49]
Ascorbate Oxidase						
Cucurbita pepo medullosa	552	61.703	140[c]	3	6.8[b]	[33, 50, 51]
Cucumis sativus	554	62.247	132[c]	n.d.	7.3[b]	[31, 52]
Ceruloplasmin						
Human	1046	129.09	134	10	4.4	[26, 53]

Notes: [a] n.d., not determined.
[b] Calculated from amino acid sequence without carbohydrate contribution.
[c] The protein exists as a homodimer in solution.

Table 2. Spectroscopical Properties of Representative Members of Blue Copper Oxidases

Blue Copper Oxidase	*Number and Type of Copper Center*	*EPR Parameters* $g_{\parallel}$	$g_{\perp}$	$A_{\parallel} \times 10^{-4}$ (cm^{-1})	*Absorption Band nm* $(\varepsilon, M^{-1}\ cm^{-1})$	*Ref.*
Laccase Fungal:						
Polyporus versicolor A	1 type 1	2.19	2.03	90	610 (4,900)	[10]
	1 type 2	2.24	2.04	194		
	1 type 3				330 (2,700)	
Neurospora crassa	1 type 1	2.19	2.04	90	595 (3,300)	[44]
	1 type 2	2.23	2.04	185		
	1 type 3				330 (2,400)	
Phlebia radiata	1 type 1	2.19	—	90	605 (n.d.)	[54]
	1 type 2	2.25	—	>170		
	no type 3[a]					
Tree: *Rhus vernicifera*	1 type 1	2.23	2.05	43	615 (5,700)	[10]
	1 type 2	2.24	2.05	200		
	1 type 3				330 (2,800)	
Ascorbate Oxidase *Cucurbita pepo medullosa*	2 type 1	2.227	2.058, 2.036	56	610 (9,700)	[55]
	2 type 2	2.242	2.053	190		
	2 type 3				330 (3,050)	
Cucumis sativus	2 type 1	2.23	2.05	63	607 (10,400)	[56]
	2 type 2	2.25	2.05	200		[57]
	2 type 3				330 (8,600)[b]	[56]
Ceruloplasmin Human	1st type 1	2.215	2.06	92	610 (10,000)	[26]
	2nd type 1	2.206	2.05	72	610 (10,000)	
	3rd type 1[c]					
	1 type 2	2.247	2.06	189		
	1 type 3				330 (3,300)	

Notes: [a] This laccase is supposed to contain only two coppers and one PQQ per mole.
[b] Seems to be incorrectly estimated.
[c] No distinct EPR signal observed. The nature of this type-1 copper will be discussed in the text.

polypeptide chain of about 550 residues. The carbohydrate content of 3% is very low. The isoelectric points calculated from the amino acid sequences without carbohydrate contribution are in the neutral pH region.

The spectral properties of the blue oxidases are very similar. The copper ions of copper proteins have been classified corresponding to their distinct spectroscopic properties [8]. Type-1 Cu^{2+} shows high absorption in the visible region ($\varepsilon > 3000$ $M^{-1}\ cm^{-1}$ at 600 nm) and an EPR spectrum with $A_{\parallel} < 95 \times 10^{-4}\ cm^{-1}$. Type-2 or normal Cu^{2+} has undetectable absorption and the EPR line shape of the usual low-molecular-mass copper complexes ($A_{\parallel} > 140 \times 10^{-4}\ cm^{-1}$). Type-3 ($Cu^{2+}$) is characterized by a strong absorption in the near-ultraviolet region (λ = 330 nm) and by the absence of an EPR signal. The type-3 center consists of a pair of copper ions that are anti-ferromagnetically coupled. The above mentioned signals disappear upon reduction.

The number and type of copper centers, EPR parameters, and the two relevant absorption bands in the visible region for several representative members of the blue oxidases are listed in Table 2. All laccases except that of the *Phlebia radiata* enzyme contain four copper ions per molecule with one type-1, one type-2, and one type-3 copper center. The EPR and absorption parameters closely resemble each other. *Phlebia radiata* laccase is supposed to contain only two copper ions with one type-1 and one type-2 and one PQQ per mole [54]. This is quite unusual and it is possible that the copper content determined for this enzyme is inaccurate. Ascorbate oxidases have eight copper ions per homodimer with two type-1, two type-2, and two type-3 copper centers. Ceruloplasmin typically contains six to seven copper ions per molecule with three type-1, one type-2, and one type-3 copper centers. It has also been proposed that there are only two type-1 copper ions and a new type-4 copper ion that is presumed to exhibit no EPR signal. In addition there is a variable content of chelatable copper. It is responsible for copper contents exceeding six copper ions per molecule but does not seem to be required for catalysis. It is now generally accepted that ceruloplasmin has three type-1 copper centers, the reason for which will be discussed below.

It turns out from the recently determined X-ray structure of ascorbate oxidase [58, 59] that the nonblue, EPR-active type-2 copper together with the type-3 copper pair is an integral part of a trinuclear active copper center. There has been experimental evidence in earlier studies that the type-2 copper is close to the type-3 copper and forms a trinuclear active copper site [60–65]. Solomon and co-workers, based on spectroscopic studies of azide binding to tree laccase [64, 65], classified this metal-binding site as a trinuclear active copper site.

4. X-RAY STRUCTURE OF ASCORBATE OXIDASE

4.1 Crystallization

The first crystals of ascorbate oxidase could be grown by Ladenstein et al. [66] in a final 1.2 M sodium–potassium phosphate buffer, pH 7.0. They were orthorhombic, space group $P2_12_12_1$ with a = 190.7 Å, b = 125.2 Å, c = 112.3 Å, and two molecules (four subunits) per asymmetric unit. Some years later Bolognesi et al. [67] obtained a different crystal form of ascorbate oxidase with 2-methyl-2,4-pentane-diol as precipitant. The crystals were orthorhombic as well, space group $P2_12_12$ with a = 106.7 Å, b = 105.1 Å, c = 113.5 Å, and one molecule (two subunits) per asymmetric unit. In both cases, the protein material was prepared from the peels of green zucchini squash (*Cucurbita pepo medullosa*). The preliminary three-dimensional X-ray structure of ascorbate oxidase based on the analysis of both crystal forms was published in 1989 by Messerschmidt et al. [58] and the polypeptide fold and the coordination of the mononuclear blue copper site and the unprecedented trinuclear copper site were described. The structure has now been refined to 1.9 Å resolution and its detailed description has been the subject of a further publication [59].

4.2 Overall Description of the Structure

The subunits are arranged in the crystals as homotetramers with D_2 symmetry. The structure of a subunit is shown schematically in Figure 1 [68]. Each subunit of 552 amino acid residues has a globular shape with dimensions of 49 × 53 × 65 Å and is built up by three domains arranged sequentially on the polypeptide chain, tightly associated in space. The folding of all three domains is of a similar β-barrel type. It is distantly related to the small blue copper proteins like plastocyanin or azurin. Domain 1 is made up of two four-stranded β-sheets (Figure 1b) which form a β-sandwich structure. Domain 2 consists of a six-stranded and a five-stranded β-sheet. Finally, domain 3 is built up of two five-stranded β-sheets that form the β-barrel structure and a four-stranded β-sheet that is an extension at the N-terminal

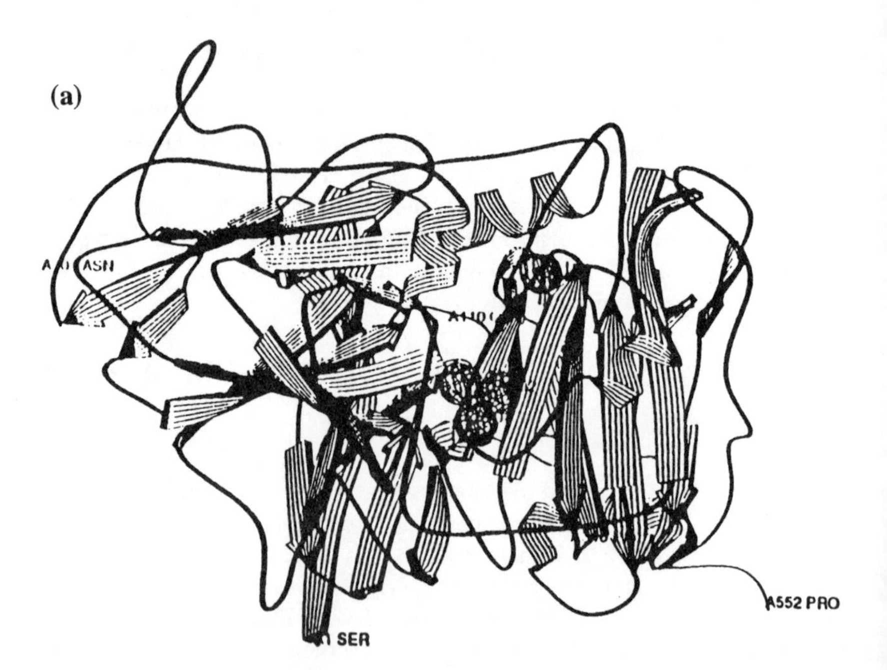

Figure 1. A schematic representation of the monomer structure of ascorbate oxidase. (**a**) Monomer plus copper ions. (**b**) Assignment of the secondary structure elements. β-sheets are represented by arrows, α-helices by helical ribbons. The figures were produced by the program RIBBON [69].

part of this domain. A topology diagram of ascorbate oxidase for all three domains and of the related structures of plastocyanin and azurin is shown in Figure 2. Ascorbate oxidase contains seven helices. Domain 2 has a short α-helix (α_1) between strands A2 and B2. Domain 3 exhibits five short α-helices that are located between strands D3 and E3 (α_3), I3 and J3 (α_4), M3 and N3 (α_5), as well as at the C-terminus (α_6, α_7). Helix α_2 connects domain 2 and domain 3.

A comparison of the different variants of the β-barrel domain structure in Figure 1 shows that domain 1 of ascorbate oxidase has the simplest β-barrel with only two four-stranded β-sheets. Plastocyanin and azurin are quite similar but between strands 4 (E1) and 6 (F1) they have insertions of one strand (plastocyanin) or one strand and an α-helix (azurin). Domain 2 has one additional strand, H2, in sheet D next to strand E2 (sheet B and strand E1 in domain 1) and two additional strands, F2 and G2, in sheet C next to strand I2 (sheet A and strand F1 in domain 1). Domain 3 resembles domain 2 except for the insertion of the short α-helices and the addition of the four-stranded β-sheet at its N-terminus.

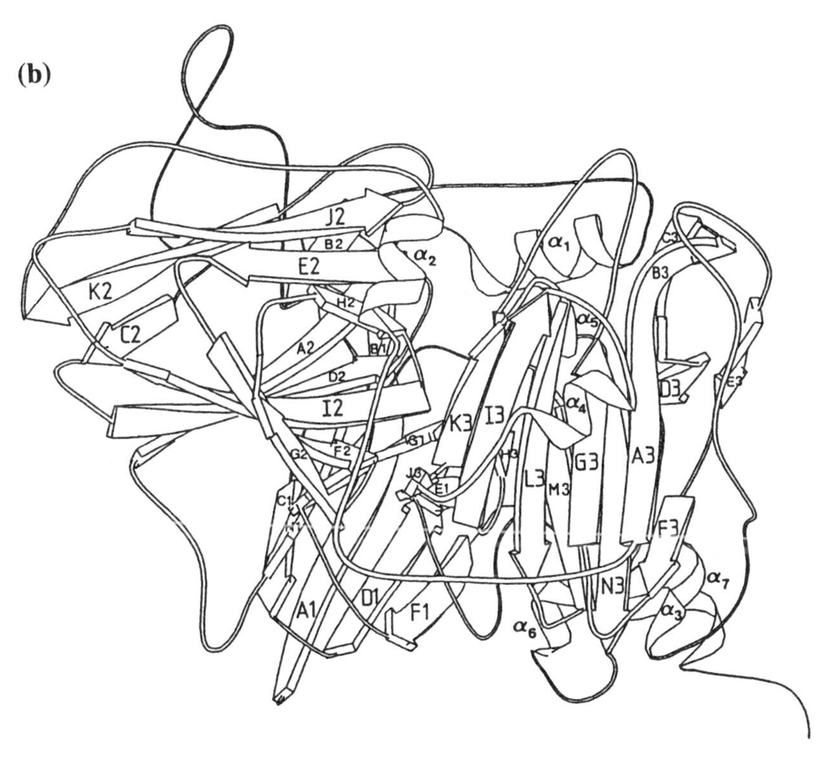

Figure 1. (Continued)

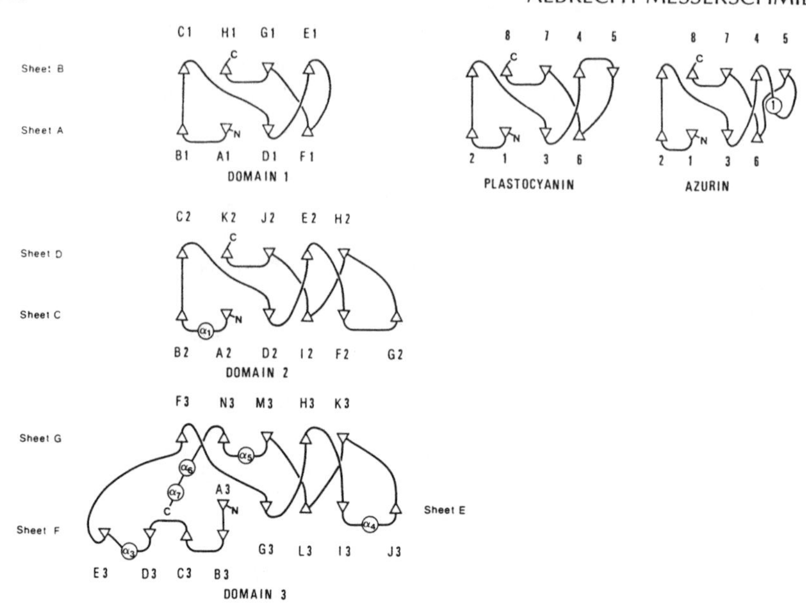

Figure 2. A topology/packing diagram of the domains of the ascorbate oxidase monomer compared with plastocyanin and azurin. Each β-strand is represented by a triangle whose apex points up or down whether the strand is viewed from C- or N-terminus. α-Helices are represented by circles.

The mononuclear copper site is located in domain 3 and the trinuclear copper species is bound between domain 1 and domain 3 (see Figure 1a). The copper site geometries will be discussed later.

Each monomer exhibits three disulfide bridges. These are between domain 1 and domain 2 (Cys-19–Cys-201), domain 1 and domain 3 (Cys-81–Cys-538) and within domain 2 (Cys-180–Cys-193). Three putative attachment sites for *N*-glycosidicly-linked carbohydrate moieties are present in the amino acid sequence of ascorbate oxidase from zucchini (Asn-92–Phe-93–Thr-94; Asn-325–Phe-326–Thr-327; Asn-440–Leu-441–Ser-442), but only Asn-92 shows density for an *N*-acetyl-glucosamine group.

4.3 Secondary Structure and Tetramer Contact Surface Areas

The secondary structure was assigned on the basis of main-chain hydrogen bonding using the algorithm of Kabsch and Sander [69] with a cutoff energy of 0.5 kcal/mol for the definition of a hydrogen bond interaction. Ramachandran plots of the refined models for subunit A and B show that all non-glycine residues lie in or close to energetically allowed regions. All residues with positive Φ-values are located in turns of the polypeptide chain. There are six α-helices in the monomer

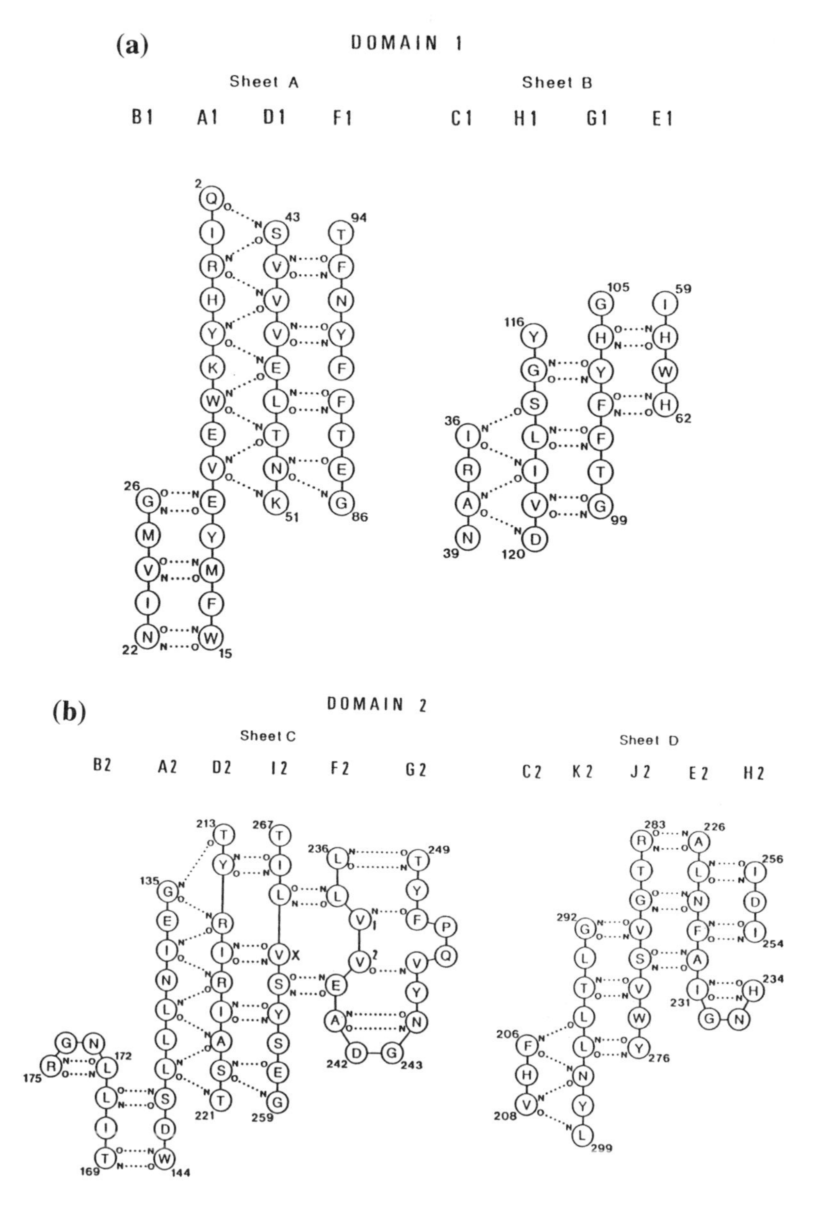

Figure 3. Main-chain hydrogen bonding scheme of the 7 β-sheets in ascorbate oxidase. (**a**) Domain 1. (**b**) Domain 2, wide β-buldge: x → 264; 1 → 238; 2 → 239. (**c**) Domain 3, classic β-buldge: x → 492; 1 → 457; 2 → 458. The amino acid residues are indicated by 1-letter code.

(c)

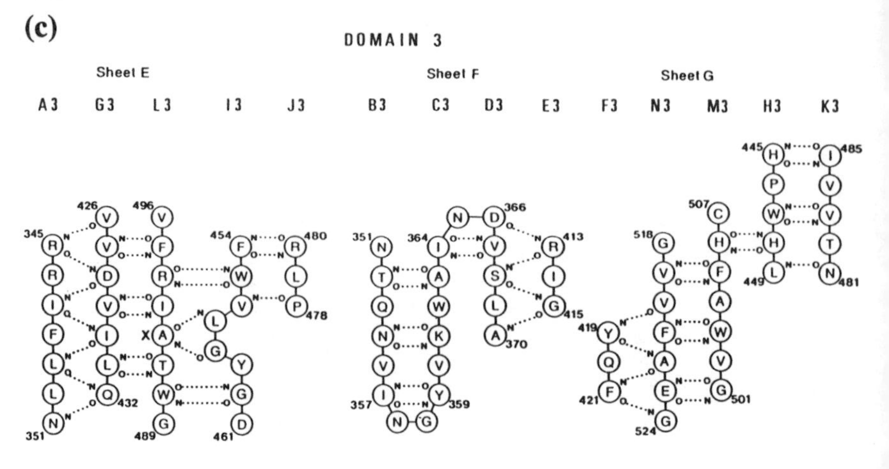

Figure 3. (Continued)

of ascorbate oxidase involving a total of 37 residues or 6.7% of the polypeptide. In addition there are one longer and three short 3_{10} helices. The average conformation angles of these secondary structures for subunit A and B have been calculated and are for both α-helices and 3_{10} helices close to those reported by Barlow and Thornton [70]. Exceptions are α_3 and α_5, which have one $i \rightarrow i + 5$ bond at their C-terminal end to give a short length π-helix.

All turns have been classified according to Crawford et al. [71] and Richardson [72]. One subunit exhibits 26 reverse turns (10 type I, 5 type I′, 10 type II, 1 type II′), three reverse turns associated with α-helices, three near-reverse turns (1 type I, 2 type II), and three open turns. The preference of glycine at position 3 in the turn is very pronounced. There are three *cis*-proline turns [73] and one Asx turn [74] which displays conventional geometry. The *cis*-prolines have no unusual function in the structure.

The residues involved in the seven β-sheet structures and the hydrogen bonding patterns are shown in Figure 3. All sheets exhibit the characteristic right-handed twist when viewed along the strand direction. There is a "wide" β-bulge [72] in β-sheet C of domain 2 (Figure 3b) at residues Val-238 and Val-239. β-Sheet E of domain 3 (Figure 3c) contains a classic β-bulge [72] at residues Leu-457 and Gly-458.

Contact surface areas for the monomer–monomer interactions within the homotetramer present in both crystal forms were calculated using the algorithm of Lee and Richards [75]. It is evident from these calculations [59] that the contact surface areas between the two dimers related about the crystallographic dyad are by far the largest. Ascorbate oxidase from zucchini occurs in solution as a dimer, which is likely the dimer mediated by the crystallographic dyad. The other dimers

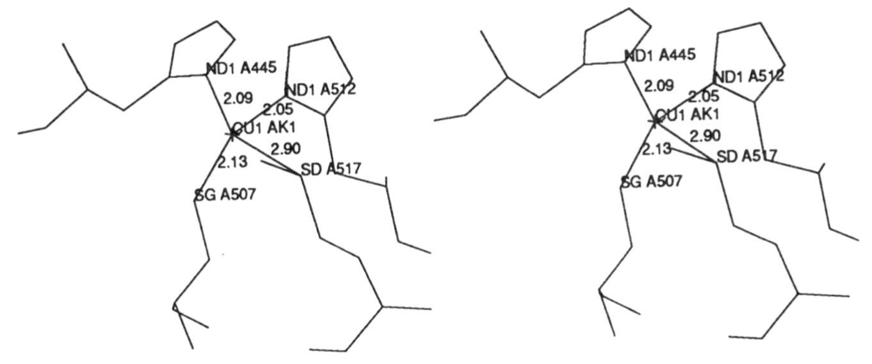

Figure 4. Stereo drawing of the type-1 copper site in domain 3. The displayed bond distances are for subunit A.

are related by the local dyad and are nearly identical. The tetramer is a dimer of dimers, which is related by a local dyad.

4.4 Copper Site Geometries

The mononuclear copper site is located in domain 3 and has the four canonical type-1 copper ligands (His, Cys, His, Met) also found in plastocyanin and azurin. It is coordinated to the ND1 atoms of His-445 and His-512, the SG atom of the Cys-507, and the SD atom of Met-517 in a distorted trigonal pyramidal geometry. The SD atom is at the long apex (see Figure 4). Bond lengths of the type-1 copper for both subunits are displayed in Table 3. They are compared with oxidized poplar plastocyanin [76] and azurin from *Pseudomonas aeruginosa* [77]. Figure 5 shows an overlay of the type-1 copper site in azurin, plastocyanin, and ascorbate oxidase.

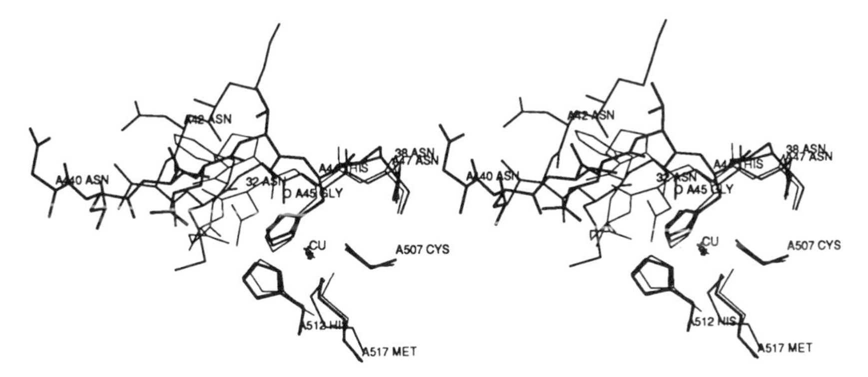

Figure 5. Stereo plot of the overlay of the type-1 copper sites for poplar plastocyanin (thin line), *Pseudomonas aeruginosa* azurin (medium thick line), and ascorbate oxidase (thick line).

Table 3. Copper–copper and Copper–Ligand Distances in Ascorbate Oxidase Compared to Copper–Ligand Distances for the Type-1 Copper Site in Plastocyanin and Azurin

		Distance (Å)		*Azurin*[a]		
Atom 1	*Atom 2*	*A*	*B*	*wtp5*	*wtp9*	*Plastocyanin*[b]
CU1 K1	CU2 K3	12.20	12.20			
CU1 K1	CU3 K3	12.73	12.65			
CU1 K1	CU4 K3	14.87	14.86			
		Type-1 Copper Site				
CU1 K1	ND1 445[c]	2.10	2.12	2.11	2.09	2.04
CU1 K1	SG 507	2.13	2.03	2.25	2.26	2.13
CU1 K1	ND1 512	2.05	2.11	2.03	2.04	2.10
CU1 K1	SD 517	2.90	2.83	3.15	3.12	2.90
CU1 K1	O 444	4.78	4.83	2.97	2.95	3.82
	Trinuclear Copper Site					
CU2 K3	CU3 K3	3.68	3.73			
CU2 K3	CU4 K3	3.78	3.90			
CU3 K3	CU4 K3	3.66	3.69			
CU2 K3	NE2 106	2.16	2.17			
CU2 K3	NE2 450	2.06	2.09			
CU2 K3	NE2 506	2.07	2.09			
CU3 K3	ND1 62	1.98	2.13			
CU3 K3	NE2 104	2.19	2.14			
CU3 K3	NE2 508	2.14	2.08			
CU4 K3	NE2 60	2.00	2.02			
CU4 K3	NE2 448	2.09	2.05			
CU2 K3	OH1 K3	2.00	2.00			
CU3 K3	OH1 K3	2.06	2.00			
CU4 K3	OH3 K3	2.02	2.03			

Notes: [a] The wtp5 and wtp9 recombinant wild-type structures are from *Pseudomonas aeruginosa* azurin at pH 5.5 and 9.0. The data are from Nar et al. [77].

[b] Plastocyanin is from poplar leaves. The data are from Guss and Freeman [76].

[c] Ascorbate oxidase numbering.

The copper is penta-coordinated in azurin (glycine) with the main chain carbonyl oxygen of the residue preceding the first histidine copper ligand on the polypeptide chain. This carbonyl oxygen forms the fifth ligand. In poplar plastocyanin, the homologous peptide is a proline which is the beginning of a turn causing the carbonyl oxygen to be removed to a distance of 3.8 Å from the type-1 copper ion. In ascorbate oxidase, the corresponding strand is extended moving the carbonyl oxygen to a distance of 4.8 Å away from the type-1 copper ion. This extended strand contributes Glu-443 to the formation of the binding site of the reducing substrate [59].

The trinuclear copper site (see Figure 6) has eight histidine ligands symmetrically supplied by domain 1 and domain 3 and two oxygen ligands. Seven histidines are ligated by their NE2 atoms to the copper ions, whereas His-62 is ligated to CU3 K3 by its ND1 atom. In the preliminary structural report on ascorbate oxidase [58] based on lower resolution data, all eight histidine residues were modeled with their NE2 atoms as ligands to the copper atoms of the trinuclear site. The high-resolution data allows an unequivocal interpretation. His-62 is part of a β-sheet while His-450 is not. An overlay of the relevant parts of domain 3 onto domain 1 shows that His-62 comes closer with its main chain atoms to the copper ion CU3 than the corresponding atoms of His-450. As a consequence, the side chain of His-62 has to adopt a conformation in which ND1 of the imidazole ring is ligated to the copper ion CU3.

The trinuclear copper site may be subdivided into a copper pair (CU2 K3, CU3 K3) with six histidine ligands in a trigonal prismatic arrangement. The pair is bridged by an OH^- or O^{2-}, which leads to a strong antiferromagnetic coupling and makes this copper pair EPR silent. The pair very likely represents the spectroscopic type-3 copper. The remaining copper (CU4 K3) has two histidine ligands and an OH^- or H_2O ligand and probably represents the spectroscopic type-2 copper. Both oxygen ligands could be clearly detected from difference Fourier maps. An oxygen ligand in the center of the three copper ions could not be detected.

Cole et al. [78] studied the electronic structure of the laccase trinuclear copper active site by the use of absorption, circular dichroism, and low-temperature magnetic circular dichroism spectroscopies. The assigned ligand field transition energies indicated that all three copper ions have tetragonal geometries and that the

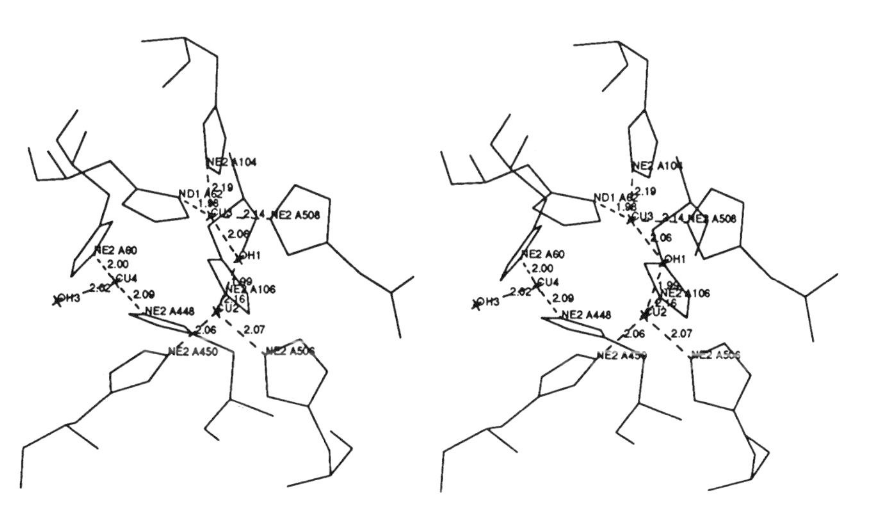

Figure 6. Stereo drawing of the trinuclear copper site. The displayed bond distances are for subunit A.

two type-3 copper ions are inequivalent. The latter interpretation fits well with our structural model as CU3 K3 of the the copper pair is ligated to one ND1 and two NE2 atoms, but the former interpretation, tetragonal coordination geometries for all three copper ions, is not consistent with the structure. The copper ions of the pair are both tetrahedrally coordinated, whereas the type-2 copper has three ligands. The existence of a central oxygen ligand would give rise to a penta-coordination of both copper pair atoms (but not a tetragonal–pyramidal coordination) and a square–planar coordination for the spectroscopic type-2 copper. Messerschmidt et al. [59] examined by modeling whether, by side chain torsion angle rotations, His-62 and His-450 could become bridging ligands between CU4 K3 and CU2 K3 or CU3 K3 with both ring nitrogens as ligating atoms, but found this stereochemically unreasonable. The copper–copper and copper–ligand distances of the trinuclear copper site are displayed in Table 3 for both subunits. Their mean values [Cu–N, 2.09 Å (σ = 0.06 Å); Cu–O, 2.02 Å (σ = 0.02 Å)] are comparable to those of binuclear copper model compounds with nitrogen and oxygen copper ligands [79, 80] in the accuracy range of 0.1 Å for the copper–ligand bonds. There is no long Cu–N bond (about 2.7 Å) as found in the binuclear copper center of hemocyanin from *Panulirus interruptus* determined by X-ray crystallography at 3.2 Å resolution [81]. The binuclear copper center in hemocyanin has a trigonal antiprismatic arrangement of the six histidine ligands, whereas the type-3 copper ions in ascorbate oxidase show the trigonal prismatic coordination. The presence of the type-2 copper and its ligands would not allow the former arrangement.

The average copper–copper distance in the trinuclear copper site of ascorbate oxidase is 3.74 Å (σ = 0.08Å) and the individual distances do not deviate by more

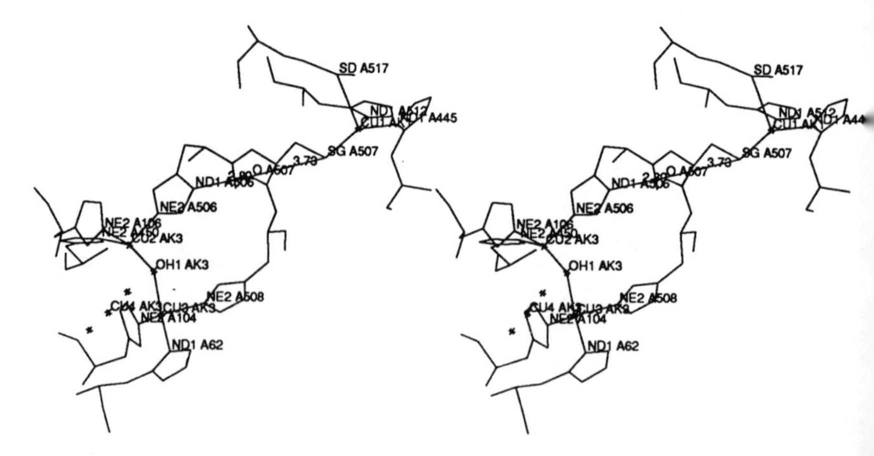

Figure 7. Stereo drawing of the region of the atomic model containing the type-1 copper center in domain 3 and the trinuclear copper center between domain 1 and domain 3.

than 0.16 Å from this mean value. The average copper–copper distance in hemocyanin is 3.54 Å [81]. The copper–copper distances are too long for copper–copper bonds but magnetic interactions are possible.

The shortest distance between the type-1 copper center and the trinuclear copper center is 12.2 Å. The His-506–Cys-507–His-508 amino acid sequence segment links the type-1 copper center and the type-3 copper center as a bridging ligand (see Figure 7). Aspects of the intramolecular electron transfer between the two redox center will be discussed in Section 10.

5. ASCORBATE OXIDASES, FUNGAL LACCASES, AND RELATED PROTEINS

A structurally based amino acid sequence alignment of ascorbate oxidase with laccase from *Neurospora crassa* and human ceruloplasmin has been carried out [82]. This alignment strongly suggests a three-domain structure for laccase, closely related to ascorbate oxidase, and a six-domain structure for ceruloplasmin. The relationship suggests that laccase, like ascorbate oxidase, has a mononuclear blue copper in domain 3 and a trinuclear copper between domain 1 and domain 3, and ceruloplasmin has mononuclear copper ions in domains 2, 4, and 6 and a trinuclear copper center between domain 1 and domain 6.

The six-domain structure and copper site model for ceruloplasmin has got further support by the elucidation of the X-ray structure of a copper-containing nitrite reductase from *Achromobacter cycloclastes* [83]. This nitrite reductase occurs in solution and in the crystal as homotrimer. The trimer shows C_3 symmetry. One monomer is built up by two domains of a similar β-barrel fold as found in the small blue copper proteins or in ascorbate oxidase. The enzyme contains two mononuclear copper centers. A type-1 copper site with the canonical ligands (His, Cys, His, Met) is located in domain 1 and a type-2 copper with three histidines and one water molecule as ligands is bound between domain 1 and domain 2 of the adjacent symmetry-related molecule. The copper-containing domains of ascorbate oxidase can be superimposed to the domains of nitrite reductase. The type-1 copper and its ligands in nitrite reductase fall at the type-1 center of ascorbate oxidase, and all three histidines of the type-2 site, as well as the copper, fall in the same site as the trinuclear cluster of ascorbate oxidase [83]. Sequence comparison [84] shows that two histidines from domain 1 (residues 100 and 135) and two from domain 2 (residues 255 and 306) correspond to four of the eight histidine ligands of the trinuclear copper cluster in ascorbate oxidase. In nitrite reductase only histidine residues 100, 135, and 306 are ligands. These comparisons suggest that the type-2 copper of nitrite reductase corresponds to one of the type-3 copper pair in ascorbate oxidase.

In the trimer of nitrite reductase a six-domain structure is realized which is reminiscent of the six-domain structure of ceroluplasmin [85], which was deduced from the amino acid sequence alignment with the other blue oxidases [82].

However, the arrangement of the six gene segments in ceruloplasmin is not simply a triplication of an ancestral nitrite reductase gene coding for two domains, but the triplication of a gene where the segments coding for the individual domains were inverted. This is due to the finding that the type-1 copper sites in ceruloplasmin are found in domains 2, 4, and 6 rather than in domain 1 and symmetrical counterparts in nitrite reductase. The X-ray structure of human ceruloplasmin has recently been solved [85a] and confirms the proposed structural model.

There are now known several amino acid sequences of fungal laccases and three of ascobate oxidases. Furthermore, amino acid sequences of seven different proteins that are related to the blue oxidases have been reported. The first of them is phenoxazinone synthase from *Streptomyces lividans* [86]. This enzyme is a copper-containing oxidase that catalyzes the coupling of 2-aminophenols to form the 2-aminophenoxazinone chromophore [87]. This reaction constitutes the final step in the biosynthesis of the potent antineoplastic agent, actinomycin. The next group comprises the gene product of a copper resistance gene from *Pseudomonas syringae* pv. *tomato* [88] and from *Xanthomonas campestris* pv. *juglandis* [88a]. These proteins are believed to function as copper-binding proteins thus making a contribution to the copper resistance. Furthermore, the *FET3* gene of *Saccharomyces cerevisae* encodes a multicopper oxidase that is required for ferrous iron uptake and

Table 4. Amino Acid Sequence Alignment of the Parts Around the Putative Copper Ligands for Fungal Laccases, Ascorbate Oxidases and Related Proteins[a]

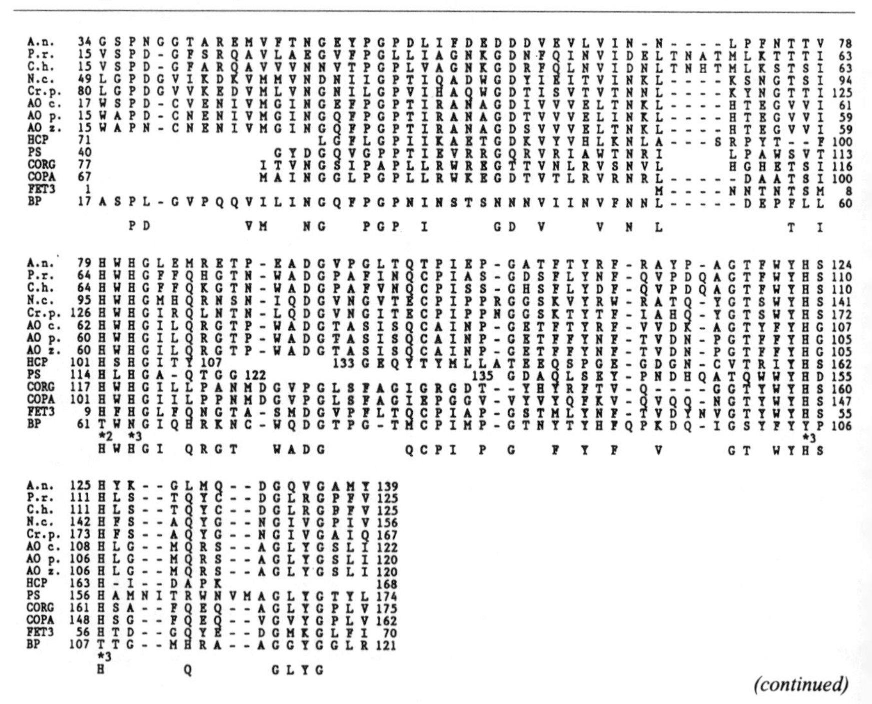

```
A.n.   34 G S P N G G T A R E M V F T N G E Y P G P D L I F D E D D D V E V L V I N - N - - - - L P F N T T V 78
P.r.   15 V S P D - G F S R Q A V L A E G V F P G L L I A G N K G D N F Q I N V I D E L T N A T M L K T T T I 63
C.h.   15 V S P D - G F A R Q A V V V N N V T P G P L V A G N K G D R F Q L N V I D N L T N H T M L K S T S I 63
N.c.   49 L G P D G V I K D K V M M V N D N I I G P T I Q A D W G D Y I E I T V I N K L - - - - K S N G T S I 94
Cr.p.  80 L G P D G V V K E D V M L V N G N I L G P V I H A Q W G D T I S V T V T N N L - - - - K Y N G T T I 125
AO c.  17 W S P D - C V E N I V M G I N G E F P G P T I R A N A G D I V V V E L T N K L - - - - H T E G V V I 61
AO p.  15 W A P D - C N E N I V M G I N G Q F P G P T I R A N A G D T V V V E L I N K L - - - - H T E G V V I 59
AO z.  15 W A P N - C N E N I V M G I N G Q F P G P T I R A N A G D S V V V E L T N K L - - - - H T E G V V I 59
HCP    71 L G F L G P I I K A E T G D K V Y V H L K N L A - - - S R P Y T - - F 100
PS     40 G Y D G Q V G P P T I E V R R G Q R V R I A W T N R I L P A W S V T 113
CORG   77 I T V N G S I P A P L L R W R E G T T V N L R V S N V L H G H E T S I 116
COPA   67 M A I N G G L P G P L L R W K E G D T V T L R V R N R L - - - - - D A A T S I 100
FET3    1 M - - - - N N T N T S M 8
BP     17 A S P L - G V P Q Q V I L I N G Q F P G P N I N S T S N N N V I I N V F N N L - - - - - D E P F L L 60

          P D  V M  N G  P G P  I  G D  V  V  N  L  T  I

A.n.   79 H W H G L E M R E T P - E A D G V P G L T Q T P I E P - G A T F T Y R F - R A Y P - A G T F W Y H S 124
P.r.   64 H W H G F F Q H G T N - W A D G P A F I N Q C P I A S - G D S F L Y N F - Q V P D Q A G T F W Y H S 110
C.h.   64 H W H G F F Q K G T N - W A D G P A F V N Q C P I S S - G H S F L Y D F - Q V P D Q A G T F W Y H S 110
N.c.   95 H W H G M H Q R N S N - I Q D G V N G V T E C P I P P R G G S K V Y R W - R A T Q - Y G T S W Y H S 141
Cr.p. 126 H W H G I R Q L N T N - L Q D G V N G I T E C P I P P N G G S K T Y T F - I A H Q - Y G T S W Y H S 172
AO c.  62 H W H G I L Q R G T P - W A D G T A S I S Q C A I N P - G E T F T Y R F - V V D K - A G T Y F Y H G 107
AO p.  60 H W H G I L Q R G T P - W A D G T A S I S Q C A I N P - G E T F F Y N F - T V D N - P G T F F Y H G 105
AO z.  60 H W H G I L Q R G T P - W A D G T A S I S Q C A I N P - G E T F F Y N F - T V D N - P G T F F Y H G 105
HCP   101 H S H G I T Y 107  133 G E Q Y T Y M L L A T E E Q S P G E - G D G N - C V T R I Y H S 162
PS    114 H L H G A - Q T G G 122  135 G D A Q L S E Y - P N D H Q A T Q W W Y H D 155
CORG  117 H W H G I L L P A N M D G V P G L S F A G I G R G D T - - Y H Y R F T V - Q - - - - G G T Y W Y H S 160
COPA  101 H W H G I I L P P N M D G V P G L S F A G I E P G G V - V Y V Y Q F K V - Q V Q Q - N G T Y W Y H S 147
FET3    9 H F H G L F Q N G T A - S M D G V P F L T Q C P I A P - G S T M L Y N F - T V D Y N V G T Y W Y H S 55
BP     61 T W N G I Q H R K N C - W Q D G T P G - T M C P I M P - G T N Y T Y H F Q P K D Q - I G S Y F Y Y P 106
          *2  *3                                                                              *3
          H W H G I  Q R G T  W A D G  Q C P I  P  G  F  Y  F  V  G T  W Y H S

A.n.  125 H Y K - - G L M Q - - D G Q V G A M Y 139
P.r.  111 H L S - - T Q Y C - - D G L R G P F V 125
C.h.  111 H L S - - T Q Y C - - D G L R G P F V 125
N.c.  142 H F S - - A Q Y G - - N G I V G P I V 156
Cr.p. 173 H F S - - A Q Y G - - N G I V G A I Q 167
AO c. 108 H L G - - M Q R S - - A G L Y G S L I 122
AO p. 106 H L G - - M Q R S - - A G L Y G S L I 120
AO z. 106 H L G - - M Q R S - - A G L Y G S L I 120
HCP   163 H - I - - D A P K 168
PS    156 H A M N I T R W N V M A G L Y G T Y L 174
CORG  161 H S A - - F Q E Q - - A G L Y G P L V 175
COPA  148 H S G - - F Q E Q - - V G V Y G P L V 162
FET3   56 H T D - - G Q Y E - - D G M K G L F I 70
BP    107 T T G - - M H R A - - A G G Y G G L R 121
          *3
          H  Q  G L Y G
```

(continued)

exhibits ferroxidase activity [89]. Another one is a product of the *Bp10* gene of *Brassica napus* (rape) [90]. The expression of the *Bp10* gene family is pollen-specific and developmentally regulated. The blood coagulation factors VIII [91] and V [92] have structural domains that are homologous to ceruloplasmin and thus to the whole family we are dealing with.

A multiple amino acid sequence alignment of the stretches around the putative copper ligands for the whole family of proteins has been carried out and is shown in Table 4. The two blood coagulation factors were omitted because these proteins have a multidomain structure with different functions. All protein copper ligands

Table 4. (Continued)

	Start	Sequence	End
A.n.	503	L I H - P P H P I H K H G N R A Y I I G N G V G K F R W E N	531
P.c.	376	- Y A L P S N A T I E L - S L P A G - - A L G - G P H P F H L H G H T F S V V R P A G S T T Y N Y V -	420
C.h.	372	- Y S L P S N A D I E I - S F P A T A A A P G - A P H P F H L H G H A F A V V R S A G S T V Y N Y D -	418
N.c.	405	- V E G V N Q W K Y W L - I E N D P D G A F S - L P H P I H L H G H D F L I L G R S P D V T A I S N T	452
Cr.p.	439	- V E E A N Q W A Y W L - I E N D P T A T G N A L P H P I H L H G H D F V V L G R S P N V S P T A - -	485
AO c.	424	- N M G E T V D V I L Q - N A N M L N P N M S - E I H P W H L H G H D F W V L G Y G E G K F Y A P E D	471
AO p.	422	- K I G E I V D V I L Q - N A N M M K E N L S - E I H P W H L H G H D F W V L G Y G D G K F - T A E E	468
AO c.	422	- K I G E V V D V I L Q - N A N M M K E N L S - E I H P W H L H G H D F W V L G Y G D G K F - S A E E	468
HCP	956	- H V G D E V N W Y L M - G M - - - - G N E I - D L H T V H F H G H S F Q Y K H R G V	991
PS	458	- T I G E G T H E Q W T - F L - - - - - N L S P I L H P M H I H L A D F Q V L G R D A	493
CORG	550	L N Y G E R L R I V L V - N D T - - - - - - - - M M S H P I H L H G M W S D L E D	581
COPA	525	L K Y G E R V R I V L V - N D T - - - - - - - - M M T H P I H L H G M W S D L E D	556
FET3	323	L E K D E I V E I V L N N Q D T - - - - - - - - G T H P F H L H G H A F A T I Q R D R	357
BP	402	- T H R T F V E V V F E - N H E - - - - - - - K - S V Q S W H L D G Y S F F S V A V E P G T W - T P E K	442
		*1 *2 *3	
		G HP HLHGHDF VLG	
A.n.	558	D F F D S R L - M D G A W I V I R Y F V Q D - K F P S I L H	485
P.c.	421	- - - - - - - - - - - - - - - - - - N P V Q R D V V S I G - - N T G D N V T I R F D T N N - P G P W F L H	452
C.h.	419	- - - - - - - - - - - - - - - - - - N P I F R D V V S T G T P A A G D N V T I R F R T D N - P G P W F L H	452
N.c.	453	R Y V F D P A V D M A R L N G N N P T R R D T A M L P - - A K G - W L L I A F R T D N - P G S W L M H	499
Cr.p.	486	- Q T P Y T F T S S D V S S L N G N N P I R R D V V M L - P P K G W L L I A F Q T T N - P G A W L M H	533
AO c.	472	E K K L N L K - - - - - - - - - - - N P P L R N T V V T F - - P Y G - W T A I R F V A D N - P G V W A F H	509
AO p.	469	E S S L N L K - - - - - - - - - - - N P P L R N T V V T F - - P Y G - W T A I R F V A D N - P G V W A F H	506
AO z.	469	E S S L N L K - - - - - - - - - - - N P P L R N T V V T F - - P Y G - W T A I R F V A D N - P G V W A F H	506
HCP	992	Y S S D V F D I - - F P G - T Y Q T L E M F P R T P G I W L L H	1020
PS	538	G - L R V M G K F D G A - Y G R F M Y H	555
CORG	596	P P G S K R T Y R V R A D A - L G T W A Y H	616
COPA	571	P P G S K R S Y R V T A D A - L G R W A Y H	591
FET3	381	Y P M R R D T - L Y V - - R P Q S N F V I R F K A D N - P G V W F F H	411
BP	443	R K N Y N L L - - - - - - - - - - - D A V S R H T V Q V Y - - P K C - W A A I L L T F D N - C G M W N V R	480
		*3	
		NP R V P G W IRF DN PG W H	
A.n.	586	C H I A S H Q M G G M A L - - - A L L D G V D V W D S *	609
P.c.	453	C H I D W H L E A A L P L - - - S S L R T S L T L R P - - L T L S P R - - - T G P T C A L S T T L W	494
C.h.	453	C H I D F H L E A G F A V - - - V F A E D I P D V A S - - A N P V P Q - - A W S D L C P I Y D A L D	495
N.c.	500	C H I A W H V S G G L S N - - - Q F L E R A Q D L R N - - S I S P A D K K A F N D N C D A W R A Y F	444
Cr.p.	534	C H I A W H V S A G L G N - - - T F L E Q P S A F V A - - G L N T N D V N Q L N S Q C K S W N A Y Y	578
AO c.	510	C H I E P H L H M G M G V - - - V F A E G V H M V G M - - I P P K A - - - - - - L A C G S T A L V K	548
AO p.	507	C H I E P H L H M G M G V - - - V F A E G V E K V G R - - I P T K A - - - - - - L A C G G T A K S L	545
AO z.	507	C H I E P H L H M G M G V - - - V F A E G V E K V G R - - I P T K A - - - - - - L A C G G T A K S L	545
HCP	1021	C H V T D H I H A G M E T - - - T T Y T V L Q N E D T - - K S G *	1046
PS	556	C H L L E H E D M G M M R - - - P F V V	572
CORG	617	C H L L Y H M E A G M M R - - - A V K V E E *	635
COPA	592	C H L L Y H M E M G M F R - - - E V R V E E *	610
FET3	412	C H I E W H L L Q G L G L - - - V L V E D P F G I Q D - - A H S Q Q L S E N H L E V C Q S C V A T E	456
BP	481	S E N T E R R Y L G Q Q L Y A S V L S P E K S L R D E Y N M P E T S - - - - - - - L Q C G - - - L V K	521
		*1*3 *1 *1	
		CHI H GM F E C	

Note: [a]A.n., laccase from *Aspergillus nidulans* [30a], P.r., laccase from *Phlebia radiata* [46]; C.h., laccase from *Coriolus hirsutus* [30c]; N.c., laccase from *Neurospora crassa* [45]; Cr.p., laccase gene from *Cryphonectria parasitica* [30e]; AO c., ascorbate oxidase from cucumber [31]; AO p., ascorbate oxidase from pumpkin [32]; AO z., ascorbate oxidase from zucchini [33]; HCP, human cerulplasmin [53]; PS, phenoxazinone synthase from *Streptomyces lividans* [86]; CORG, copper resistance gene from *Xanthomonas campestris pv. juglandis* [88a]; COPA, copper binding protein from copper resistance gene from *Pseudomonas syringae pv. tomato* [88]; FET3, Multicopper oxidase from FET3 gene of *Saccharomyces cerevisiae* [89]; BP, brassica napus Bp10 gene product [90]. Alignment positions with greater equal six identities are indicated. The secondary structure elements, known from the three-dimensional structure of ascorbate oxidase, are given at the bottom of each alignment block. *1, ligand to type-1 copper; *2, ligand to type-2 copper; *3, ligand to type-3 copper. Parts of sequences were omitted from the alignment were no relationship could be detected.

characteristic of the blue oxidases are conserved with two principal exceptions. The rape *Bp10* gene product has only one histidine of the copper ligands conserved, meaning that this protein will bind no copper and therefore will have no oxidase activity [90]. The second exception concerns the position for a possible fourth ligand of the type-1 copper. As mentioned above, there is very often a methionine at this place in small blue copper proteins as well as in blue oxidases. This methionine serves as a weak ligand to the type-1 copper. In *Aspergillus nidulans* laccase, the three ascorbate oxidases, phenoxazinone synthase, and the copper-binding proteins from the *Pseudomonas syringae* and *Xanthomonas campestris* copper resistance gene, there is a methionine at this position. Similar to *Neurospora crassa* laccase, *Phlebia radiata* and *Cryphonectria parasitica* laccase, and product of the *FET3* gene from *S. cerevisae* also contain a leucine at this position. This leucine is isosteric to methionine but no copper ligand. Recently, it has been shown, using a recombinant Met-121-Leu mutant of azurin from *P. aeruginosa*, that the methionine of the type-1 copper site is not required to generate the properties of this metal site [93]. *Coriolus hirsutus* laccase holds a phenylalanine at this position. This residue also cannot serve as a copper ligand. The role of the residue located in this position seems to tune the redox potential of the type-1 copper site which is necessary for the function of the enzyme.

An inspection of this amino acid sequence alignment shows that the sequences align very well in the segments of the conserved copper ligands. This is also valid for segments of defined secondary structure elements of the ascorbate oxidase structure (not included in the alignment). The rape *Bp10* gene product aligns very well on ascorbate oxidase although it is no copper oxidase.

6. OXIDATION–REDUCTION POTENTIALS

Redox potentials for the different copper centers in the blue oxidases have been determined for all members of the group, but in each case only for a limited number of species. The available data is summarized in Table 5 [94, 95]. The redox potentials for the type-1 copper of tree laccase and ascorbate oxidase are in the range of 330–400 mV and comparable to the values determined for the small blue copper proteins plastocyanin, azurin, and cucumber basic protein (for redox potentials of small blue copper proteins see the review of Sykes [96]). The high potential for the fungal *Polyporus* laccase is probably due to a leucine or phenylalanine residue at the fourth coordination position which have been observed in the amino acid sequences of fungal laccases from other species (see Table 4 and Section 5). Two different redox potentials for the type-1 copper were observed for human ceruloplasmin [97]. The 490 mV potential can be assigned to the two type-1 copper sites with methionine ligand and the 580 mV potential to the type-1 center with the isosteric leucine at this position (the amino acid stretches for the two other putative type-1 copper sites in ceruloplasmin were not included in Table 4). The redox potentials for the type-3 centers are equal or somewhat higher than the respective

Table 5. Oxidation–Reduction Potentials of the Copper Centers of Blue Copper Oxidases

Protein	pH	Potential E° (mV) Type-1	Type-2	Type-3	Ref.
Laccase					
Fungal: Polyporus	5.5	785	—	782	[94]
+ 1 mM NaF	5.5	780	—	570	
Tree: Rhus	7.5	394	365	434	[94, 95]
+ 10 mM NaF	7.5	390	390	390	
+ $Fe(CN)^{4-}$	7.5	434	—	483	
Ascorbate Oxidase					
Cucurbita pepo medullosa	7.0	344[a]	—	344[a]	[18]
	7.0	327[b]	—	357[b]	
Cucumis sativus	6.0	350		—	[56]
Ceruloplasmin					
Human	5.5	490, 580	—	—	[97]

Notes: [a] at 25°C
[b] at 10°C

type-1 potentials. For ascorbate oxidase a difference of +30 mV between the type-3 and type-1 potentials was observed when they were determined at 10 °C in contrast to 25 °C where no distinction was possible [18].

It is intriguing that the type-3 potential in fungal laccase is also greater, which is necessary for its enzymatic function (the reason for this will be discussed later). It will be interesting to learn what structural changes, compared to tree laccase or ascorbate oxidase, are the reason for this elevation of redox potential once the X-ray structures of a fungal and tree laccase have been determined. Redox potentials for the type-2 site have been only reported for tree laccase. They had to be determined from a redox titration of the type-2 EPR signal. The values obtained do not differ much from those of the other copper centers. Taking into account that an integral trinuclear copper site is present in the blue oxidases and that there are no separated type-2 and type-3 centers makes it more plausible to define a unique redox potential for this trinuclear copper species. The type-2 EPR signal as well as the 330-nm band are both spectroscopic signals of this copper species. Redox titrations using one of this signals should give the same results in redox potential.

7. KINETIC PROPERTIES OF LACCASE AND ASCORBATE OXIDASE

The numerous studies concerning the catalytic and binding properties of the blue oxidases are well documented in the formerly mentioned reviews (see for example, for laccase [10], for ascorbate oxidase [19] and for ceruloplasmin [26]). Important data, concentrating on laccase and ascorbate oxidase only, are summarized here.

Results from steady-state kinetics for laccase and ascorbate oxidase are given in Table 6. Listed are k_{cat}, K_m, and k_{cat}/K_m values at different pH values. The k_{cat}/K_m values indicate the low limit of the bimolecular enzyme–substrate reaction rate. The most reactive reducing substrate for tree laccase is hydroquinone with $k_{cat} = 2.4 \times 10^2\ s^{-1}$, $K_m = 2.0 \times 10^{-1}$ M, and $k_{cat}/K_m = 1.2 \times 10^3\ M^{-1}s^{-1}$, and for ascorbate oxidase ascorbate with $k_{cat} = 7.5 \times 10^3\ s^{-1}$, $K_m = 2.0 \times 10^{-4}$ M and $k_{cat}/K_m = 3.8 \times 10^7\ M^{-1}s^{-1}$. The data do not show a remarkable pH dependence.

A "ping-pong di Theorell–Chance" mechanism has been deduced for tree laccase from steady-state kinetics [98]. This mechanism is characterized by the sequential entry of the two substrates and the immediate release of the respective products. The actual catalytic cycle will consist of very different steps, but the finding that the two substrates react in a ping-pong mechanism is important.

Transient kinetics such as stopped flow, pulse radiolysis, or laser-flash photolysis has been applied to monitor the different steps during the catalytic cycle. In the presence of both substrates, complex patterns that are difficult to interpret have been obtained. Therefore, transient kinetics was carried out in the presence of only one substrate in each case. One process that has been investigated is the anaerobic reduction of the enzyme. A second is the reoxidation of the reduced enzyme by dioxygen. Results of the anaerobic reduction of laccase and ascorbate oxidase are given in Table 7. The reduction of the type-1 copper, monitored by the bleaching of the 600-nm band, is a bimolecular second-order reaction. The reaction is biphasic for tree laccase and ascorbate oxidase with hydroquinone and reductate as substrates, respectively, and it is monophasic in the other cases. The rate constants for tree laccase with hydroquinone as a substrate are in the range of 2.0×10^2 to $2 \times 10^4\ M^{-1}\ s^{-1}$ with ascending values from pH 6.5 to pH 8.5 [99]. The values for fungal laccase are three to four orders of magnitudes higher [100]. The rates for ascorbate

Table 6. Steady-State Kinetics of Laccase and Ascorbate Oxidase

Enzyme	*Substrate*	*pH*	k_{cat} *(s^{-1})*	K_m *(M)*	k_{cat}/K_m *($M^{-1}\ s^{-1}$)*	*Ref.*
Laccase	Ascorbate	6.0	1.5	1.0×10^{-2}	1.5×10^2	[19]
(*Rhus vernicifera*)	$Fe(CN)_6^{4-}$	6.0	2.0×10^1	5.0×10^{-2}	4.0×10^2	[19]
	Ascorbate	7.5	8.2	5.9×10^{-2}	1.4×10^2	[19]
	Hydroquinone	7.5	2.4×10^2	2.0×10^{-1}	1.2×10^3	[98]
	Oxygen	7.5	1.3×10^2	2.1×10^{-5}	6.1×10^6	[98]
Laccase	Ascorbate	6.0	3.0×10^1	1.7×10^{-2}	1.8×10^3	[19]
(*Rhus succedanea*)	$Fe(CN)_6^{4-}$	6.0	2.2×10^2	2.2×10^{-2}	1.0×10^4	[19]
	Ascorbate	7.5	8.8×10^1	6.0×10^{-3}	1.5×10^4	[19]
	$Fe(CN)_6^{4-}$	7.5	7.8×10^1	1.1×10^{-2}	7.1×10^3	[19]
Ascorbate oxidase	Ascorbate	6.0	7.5×10^3	2.0×10^{-4}	3.8×10^7	[19]
(*Cucurbita pepo*	$Fe(CN)_6^{4-}$	6.0	2.5×10^1	3.0×10^{-3}	8.3×10^3	[19]
medullosa)	Ascorbate	7.5	unaffected by pH			[19]
	$Fe(CN)_6^{4-}$	7.5	1.7	5.0×10^{-3}	8.5×10^2	[19]

Table 7. Anaerobic Reduction of Laccase and Ascorbate Oxidase

Enzyme	*Kind of Type-1 Copper Reduction and Rate (~610 nm)*					*Kind of Type-2 and Type-3 Copper Reduction and Rate (~330 nm and/or Reappearance of 610 nm Band)*						
	Bimolecular, Second Order											
	Substrate	*pH*	K_{init} $M^{-1}s^{-1}$	K_2 $M^{-1}s^{-1}$	*Ref.*	*Substrate*	*pH*	K_{init} $M^{-1}s^{-1}$	K_1 s^{-1}	K_2 s^{-1}	K_3 s^{-1}	*Ref.*
Fungal laccase (*Polyporus versicolor*)	Hydroquinone	5.5	1.7×10^7	—	[100]	Hydroquinone	5.5	—	—	—	1.0	[100]
	$Fe(CN)_6^{4-}$	5.5	1.5×10^6	—	[100]	Ascorbate	5.5	—	—	—	1.0	[100]
						Inactive form already at pH 5.5, OH^- binds to type-2 copper						
Tree laccase (*Rhus vernicifera*)	Hydroquinone	6.5	3.6×10^2	1.6×10^2	[99]	Hydroquinone	6.5	4.0×10^2	—	—	—	[99]
	Hydroquinone	7.4	1.6×10^3	8.0×10^2	[99]	Hydroquinone	7.4	1.5×10^3	—	—	0.4	[99]
	Hydroquinone	8.5	1.8×10^4	4.0×10^3	[99]	Hydroquinone	8.5	8.0×10^3	—	—	0.25	[99]
	$ArNO_2^{\bullet-}$	7.4	$< 2\times10^6$	—	[101]	$CO_2^{\bullet-}$	7.0	—	—	—	1.0	[107]
						Inactive form at high pH, OH^- binds to type-2 copper						
Ascorbate oxidase (*Cucurbita pepo medullosa*)	Reductate	6.1	2.2×10^4	1.2×10^4	[18]	Reductate	6.1	—	100	—	—	[18]
	Reductate	7.0	1.7×10^4	1.0×10^4	[18]	Reductate	7.0	—	100	—	—	[18]
	Reductate	7.8	1.4×10^4	0.8×10^4	[18]	Reductate	7.8	—	100	—	—	[18]
	Ascorbate	much faster, reaction within the deadtime of instrument			[18]	Rapid-freeze-quench EPR experiments indicate, that type-2 copper is reduced more slowly						[18]
	$ArNO_2^{\bullet-}$	7.5	$1\text{–}3\times10^7$	—	[102]							
	Lumiflavin	7.0	2.7×10^7	—	[103]							
		7.0	3.8×10^7	—	[94]	Lumiflavin	7.0	—	97	—	2.4	[104]
	$CO_2^{\bullet-}$	7.0	1.2×10^9	—	[105]	$CO_2^{\bullet-}$	7.0	—	201	20	2.3	[105]
		7.0	1.1×10^9	—	[104]		7.0	—	120	—	2.0	[104]
	$MV^{\bullet+}$	7.0	2.6×10^7	—	[104]	$MV^{\bullet+}$	7.0	—	127	—	2.3	[104]
	Deazaflavin	7.0	1.5×10^8	—	[104]	Deazaflavin	7.0	—	121	—	2.5	[104]

oxidase with reductate as substrate are about 10^4 M^{-1} s^{-1} and exhibit no remarkable pH dependence [18]. Reaction of one electron-reduced nitroaromatics ($ArNO_2^{\bullet-}$) gives rate constants of < 2×10^6 M^{-1} s^{-1} for tree laccase [101] and 1–3 × 10^7 M^{-1} s^{-1} [102] for ascorbate oxidase. Ascorbate as substrate for ascorbate oxidase should be at least as fast as the given rate constants since the reaction is completed within the deadtime of the stopped-flow instrument (3 to 5 ms [18]). Recent studies on ascorbate oxidase using laser-flash photolysis [103] or pulse radiolysis [104, 105] reveal rate constants for the radicals of lumiflavin, deazaflavin, CO_2^- and $MV^{\bullet+}$ between 10^7 and 10^9 M^{-1} s^{-1}. The pH dependence was not determined. The data suggests that the type-1 copper is the primary electron receptor from the reducing substrate. In the case of ascorbate oxidase with ascorbate as substrate, it has been shown that one electron is transferred from ascorbate and the generated semidehydroascorbate spontaneously dismutates in solution [106].

The anaerobic reduction of the trinuclear copper species monitored by the bleaching of the 330-nm band, the reappearance of the 600-nm band, and the disappearance of the type-2 EPR-signal appear to be multiphasic processes in most cases. For fungal laccase, the process is monophasic with a unimolecular rate constant of 1.0 s^{-1} [100]. Tree laccase with hydroquinone as substrate [99] displays a pH dependence with a bimolecular reaction ($K_{init} = 4.0 \times 10^2$ $M^{-1}s^{-1}$) at pH 6.5 and a biphasic behavior at pH 7.4 and 8.5. The first reaction is a bimolecular second-order reaction with increasing rate constants at ascending pH. The second one is a unimolecular reaction with rate constants of about 0.3 s^{-1}. Pulse radiolysis studies on tree laccase with CO_2^- radical as a substrate at pH 7.0 reported a monophasic reaction (K = 1.0 s^{-1}) [107]. The disappearance of the type-2 EPR-signal has been determined for tree laccase by rapid freeze quench experiments [108]. It is as fast as the reduction of the type-1 copper. Tree laccase is inactive at high pH. As the type-2 EPR signal is altered it has been concluded that an OH^- binds to the type-2 copper, causing the inactivation of the enzyme.

The anaerobic reduction of the trinuclear copper center for ascorbate oxidase with different substrates presents a distinct picture. The reaction with reductate is monophasic with a unimolecular rate constant of 100 s^{-1} [18], independent of pH. Rapid freeze quench EPR experiments indicate that the type-2 EPR-signal vanishes more slowly [18]. The pulse radiolysis studies of the radicals of lumiflavin, deazaflavin, CO_2^-, and $MV^{\bullet+}$ at pH 7.0 [104, 105] showed a biphasic behavior with an initial, faster reaction (K = 97–127 s^{-1}) and a final, slower reaction (K ~2 s^{-1}) [104]. Different results have been obtained by Farver and Pecht [105] with $CO_2^{\bullet-}$ as a substrate. They found a triphasic reaction with unimolecular rate constants $K_1 = 201\ s^{-1}$, $K_2 = 20\ s^{-1}$, and $K_3 = 2.3\ s^{-1}$. The final constant is twice that in a study by Kyritsis et al. [104], whereas the third constant is identical. The second constant was not observed in the study.

The unimolecular reduction of the trinuclear copper center is caused by the intramolecular electron transport from the type-1 copper to the trinuclear copper species. The turnover numbers represented by the k_{cat} values (see Table 6) are one

Table 8. Reoxidation of Reduced Enzyme

Enzyme	*Oxidant*	*Signal of Intermediate*	*Putative Intermediate*	*Rate of Reoxidation*	*Rate of Decay of Intermediate*	*Ref.*
Tree laccase (*Polyporus versicolor*)	O_2	Absorption maximum at 360 nm, new EPR signal at low temperature (10K)	$O^{\bullet -}$ radical	615 nm band 5×10^6 (M^{-1} s^{-1}), 330 nm band 5×10^6 (M^{-1} s^{-1}), 360 nm absorption maximum 5×10^6 (M^{-1} s^{-1}), type-2 copper EPR signal, halftime 20s	Absorption maximum at 360 nm, halftime 20s, first order reaction	[109]
T1Hg-derivative	O_2	Absorption bands at 340 and 470 nm, CD changes in 500–100 nm range, disappearance of MCD signal at 730 nm	HO_2^-	340 nm and 470 nm band$^{\$}$ 2×10^6 (M^{-1} s^{-1})	Formation is irreversible	[113]
	$H_2O_2^*$	CD changes in 250–300 nm range	O_2^{2-}	CD changes, formation in < 10 minutes	New CD spectrum stable > 6 h	[111]
Ascorbate oxidase (*Cucurbita pepo medullosa*)	O_2	No data	No data	610 nm band 5×10^6 (M^{-1} s^{-1}) 330 nm band 5×10^6 (M^{-1} s^{-1})	No data	[110]
	H_2O_2	Absorption band at 350 nm, pH 7.6	O_2^{2-}	350 nm band formation in several minutes	350 nm decay over several hours	[112]

Notes: * addition to oxidized enzyme in equimolar amount

$ measured at 3°C

to two orders of magnitudes greater than the observed intramolecular electron transfer rates. They should be at least as large as the turnover numbers. These low rates seem to imply that the enzymes are in an inactive form under the conditions of the anaerobic reduction experiments. It will be shown later that there is experimental evidence for this assumption.

The reoxidation studies on laccase and ascorbate oxidase are listed in Table 8. The reoxidation of the type-1 copper and of the trinuclear copper site occurs at a rate of 5×10^6 M^{-1} s^{-1} both for tree laccase [109] and for ascorbate oxidase [110]. During reoxidation with H_2O_2, an O_2^{2-}-intermediate is formed in several minutes, which is documented for tree laccase by changes in the CD-spectrum [111] and for ascorbate oxidase in the formation of an absorption band at 350 nm [112]. The intermediates are stable for several hours in both enzymes. An interesting intermediate was found during reoxidation of fully reduced tree laccase by dioxygen [109]. This intermediate caused the rapid formation of an absorption maximum at 330 nm and also affected the type-2 copper EPR signal. It decayed in a first-order reaction with a half-time of 20 s. A new EPR signal at low temperature (10 K) due to this radical could be detected. This radical has been described as an $O^{\bullet-}$ radical. Recently, Solomon and co-workers [78, 113, 114] have identified and spectroscopically characterized an oxygen intermediate during the reaction of either fully reduced native tree laccase or T1Hg-laccase with dioxygen. The intermediate has been described as hydroperoxide binding as 1,1-μ between either CU2 and CU4 or CU3 and CU4.

From these data, it follows that the dioxygen binds to the trinuclear copper site. This species may store three electrons and transfer them to the bound dioxygen followed by a final one-electron transfer.

From the smoothed van der Waals surface (Conolly surface) calculated for the X-ray structure model of ascorbate oxidase, a binding pocket for the reducing substrate near the type-1 copper and two channels providing access to the trinuclear copper species could be identified [59]. The binding pocket for the reducing substrate is complementary to an ascorbate molecule. The channel giving access to copper atoms CU2 and CU3 of the trinuclear site is much broader than the channel leading to CU4. The former is the main channel for the entrance of the dioxygen and protons and the release of the water molecules. Figures and a detailed description of the pocket and the channels are given in Messerschmidt et al. [59].

8. FUNCTIONAL DERIVATIVES

8.1 General Remarks

The blue oxidases may be inhibited by a number of different substances. For ceruloplasmin, there are systematic studies showing that two groups of more specific inhibitors exist: (1) halides and inorganic anions, and (2) carboxylate ions [115–118]. The rank of inhibition effects of anions with ceruloplasmin is:

$CN^- > N_3^- > OCN^- > SCN^- > SeCN^- > F^- > I^- \simeq NO_3^- \simeq Cl^- > Br^- >$ (ClO_4^-, tetraborate, borate, phosphate, sulfate, and cacodylate) [116]. Inhibition by CN^-, N_3^-, and F^- has been investigated predominantly with laccase [108, 119, 120] and ascorbate oxidase [121]. The order of inhibitory action is the same as that observed for ceruloplasmin. The inhibition by these anions appears to be understood quite satisfactory. Strong binding to the spectroscopic type-2 copper is indicated by a perturbation of the EPR signal of this site.

Laccases also become inhibited at higher pH values. Tree laccase is inhibited above pH 6.5. It appears that at pH 7.4 ~ 50% of the enzyme molecules are inhibited due to the binding of OH^- to the type-2 copper. The nature of this OH^--binding to the resting form of ascorbate oxidase will be demonstrated below.

The reaction of nitric oxide with laccase [61] and ascorbate oxidase [122] has been studied as well. Nitric oxide fully reduces fungal and tree laccase when it is added to the oxidized enzyme under anaerobic conditions. In addition the binding of one NO molecule to laccase can be detected. This is characterized by a new EPR signal and has been described as coordinated with the type-2 copper [61]. Only the reduction of the type-1 copper has been observed when NO has been added to ascorbate oxidase under anaerobic conditions.

Other functional derivatives have been prepared. Two of them deserve special mention, namely the type-2 copper depleted (T2D) enzyme and a derivative where the type-1 copper was replaced by mercury (T1Hg). The T2D-derivative will be dealt with in a special paragraph. The T1Hg-form was originally prepared by McMillin and co-workers [123] from tree laccase. This derivative lacks the type-1 copper EPR signal and makes possible the recording of the type-2 EPR signal alone. It has been used to study the reaction of dioxygen with the reduced derivative [113]. As already mentioned, the formation of an oxygen intermediate could be observed. This intermediate was bound irreversibly to the trinuclear copper site due to the abortion of further reduction of the dioxygen caused by the lack of the type-1 copper redox center. The preparation of the apo-enzyme and reconstitution are possible for all three blue oxidases (see for laccase [10], for ascorbate oxidase [19], and for ceruloplasmin [26]).

X-ray crystal structures of four functional derivatives of ascorbate oxidase were determined [124, 125]. The results of these investigations and implications for the catalytic mechanism of the blue oxidases will be outlined in the next section.

8.2 X-Ray Structure of the Type-2 Depleted (T2D) Form of Ascorbate Oxidase

It has been demonstrated by several groups that copper can be selectively removed from the blue oxidases, causing a disappearance of the type-2 EPR signal. Several methods have been described for laccase from *Polyporus versicolor* [126, 127] and from the Japanese lacquer tree *Rhus vernicifera* [128–131], and for ascorbate oxidase from zucchini [131–133]. All procedures except one involve working under reducing conditions with metal chelating reagents, such as EDTA, dimethyl glyoxime, bathocuproine disulfonate, or nitrilotriacetate. Reaction of *N,N*-diethyldithiocarbamate with ascorbate oxidase under aerobic conditions in solution gave the type-2-depleted enzyme [133]. Many experiments were carried out on T2D multicopper oxidases in the past and to interpret these experiments it is important to know the actual occupation of the copper sites in the depleted enzyme. An X-ray structure analysis of the depleted enzyme of ascorbate oxidase [124] provided new information on this point.

Crystals of native oxidized ascorbate oxidase were anaerobically dialyzed in microcells against 50 mM sodium phosphate buffer, pH 5.2, containing 25% (v/v) methylpentanediol (MPD), 1 mM EDTA, 2 mM dimethylglyoxime (DMG), and 5 mM ferrocyanide for 7 and 14 h. Thereafter, crystals were brought back to the aerobic 25% MPD solution, buffered with 50 mM sodium phosphate, pH 5.5. This procedure is based on Avigliano et al.'s [132] method of preparing T2D ascorbate oxidase in solution, and was modified by Merli et al. [134] for use with ascorbate oxidase crystals. The 2.5 Å resolution X-ray structure analysis by difference-Fourier techniques and crystallographic refinement shows that about 1.3 copper ions per ascorbate oxidase monomer are removed. The copper is lost from all three copper sites of the trinuclear copper species, whereby the EPR-active type-2 copper is the most depleted (see Figure 8). Type-1 copper is not affected. The EPR spectra from polycrystalline samples of the respective native and T2D ascorbate oxidase were recorded. The native spectrum exhibits the type-1 and type-2 EPR signals in a ratio of about 1:1, as expected from the crystal structure. The T2D spectrum reveals the characteristic resonances of the type-1 copper center, also observed for T2D ascorbate oxidase in frozen solution, and the complete disappearance of the spectroscopic type-2 copper. This observation indicates preferential formation of a

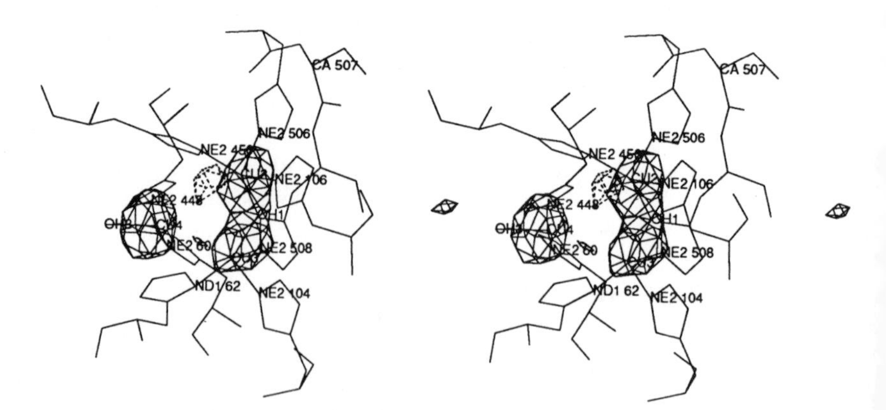

Figure 8. Averaged FO_{T2D}–FC_{T2D}-difference electron density map plus atomic model around the trinuclear copper site. Contour levels: –18.0 solid line, 18.0 dashed line, magnitudes of hole less than –35.0.

copper-depleted form with the hole equally distributed over all three copper sites. Each of these copper-depleted species may represent an anti-ferromagnetically coupled copper pair that is EPR-silent and could explain the disappearance of the type-2 EPR signal.

8.3 X-ray Structure of the Reduced Form of Ascorbate Oxidase

Crystals of the reduced form (REDU) of ascorbate oxidase [125] had to be prepared and mounted in the glass capillary in a glove box that was flushed with argon gas and operated with a slight overpressure of argon. The degassed buffer solution was stored in the glove box. Dithionite was added to the buffer solution to a concentration of 10 mM. The siliconized X-ray capillary was washed with the buffer. The crystals were soaked in the reducing buffer for 30 min. After 15 min, the crystals lost their blue color. They were mounted in the X-ray capillary and carefully sealed with wax that had been degassed in the desiccator. Crystals mounted this way remained colorless and reduced over weeks.

The 2.2 Å resolution X-ray structure analysis by difference-Fourier techniques and crystallographic refinement delivered the following results [125]. The geometry at the type-1 copper remains much the same compared to the oxidized form. The mean copper–ligand bond lengths of both subunits are increased by 0.04 Å in average, which is insignificant but may indicate a trend. Similar results have been obtained for the reduced forms of poplar plastocyanin at pH 7.8 [135], azurin from *Alcaligenes denitrificans* [136], and azurin from *Pseudomonas aeruginosa* [137]. In reduced poplar plastocyanin at pH 7.8 a lengthening of the two Cu–N(His) bonds by about 0.1 Å is observed. In reduced azurin, pH 6.0, from *Alcaligenes denitrificans*, the distances from copper to the axial methionine and the carbonyl oxygen each increase by about 0.1 Å. The same shifts are found in the refined structures of reduced azurin from *Pseudomonas aeruginosa* determined at pH values of 5.5 and 9.0 [137]. The estimated accuracy of the copper–ligand bond lengths in the high-resolution structures of the above mentioned small blue copper proteins is about 0.05 Å. The type-1 copper sites in the small blue copper proteins as well as in ascorbate oxidase require little reorganization in the redox process.

A schematic drawing of the reduced form of ascorbate oxidase is shown in Figure 9. The structural changes are considerable at the trinuclear copper site. Thus, on reduction, the bridging oxygen ligand OH1 is released and the two coppers, CU2 and CU3, move towards their respective histidines and become three coordinate, a preferred stereochemistry for Cu(I). The copper–copper distances increase from an average of 3.7 Å to 5.1 Å for CU2–CU3, 4.4 Å for CU2–CU4, and 4.1 Å for CU3–CU4. The mean values of the copper–ligand distances of the trinuclear copper site are comparable to those of native oxidized ascorbate oxidase and binuclear copper model compounds with nitrogen and copper ligands [79, 80]. CU4 remains virtually unchanged between reduced and oxidized forms. Coordinatively unsaturated copper(I) complexes are known from the literature. Linear two-coordinated

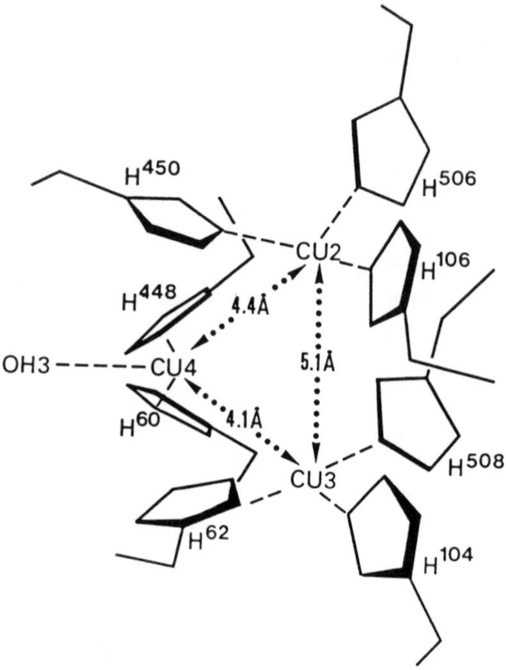

Figure 9. Schematic drawing of the reduced form of ascorbate oxidase around the trinuclear copper site. The included copper–copper distances are the mean values between both subunits.

[138] and T-shaped three-coordinated [139] copper(I) compounds have been reported. The copper nitrogen distances for both linearly arranged nitrogen ligands are about 1.9 Å, about 0.1 Å shorter than copper–nitrogen bond lengths in copper(II) complexes. In T-shaped copper(I) complexes, the bond length of the third ligand is increased. Copper ion CU4 is in a T-shape threefold coordination not unusual for copper(I) compounds. The structure of the fully reduced trinuclear copper site is quite different therefore from that of the fully oxidized resting form of enzyme, and its implications for the enzymatic mechanism will be discussed below.

8.4 X-ray Structure of the Peroxide Form of Ascorbate Oxidase

Native crystals of ascorbate oxidase were soaked in harvesting buffer solution (50 mM sodium phosphate, 25% MPD, pH 5.5) containing different H_2O_2 concentrations [125]. H_2O_2 concentrations greater than 20 mM caused the crystals to crack, to become brownish, and finally to decompose within several hours depending on the concentration of H_2O_2 used. At 10 mM H_2O_2, the crystals remained blue and did not crack at 1-1/2 to 2 days. A native crystal was therefore soaked for 2 h in 10

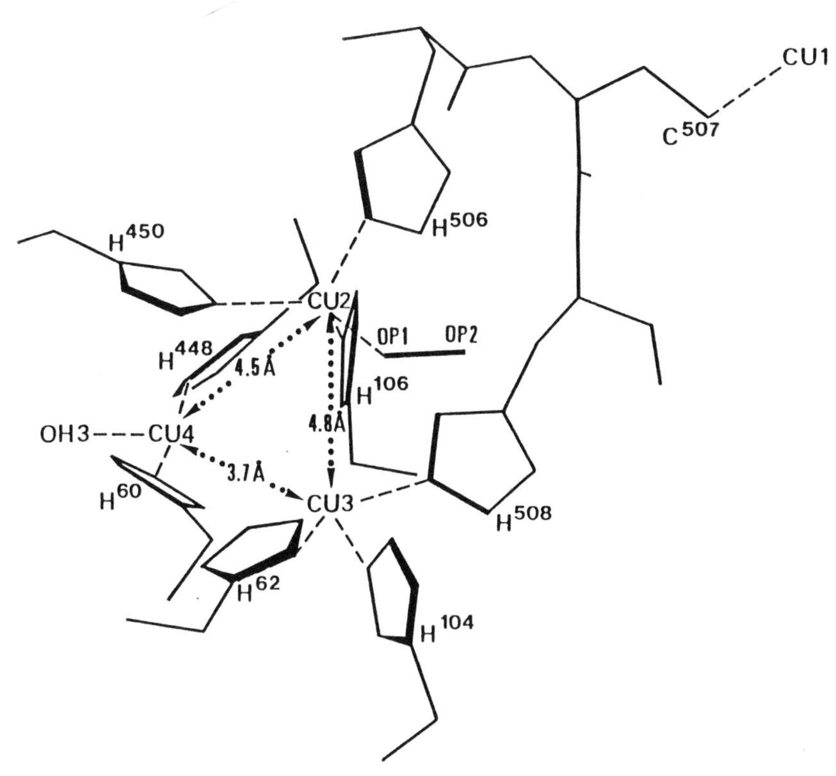

Figure 10. Schematic drawing of the peroxide form of ascorbate oxidase around the trinuclear copper site. The included copper–copper distances are the mean values between both subunits.

mM H_2O_2 containing harvesting buffer solution, mounted in the X-ray capillary, and immediately used for the X-ray intensity measurements.

The 2.6 Å resolution X-ray structure analysis by difference-Fourier techniques and crystallographic refinement reveal the following picture illustrated in Figure 10 [125]. The geometry at the type-1 copper site is not changed compared to the oxidized form. The copper–ligand bond distances averaged for both subunits show no significant deviations from those of the oxidized form. As in the reduced form, the structural changes are remarkable at the trinuclear copper site. The bridging oxygen ligand OH1 is absent, the peroxide binds terminally to the copper atom CU2 as hydroperoxide, and the copper–copper distances increase from an average of 3.7 to 4.8 Å for CU2–CU3 and 4.5 Å for CU2–CU4. The distance CU3–CU4 remains at 3.7 Å. The mean values of the copper–ligand distances of the trinuclear copper site are again comparable to native oxidized ascorbate oxidase and corresponding copper model compounds.

Copper ion CU3 is threefold coordinated as in the reduced form but the coordination by the ligating N-atoms of the corresponding histidines is not exactly trigonal–planar and the CU3 atom is at the apex of a flat trigonal pyramid. The coordination sphere around CU4 is not affected and similar in all three forms. Copper atom CU2 is fourfold coordinated to the NE2 atoms of the three histidines, as in the oxidized form, and by one oxygen atom of the terminally bound peroxide molecule in a distorted tetrahedral geometry. Its distance to CU3 increases from 4.8 Å in the oxidized peroxide derivative to 5.1 Å in the fully reduced enzyme. The bound peroxide molecule is directly accessible from the solvent through a channel leading from the surface of the protein to the CU2–CU3 copper pair. This channel has already been described earlier and its possible role as dioxygen transfer channel has been discussed. An interesting feature is the close proximity of the imidazole ring of histidine 506 to the peroxide molecule. Histidine 506 is part of one possible electron transfer pathway from the type-1 copper to the trinuclear copper site and could indicate a direct electron pathway from CU1 to dioxygen. It may also help to stabilize important intermediate states in the reduction of dioxygen.

The strong positive peaks at CU2 in both FO_{NATI}–FO_{PEOX} and FO_{REDU}–FO_{PEOX} electron density maps could not be explained by a shift of CU2 alone. Occupancies of the copper atoms, the oxygen atom OH3, and the peroxide molecule were refined. Type-1 copper CU1 is almost not affected. Copper atoms CU3 and CU4 are only partly removed but copper atom CU2 is about 50% depleted. The oxygen ligands exhibit full occupancy. The treatment of crystals of ascorbate oxidase with hydrogen peroxide generates not only a well-defined peroxide binding but also a preferential depletion of copper atom position CU2. In the copper-depleted molecules the ligating histidine 106 adopts an alternative side chain conformation detected in the 2FO–FC-map, calculated with the final peroxide derivative model coordinates. This map shows that histidine 106 moves away when copper atom CU2 is removed and opens the trinuclear site even more. From the T2D crystal structure of ascorbate oxidase, it is apparent that copper from all three metal binding sites of the trinuclear copper species is removed in different amounts. The movement of the histidine 106 side chain could explain how this process is accomplished.

Copper depletion may also cause instability of the protein against hydrogen peroxide. Reaction of hydrogen peroxide with excess ascorbate oxidase in solution leads to a rapid degradation of the enzyme [50]. This can be monitored in the UV/visible PEOX–NATI difference spectrum by a negative band at 610 nm and a positive band at 305 nm. Adding four equivalents of hydrogen peroxide per monomer ascorbate oxidase does not lead to enzyme degradation and gives a positive peak at 305 nm indicative for the peroxide binding. Unfortunately, it was not possible to monitor a UV/visible spectrum of dissolved crystals after X-ray data collection because of the dissociation of the bound peroxide in solution.

The reaction of dioxygen with laccase or ascorbate oxidase has been reviewed in Section 7 and in Messerschmidt et al. [59], where the possible binding modes of dioxygen to binuclear and trinuclear copper centers are discussed. A novel mode of dioxygen binding to a binuclear copper complex was found in a compound synthesized by Kitajimia et al. [140]. The complex contains peroxide in the μ-η^2:η^2 mode, i.e., side-on between the two copper(II) ions. Such binding mode of dioxygen has been detected in the crystal structure of the oxidized form of *Limulus polyphemus* subunit II hemocyanin [141]. However, the binding mode of dioxygen to the trinuclear copper site in the blue oxidases appears to be different, as can be seen from the X-ray structure of the peroxide derivative of ascorbate oxidase.

During its reaction with fully reduced laccase, dioxygen binds to the trinuclear copper species and three electrons are very rapidly transferred to it, resulting in the formation of an "oxygen intermediate" with a characteristic optical absorption near 360 nm [109, 142] and a broad low-temperature EPR signal near g = 1.7 [143, 144]. The type-1 copper is concomitantly reoxidized when the low-temperature EPR signal is formed. The oxygen intermediate decays very slowly ($t_{1/2}$ ~ 1 to 15 s), correlated with the appearance of the type-2 EPR signal [145]. Solomon and co-workers [78, 113, 114] have identified and spectroscopically characterized an oxygen intermediate during the reaction of either fully reduced native tree laccase or T1Hg-laccase with dioxygen. They concluded from their spectroscopic data that the intermediate binds as 1,1-μ-hydroperoxide between either CU2 and CU4 or CU3 and CU4. Since it is unlikely that the dioxygen migrates or rearranges coordination during reduction, Messerschmidt et al. [125] proposed that the binding site and mode determined for the peroxide derivative of ascorbate oxidase is representative for all reaction intermediates of dioxygen and, by homology arguments, is in all blue oxidases. The relevance of this binding mode for the catalytic mechanism will be discussed later.

8.5 X-ray Structure of the Azide Form of Ascorbate Oxidase

The azide derivative was obtained by soaking the native crystals in the harvesting buffer solution containing 50 mM sodium azide for 24 h [125]. Binding of azide was indicated by a change of color of the crystal from blue to brownish. After X-ray intensity data collection, the crystals were dissolved in a solution containing 50 mM azide, 5 mM phosphate buffer, pH 5.5, and 8% MPD and the UV/visible spectrum was recorded at room temperature.

The results of the 2.3 Å resolution X-ray structure analysis by difference-Fourier techniques and crystallographic refinement are depicted in Figure 11 [125]. The geometry at the type-1 copper site is not changed compared with that of the native form. The copper–ligand bond distances averaged for both subunits show no significant deviations from those of the native form. Again, the structural changes are large at the trinuclear copper site. The bridging oxygen ligand OH1 and water molecule 145 have been removed, CU2 moves towards the ligating histidines, and two azide molecules bind terminally to it. The copper–copper distances increase from an average of 3.7 to 5.1 Å for CU2–CU3 and 4.6 Å for CU2–CU4. The distance CU3–CU4 is decreased to 3.6 Å. The mean values of the copper–ligand

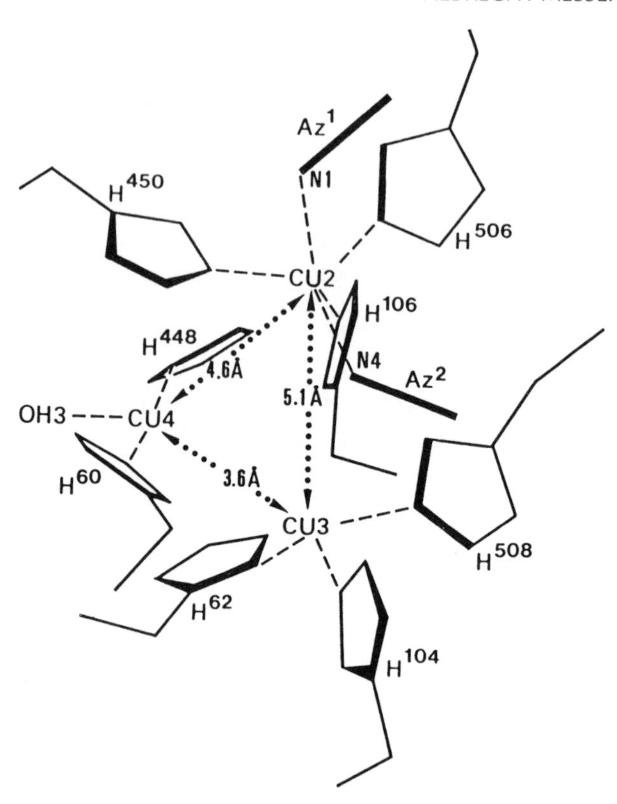

Figure 11. Schematic drawing of the azide form of ascorbate oxidase around the trinuclear copper site. The included copper–copper distances are the mean values between both subunits.

distances of the trinuclear copper site are again comparable to those of native ascorbate oxidase and corresponding copper model compounds.

The coordination of CU3 resembles that in the peroxide form. The threefold coordination by histidines is a very flat trigonal pyramid. The coordination sphere around CU4 is not affected. CU2 is fivefold coordinated to the NE2 atoms of the three histidines, as in the reduced form, and to the two azide molecules. The two azide molecules are terminally bound at the apexes of a trigonal bipyramid. Both azide molecules bind to the copper atom CU2, which is well accessible from the broad channel leading from the surface of the protein to the CU2–CU3 copper pair. It is not unexpected that the second azide molecule (az^2 in Figure 11) binds similarly to the peroxide molecule, as azide is regarded as a dioxygen analogue. There is no bound azide molecule bridging either CU2 with CU4 or CU3 with CU4.

The binding of azide to laccase as well as to ascorbate oxidase has been studied extensively by Solomon and co-workers [64, 65, 78, 146], and by Marchesini and associates [147, 148] by spectroscopic techniques. The derived spectroscopic models involve the binding of two azide molecules for laccase and three azide molecules for ascorbate oxidase with different affinities. Since the binding of the high-affinity azide molecules seems to generate spectral features related to type-2 and type-3 copper, the spectroscopic data were interpreted as the binding of at least one azide molecule as a 1,3-μ-bridge between the type-3 copper ions and the type-2 copper ion.

After the X-ray data collection of the azide derivative, the crystal was dissolved in azide-containing buffer and a UV/visible spectrum was recorded to check the spectral properties of the sample [125]. The spectrum was characterized by a broad increase of absorption in the 400–500 nm region and an intense absorption maximum at 425 nm, very similar to the results of Casella et al. [147].

There are many structural studies of copper coordination compounds with azide ligands, mainly of mononuclear and binuclear copper complexes but a few also of trinuclear copper complexes. A comprehensive review on copper coordination chemistry has been written by Hathaway [149]. Azide binds only terminally to mononuclear systems. Fivefold coordination of nitrogen ligands including azide to Cu^{2+} is frequently found arranged as a trigonal bipyramid. In binuclear systems azide may bind terminally as 1,1-μ or bridging as 1,3-μ. Similarly two azides may bind di-1,1-μ or di-1,3-μ. Interaction with all three copper ions of a trinuclear complex may be either terminal as 1,1,1-μ or bridging as 1,1,3-μ. In the X-ray crystal structure of ascorbate oxidase two azide molecules bind terminally to the type-3 CU2. Azide binding in ascorbate oxidase therefore resembles the binding of azide to an isolated copper ion. In fact there is little interaction of CU2 with CU3 and CU4, which are 5.1 Å and 4.6 Å away, respectively.

The coordination of copper ion CU4 in the native oxidized structure is of some interest. It has only three ligating atoms at close distances, forming a T-shape coordination that is known for Cu(I) complexes (see discussion of reduced form). However, the ligand field is completed if we take into account the π-electron systems of the imidazole rings of histidines 62 and 450 (see Figure 6). A ligand field with tetragonal–pyramidal symmetry around CU4 is then formed. The shortest distances of CU4 are to CD2 450 with 3.4 Å and to CG 62 with 3.6 Å. These distances are too long for strong copper–π-electron interactions but the histidines will contribute to the CU4 ligand field.

9. THE CATALYTIC MECHANISM

Catalytic reaction schemes for laccase and ceruloplasmin have been formulated on the basis of the mechanistic studies and the state of characterization of the copper redox centers at this time. They are outlined in the reviews on laccase by Reinhammar [10] and on ceruloplasmin by Ryden [26]. The degree of correctness of these reaction schemes is rather limited due to the fact that structure and spatial arrangement of the copper centers were unknown at this time.

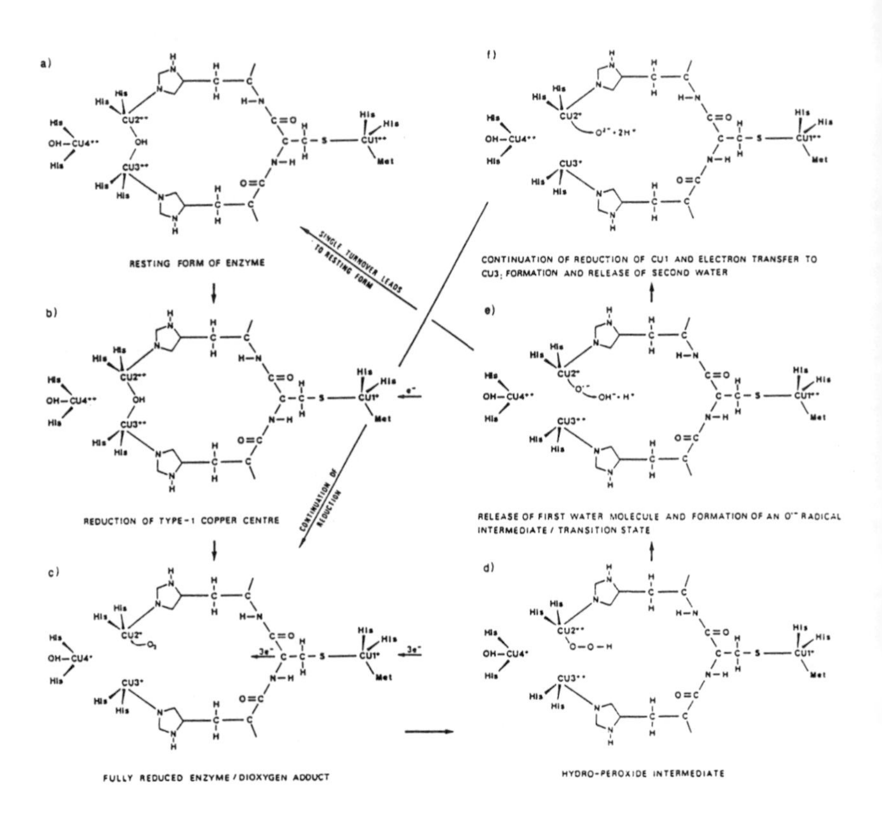

Figure 12. Proposal for the catalytic mechanism of ascorbate oxidase.

A tentative catalytic mechanism of ascobate oxidase has been proposed based on the refined X-ray structure and on spectroscopic and mechanistic studies of ascorbate oxidase and the related laccase. The results of these studies were discussed in detail [59]. The X-ray structure determinations of the fully reduced and peroxide derivatives define two important intermediate states during the catalytic cycle. A proposal for the catalytic mechanism incorporating this new information is given in [125] and presented in Figure 12. This scheme should be valid in principle also for laccase due to the close similarities of both blue oxidases.

The catalytic cycle starts from the resting form (Figure 12a) in which all four copper ions are oxidized and CU2 and CU3 are bridged by an OH^- ligand. CU2 and CU3 are most likely the spin-coupled type-3 pair of copper and CU2 is the type-2 copper. The first step is the reduction of the type-1 copper CU1 by the reducing substrate in a one-electron transfer step (Figure 12b). The electrons are transferred through the protein to either CU2 or CU3. Electron transfer may be through-bond, through-space, or a combination of both. A branched through-bond pathway is available, leading to CU2 with nine bonds (including a hydrogen bond) and to CU3 with 11 bonds, respectively. The fully reduced enzyme requires four electrons to be transferred (Figure 12c). Its structure has been described in a previous section. The hydroxyl bridge between the copper pair has been released and the distance of copper atom CU2 to CU3 has been increased to about 5.1 Å. Considerable reorganization energy will be necessary to reach this state from the resting form of enzyme. At this stage, dioxygen may bind to the enzyme at CU2, probably in the manner shown in the peroxide derivative described in the appropriate section. A transfer of two electrons from the copper pair to dioxygen leads to the formation of a hydroperoxide intermediate (Figure 12d). A third electron may be transferred from CU4 to the hydroperoxide intermediate, and a fourth electron from the type-1 copper to copper ion CU2. The O–O bond is broken at this stage and the first water molecule released (Figure 12e). An oxygen radical has been detected in laccase by EPR. The EPR spectrum indicates that the type-1 copper has been reoxidized and that the EPR signals of the oxygen radical intermediate and type-1 copper are present. The CU2 is in the reduced state, whereas the oxidized copper atoms CU3 and CU4 may be spin-coupled and EPR silent. The reduced CU2 may facilitate O–O bond breakage and release of water. The catalytic cycle is continued by a further reduction of the type-1 copper center by the reducing substrate. This electron may be transferred to CU3 of the copper pair via the 11-bond pathway. Now, the fourth electron may be transferred to the oxygen radical intermediate from copper atom CU2 and the second water molecule released (Figure 12f). In the case of only four electron equivalents, the reaction may lead to the resting form and the second water may remain bound as a bridging ligand between CU2 and CU3 concomitant with a substantial rearrangement within the trinuclear copper site and its coordinating ligands. If turnover is continued, this will not occur and the trinuclear copper site may maintain a structure very close to that found in the fully reduced form. Only minor rearrangements will take place at the trinuclear copper site during the catalytic cycle, a prerequisite for facile electron transfer reactions. The four protons required for the formation of the two water molecules from dioxygen may be supplied from bulk water through the dioxygen channel via the water molecules bound in the vicinity of CU2 and CU3.

10. ELECTRON TRANSFER PROCESSES

10.1 Electron Transfer to the Type-1 Copper Redox Center

As previously mentioned, the electron transfer from one-electron-reduced nitroaromates ($ArNO_2^{\cdot-}$), $CO_2^{\cdot-}$, methyl viologen$^{\cdot+}$, lumiflavin, or deazaflavin to the type-1 copper center (see Table 7) takes place in a bimolecular second-order reaction with rates compatible with or higher than the turnover number, with ascorbate as reducing substrate. The electron transfer from ascorbate to the type-1

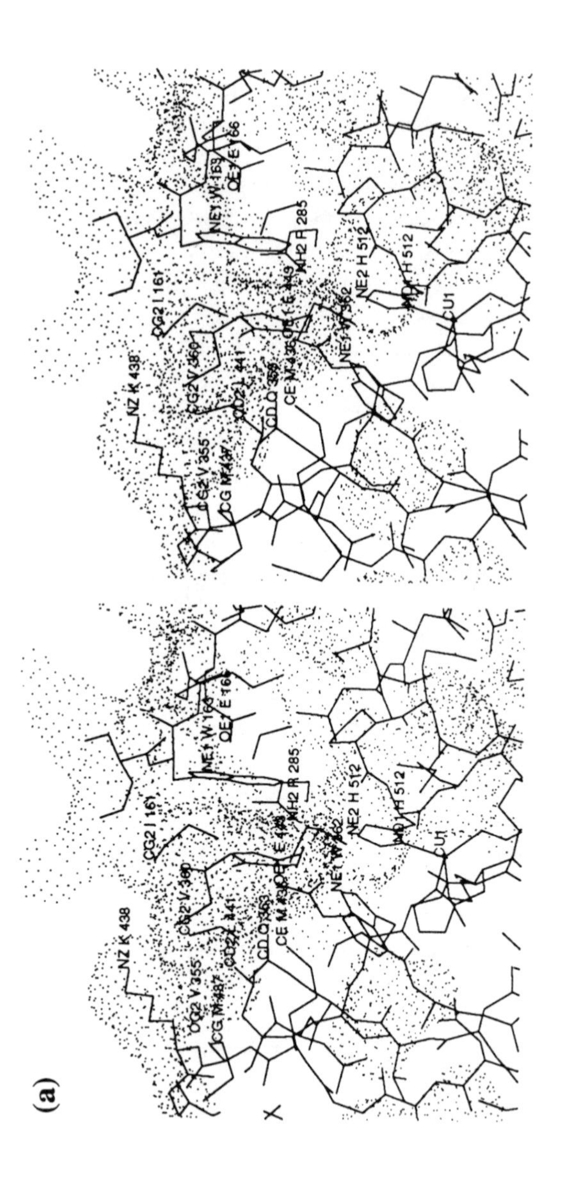
(a)
NZ K 438
CG2 I 161
CG2 V 360
CG2 V 355
CG M 437
CD2 L 441
CE M 435
NE1 W 163
NH2 R 285
NE2 H 512
ND1 H 512
CU1

(b)

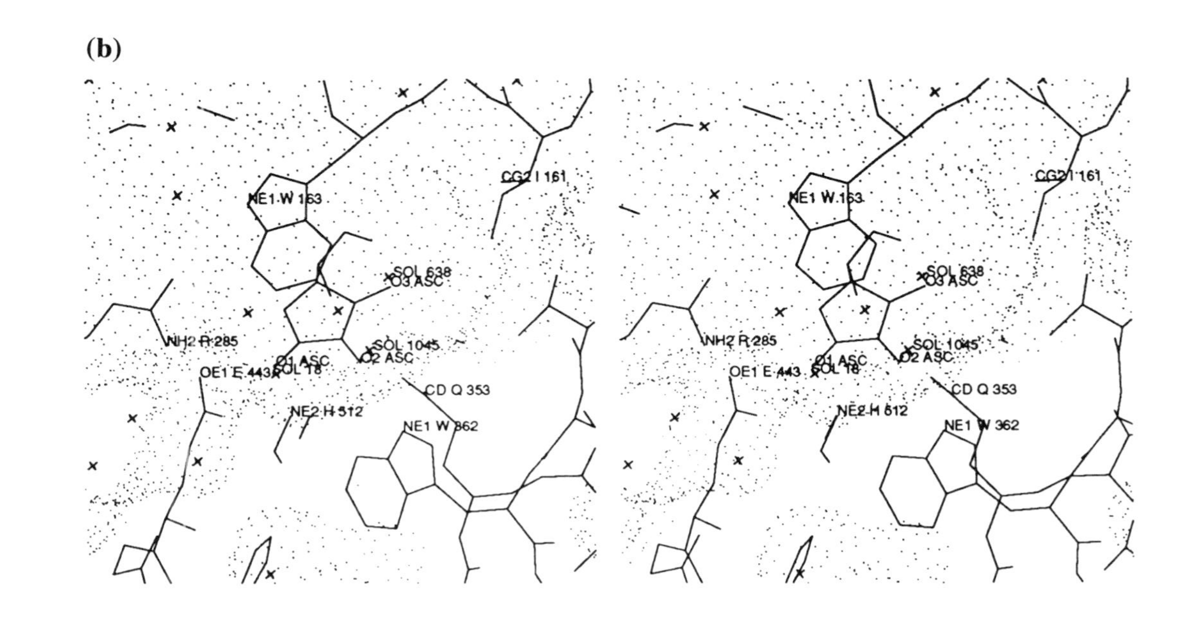

Figure 13. Stereo drawing of the binding site near the type-1 copper plus docked L-ascorbate. Atomic model plus Conolly dot surface. (**a**) Viewed perpendicular to the ND1 His512–CU1 K1 bond. (**b**) Viewed parallel to the ND1 His512–CU1 K1 bond.

copper center can be even faster and is completed within the deadtime of the stopped-flow instrument. It is therefore not the rate-limiting step in the overall reaction. The factors governing electron transfer may be described within the framework of Marcus electron transfer theory [150]. They are expressed in terms of driving force, distance of redox centers, reorganization energy, etc.

The driving force, calculated from the difference in the redox potentials (+344 mV for the type-1 copper in ascorbate oxidase (see Table 6); +295 mV for the couple ascorbate/ascorbate free radical [151], is 49 mV. In the proposed modeled encounter complex [59], there is a short distance of about 7 Å between the two redox centers (distance CU1 AK1-O1 ASC1 = 6.8 Å; distance CU1 AK1-O2 ASC1 = 7.5 Å) and an effective parallel arrangement of the rings with good overlap of the π-electron density systems facilitating a rapid electron transfer (see Figure 13).

It is well documented for small blue copper proteins, such as plastocyanin, that there are minimal structural changes on reduction and reoxidation [135]. The reorganization energy is, therefore, probably small.

10.2 Intramolecular Electron Transfer from the Type-1 Copper Center to the Trinuclear Copper Center

Long-distance intramolecular electron transfer can be described in the frame of the theory of Marcus [150]. In the formulation of Lieber et al. [152], the intramolecular electron rate constant k_{ET} can be written as,

$$k_{ET} = \nu_n \Gamma \exp(-\Delta G^*/RT) \tag{1}$$

where ν_n is the nuclear frequency factor, normally $10^{13}s^{-1}$, Γ is the electronic factor, and ΔG^* is the activation free energy for ET. The electronic factor, Γ, is at unity when the donor and acceptor are strongly coupled, but is much smaller at long donor–acceptor distances. In such cases, Γ is expected to fall off with distance (d),

$$\Gamma = \Gamma(d_0) \exp[-\beta(d - d_0)] \tag{2}$$

where d_0 is the van der Waals contact distance, normally taken as 3 Å [150], and β is the electron coupling factor, which decreases with increasing distance and depends on the nature of the intervening medium. ΔG^* depends on the reaction free energy, ΔG^o, and the nuclear reorganization energy λ [150] according to the equation:

$$\Delta G^* = (\Delta G^o + \lambda)^2 / 4\lambda \tag{3}$$

The ET rate is maximal when $-\Delta G^o = \lambda$.

The shortest distances, d, between the type-1 copper and the copper ions of the trinuclear copper site are 12.2 Å (CU1 K1-CU2 K3) and about 12.7 Å (CU1 K1-CU3 K3) (see Table 3). Figure 7 shows that His506–Cys507–His508 serves as a bridging ligand between the two redox centers, providing a bifurcated pathway for ET from the type-1 copper center to the trinuclear copper species. The difference in the redox potential of the type-1 copper center and the type-3 copper ions, the driving force, measured at 10°C, is $-\Delta G^{o} = 30$mV [18]. However, the binding of dioxygen to the partly reduced protein and the presence of reduction intermediates may affect this redox potential (a very slow equilibration was found between type-1 and type-3 copper ions in ascorbate oxidase in the absence of dioxygen [153]). For the reorganization energy, λ, and the electronic coupling factor, β, no estimates can be derived for ascorbate oxidase, but reasonable values for proteins are $\lambda = 1$ eV and $\beta = 1.2$ Å^{-1}, according to Gray and Malmström [154]. These values inserted into Eq. (1) yield $k_{ET} \sim 10^5$ s^{-1}. Changing β to 1.6 Å^{-1} gives $k_{ET} \sim 4 \times 10^3$ s^{-1}, a value closer to the observed turnover number of 8×10^3 s^{-1}.

It has been suggested that the electron transfer in proteins may not be designed for very fast intramolecular ET with the exception of light-induced ET in photosynthetic reaction centers [155]. They could even be designed to slow down these rapid rates, which might otherwise lead to biological "short circuits". Related to this point is the observation that maximal rates for intramolecular ET in organic donor–acceptor molecules with rigid spacers are significantly faster than those for Ru-labeled protein systems at similar distances [156].

In the case of laccase and ascorbate oxidase, the observed ET rates for the reduction of the type-3 copper ions (see Table 6) are lower than the observed turnover number. This can be explained only by the possibility that the enzymes are in a resting form under the experimental conditions. A considerable reorganization energy seems to be necessary to get to the reduced state of the type-3 copper ions (release of the bridging OH^- and movement of the copper CU2 and CU3). From these data it can not be decided what the rate-limiting step is in the catalytic cycle, either this intramolecular ET or the reaction of the dioxygen at the trinuclear copper site.

ET from the type-1 copper to the type-3 copper pair of the trinuclear copper site may be through-bond, through-space, or a combination of both. A through-bond pathway is available for both branches each with 11 bonds (see Figure 7). The alternative combined through-bond and through-space pathway from the type-1 copper CU1 K1 to CU2 K3 of the trinuclear center involves a transfer-from the SG atom of Cys-507 to the main-chain carbonyl of Cys-507 and through the hydrogen bond of this carbonyl to the ND1 atom of the His-506.

Kyritsis et al. [104] carried out a theoretical pathway analysis for ascorbate oxidase using an algorithm, which was developed by Beratan and Onuchic [157–159] to help identify the most favorable long-range electron transfer pathways in metalloproteins. According to this analysis, the most favored route consists of four covalent bonds and a hydrogen bond between main-chain carbonyl O of Cys-507 and ND1 atom of His-506 (hydrogen-bond length 2.9 Å) (see Figure 7), which gives an electronic coupling ε^2 value of 2.3×10^{-3}. The second most efficient pathway contains an extra covalent bond between the ND1 and the CG atoms of His-506, and gives an ε^2 value of 8.3×10^{-4}. For the second imidazole ring the best pathway consists of seven covalent bonds between SG of Cys-507 and CG of His-508, with an ε^2 value of 7.8×10^{-4}. Thus the hydrogen-bonded Cys-507–His-506 pathway gives approximately three times more efficient electronic coupling than the Cys-507–His-508 route. In the case of the type-1 blue copper protein plastocyanin, it is believed that a similar electron transfer pathway, consisting of the copper ligand Cys-84 and the adjacent highly conserved Tyr-83, is relevant and made use in the reaction with cytochrome *f* [160].

10.3 Electron Transfer within the Trinuclear Copper Site

Electron exchange within the trinuclear copper site is expected to be very fast due to the short distances between the copper atoms (from 3.7 to 5.2 Å in the reduced form), as is ET to the bound dioxygen. This fast electron exchange is necessary to include CU4 into the redox processes. CU4 is at a greater distance to the type-1 copper CU1 than CU2 and CU3.

ACKNOWLEDGMENTS

The author wishes to thank Professors R. Huber and R. Ladenstein and other colleagues involved in the X-ray structural work on ascorbate oxidase. These results would not have been possible without their expertise, cooperation, and support.

REFERENCES

[1] King, T. E., Mason, H. S. and Morrison, M. (eds.), Prog. Clin. Biol. Res., 274 (1988) 1–789.
[2] Capaldi, R. A., Annul Rev. Biochem., 59 (1990) 569–596.
[3] Malkin, R. and Malmström, B. G., Adv. Enzymol., 33 (1970) 177–243.
[4] Yoshida, H., J. Chem. Soc. (Tokyo), 43 (1883) 472.
[5] Bertrand, G. C. R., Hebd. Acad. Sci. (Paris), 118 (1894) 1215.
[6] Keilin, D. and Mann, T., Nature (London), 143 (1939) 23.
[7] Malmström, B. G., Andreasson, L.-E. and Reinhammar, B., in Boyer, P.D. (ed.), *The Enzymes*, Vol. 12B, Academic Press, New York, 1975, pp. 507–579.
[8] Fee, J. A., Struct. Bond., 23 (1975) 1–60.
[9] Reinhammar, B. and Malmström, B. G., in Spiro, Th. G. (ed.), *Copper Proteins, Metal Ions in Biology*, Vol. 3, John Wiley & Sons, New York, 1981, pp. 109–149.
[10] Reinhammar, B., in Lontie, R. (ed.), *Copper Proteins and Copper Enzymes*, Vol. 3, CRC Press, Boca Raton, FL, 1984, pp. 1–35.
[11] Farver, O., and Pecht, I., in Lontie, R. (ed.), *Copper Proteins and Copper Enzymes*, Vol. 1, CRC Press, Boca Raton, FL, 1984, pp. 183–214.
[12] Mayer, A. M., Phytochemistry, 26 (1987) 11–20.
[13] Szent-Györgyi, A., Biochem. J., 22 (1928) 1387.
[14] Szent-Györgyi, A., Science, 72 (1930) 125.
[15] Lovett-Janison, P. L., and Nelson, J. M., J. Amer. Chem. Soc., 62 (1940) 1409.
[16] Stotz, E., J. Biol. Chem., 133 (1940) c.
[17] Dawson, C. R., in Peisach, J., Aisen, P. and Blumber, W. E. (ed.), *The Biochemistry of Copper*, Academic Press, New York, 1966, pp. 305–337.

[18] Kroneck, P. M. H., Armstrong, F. A., Merkle, H. and Marchesini, A., in Seib, P. A. and Tolbert, B. M. (eds.), *Advances in Chemistry Series, No. 200. Ascorbic Acid: Chemistry, Metabolism, and Uses*, The American Chemical Society, New York, 1982, pp. 223–248.
[19] Mondovi, B. and Avigliano, L., in Lontie, R. (ed.), *Copper Proteins and Copper Enzymes*, Vol. 3, CRC Press, Boca Raton, FL, 1984, pp. 101–108.
[20] Finazzi-Agro, A., Life Chemistry Reports, 5 (1987) 199–209.
[21] Holmberg, C. G., Acta Physiol. Scand., 8 (1944) 227.
[22] Holmberg, C. G. and Laurell, C.-B., Scand. J. Lab. Clin. Invest., 3 (1951) 103.
[23] Malmström, B. G. and Ryden, L., in Singer, T. (ed.), *Biological Oxidations*, John Wiley and Sons, New York, 1968, pp. 415–438.
[24] Poulik, M. D. and Weiss, M. L., in Putnam, F. W. (ed.), *The Plasma Proteins*, Vol. 2, Academic Press, New York, 1975, pp. 51–108.
[25] Frieden, E. and Hsieh, H. S., Adv. Enzymol., 44 (1976) 187–236.
[26] Ryden, L., in Lontie, R. (ed.), *Copper Proteins and Copper Enzymes*, Vol. 3, CRC Press, Boca Raton, FL, 1984, pp. 37–100.
[27] Laurie, S. H. and Mohammed, E. S., Coord. Chem. Rev., 33 (1980) 279.
[28] Cousins, R. J., Physiol. Rev., 65 (1985) 238.
[29] Arnaud, P., Gianazza, E. and Miribel, L., Methods in Enzymology, 163 (1988) 441–452.
[30] Chichiricco, G., Ceru, M. P., D'Alessandro, A., Oratore, A. and Avigliano, L., Plant Sci., 64 (1989) 61–66.
[30a] Aramayo, R. and Timberlake, W. E., Nucl. Acids Res., 18 (1990) 3415.
[30b] Kojima, Y., Tsukuda, Y., Rawai, Y., Tsukamota, A., Sugiura, J., Sakaino, M. and Kita, Y., J. Biol. Chem., 265 (1990) 15224–15230.
[30c] Choi, G. H., Larson, T. G. and Nuss, D. I., Mol. Plant-Microbe Interactions, 5 (1992) 119–128.
[31] Ohkawa, J., Okada, N., Shinmyo, A. and Takano, M., Proc. Natl. Acad. Sci. USA, 86 (1989) 1239–1243.
[32] Esaka, M., Hattori, T., Fujisawa, K., Sakajo, S. and Asahi, T., Eur. J. Biochem., 191 (1990) 537–541.
[33] Rossi, A., Petruzzelli, R., Messerschmidt, A. and Finazzi-Agro, A., private communication. (1991).
[34] Marchesini, A., Cappalletti, P., Canonica, L., Danieli, B., and Tollari, S., Biochim. Biophys. Acta, 484 (1977) 290–300.
[35] Butt, V. S., in Davies, R. (ed.) *The Biochemstry of Plants. A Comprehensive Treatise, Metabolism and Respiration*, Academic Press, New York, 1980, pp. 85–95.
[36] Lin, L.-S. and Varner, J. E., Plant Physiol. 96 (1991) 159–165.
[37] Hayashi, R. and Morohashi, Y., Plant Physiol., 102 (1993) 1237–1241.
[38] *Biochemical Catalogue*, p. 108. Boehringer Mannheim (1991).
[39] Wimasalena, R. and Dharmasena, S., Anal. Biochem., 210 (1993) 58–62.
[40] Fahraeus, G. and Reinhammar, B., Acta Chem. Scand., 21 (1967) 2367–2378.
[41] Mosbach, R., Biochim. Biophys. Acta, 73 (1963) 204–212.
[42] Briving, C., Ph.D. Thesis, University of Göteborg, 1975.
[43] Froehner, S. C. and Eriksson, K.-E., J. Bacteriol., 120 (1974) 458–465.
[44] Lerch, K., Deinum, J. and Reinhammar, B., Biochem. Biophys. Acta, 534 (1978) 7–14.
[45] Germann, U. A., Müller, G., Hunziker, P. E., and Lerch, K. , J. Biol. Chem., 263 (1988) 885–896.
[46] Saloheimo, M., Niku-Paalova, M.-L. and Knowles, J. K. C., J. General Microbiol., 137 (1991) 1537–1544.
[47] Niku-Paalova, M.-L., Karhunen, E., Salola, P. and Raumio, V., Biochem. J., 254 (1988) 877–884.
[48] Rehmann, A. U. and Thurston, C. F., J. General Microbiol., 138 (1992) 1251–1257.
[49] Reinhammar, B., Biochim. Biophys. Acta, 205 (1970) 35–47.
[50] Marchesini, A. and Kroneck, P. M. H., Eur. J. Biochem., 101 (1979) 65–76.

[51] D' Andrea, G., Bowstra, J. B., Kamerling, J. P. and Vliegenthart, J. F. G., Glycoconjugate, 5 (1988) 151–157.
[52] Nakamura, T., Makino, N. and Ogura, Y., J. Biochem. (Tokyo), 64 (1968) 189.
[53] Takahashi, N., Ortel, T. L. and Putnam, F. W., Proc. Natl. Acad. Sci. USA, 81 (1984) 390–394.
[54] Karhunen, E., Niku-Paalova, M.-L., Viikari, L., Haltia, T., van der Meer, R. A. and Duine, J. A., FEBS Lett., 267 (1990) 6–8.
[55] deleted.
[56] Kawahara, K., Suzuki, S., Sakurai, T. and Nakahara, A., Arch. Biochem. Biophys., 241 (1985) 179–186.
[57] Sakurai, T., Suzuki, S. and Tanabe, Y., Inorg. Chim. Acta, 157 (1989) 117–120.
[58] Messerschmidt, A., Rossi, A., Ladenstein, R., Huber, R., Bolognesi, M., Gatti, G., Marchesini, A., Petruzzelli, R. and Finazzi-Agro, A., J. Mol. Biol., 206 (1989) 513–529.
[59] Messerschmidt, A., Ladenstein, R., Huber, R., Bolognesi, M., Avigliano, L., Petruzzelli, R., Rossi, A. and Finazzi-Agro, A., J. Mol. Biol., 224 (1992) 179–205.
[60] Bränden, R., and Deinum, J., FEBS Lett., 73 (1977) 144–146.
[61] Martin, C. T., Morse, R. H., Kanne, R. M., Gray, H. B., Malmström, B. G. and Chan, S. I., Biochemistry, 20 (1981) 5147–5155.
[62] Morpurgo, L., Desideri, A., and Rotilio, G., Biochem. J., 207 (1982) 625–627.
[63] Winkler, M. E., Spira, D. J., LuBien, C. D., Thamann, T. J. and Solomon, E. I., Biochem. Biophys. Res. Comm., 107 (1982) 727–734.
[64] Allendorf, M. D., Spira, D. J., and Solomon, E. I. Proc. Natl. Acad. Sci. USA, 82 (1985) 3063–3067.
[65] Spira-Solomon, D. J., Allendorf, M. D., and Solomon, E. I., J. Amer. Chem. Soc., 108 (1986) 5318–5328.
[66] Ladenstein, R., Marchesini, A., and Palmieri, S., FEBS Lett., 107 (1979) 407–408.
[67] Bolognesi, M., Gatti, G., Coda, A., Avigliano, L., Marcozzi, G., and Finazzi-Agro, A. J. Mol. Biol., 169 (1983) 351–352.
[68] Priestle, J. P., J. Appl. Crystallogr., 21 (1988) 572–576.
[69] Kabsch, W. and Sander, S., Biopolymers, 22 (1983) 2577–2637.
[70] Barlow, D. J. and Thornton, J. M., J. Mol. Biol., 168 (1983) 867–885.
[71] Crawford, J. L., Lipscomb, W. N. and Schellmann, C. C., Proc. Natl. Acad. Sci. USA, 70 (1973) 538–542.
[72] Richardson, J. S., Adv. Protein Chem., 34 (1981) 167–339.
[73] Huber, R. and Steigemann, W., FEBS Lett., 48 (1974) 235–237.
[74] Baker, E. N. and Hubbard, R. E., Prog. Biophys. Mol. Biol., 44 (1984) 97–179.
[75] Lee, B. K. and Richards, F. M., J. Mol. Biol., 55 (1971) 379–400.
[76] Guss, J. M. and Freeman, H. C., J. Mol. Biol., 169 (1983) 521–563.
[77] Nar, H., Messerschmidt, A., Huber, R., van de Kamp, M. and Canters, G. W., J. Mol. Biol., 221 (1991) 765–772.
[78] Cole, J. L., Clark, P. A. and Solomon, E. I., J. Amer. Chem. Soc., 112 (1990) 9534–9548.
[79] Karlin, K. D., Hayes, J. C., Gultneh, Y., Cruse, R. W., McKnown, J. W., Hutchinson, J. P. and Zubieta, J., J. Amer. Chem. Soc., 106 (1984) 2121–2128.
[80] Chaudhuri, P., Ventor, D., Wieghardt, K., Peters, E., Peters, K. and Simon, A., Angew. Chem., 97 (1985) 55–56.
[81] Volbeda, A. and Hol, W. G. J., J. Mol. Biol., 209 (1989) 249–279.
[82] Messerschmidt, A. and Huber, R., Eur. J. Biochem., 187 (1990) 341–352.
[83] Godden, J. W., Turley, S., Teller, D. C., Adman, E. T., Liu, M. Y., Payne, W. J. and LeGall, J., Science, 258 (1991) 438–442.
[84] Fenderson, F. F., Kumar, S., Adman, E. T., Liu, M.-Y., Payne, W. J. and LeGall, J., Biochemistry, 30 (1991) 7180–7185.
[85] Adman, E. T., Adv. Protein Chem., 42 (1991) 145–197.

[85a] Zaitseva, I., Zaitsev, V., Card, G., Moshkov, K., Bax, B., Ralph, A. and Lindley, P., J. Biol. Inorg. Chem., 1 (1996) 15–23.

[86] Villafranca, J. F., Freeman, F. C. and Kotcherar, A., in Karlin, K. D. and Tyeklar, Z. (eds.), *Bioinorganic Chemistry of Copper*, Chapman and Hall, New York, 1993, pp. 439–446.

[87] Barry, C. E., Nayar, P. G. and Begley, T. P., Biochemistry, 28 (1989) 6323–6333.

[88] Mellano, M. A. and Cooksey, D. A., J. Bacteriol., 170 (1988) 2879–2883.

[88a] Lee, Y. A., Hendson, M., Panopoulos, N. J. and Schroth, M. N., J. Bacteriol., 176 (1994) 173–188.

[89] Askwith, C., Eide, D., VanHo, A., Bernhard, P. S., Li, L., Davis-Raplan, S., Sipe, D. M. and Kaplan, J., Cell, 76 (1994) 403–410.

[90] Albani, D., Sardana, R., Robert, L. S., Altosaar, I., Arnison, P. G. and Fabijanski, S. F., Plant J., 2 (1992) 331–342.

[91] Vehar, G. A., Keyt, B., Eaton, D., Rodrigues, H., O'Brien, D. P., Rotblat, F., Oppermann, H., Keck, R., Wood, W. I., Harkins, R. N., Tuddenham, E. G., Lawn, R. M. and Capon, D. J., Nature, 312 (1984) 337–342.

[92] Kane, W.-H. and Davie, E. W., Proc. Natl. Acad. Sci. USA, 83 (1986) 6800–6804.

[93] Karlsson, G., Aasa, R., Malmström, B. G. and Lundberg, L. G., FEBS Lett., 253 (1989) 99–102.

[94] Reinhammar, B., Biochim. Biophys. Acta, 275 (1972) 245–259.

[95] Reimhammar, B. and Vänngard, T., Eur. J. Biochem., 18 (1971) 463–468.

[96] Sykes, A. G., Adv. Inorg. Chem., 36 (1991) 377–408.

[97] Deinum, J. and Vänngard, T., Biochim. Biophys. Acta, 310 (1973) 321–330.

[98] Peterson, L. C. and Degn, H., Biochim. Biophys. Acta, 526 (1978) 85–92.

[99] Andreasson, L.-E. and Reinhammar, B., Biochim. Biophys. Acta, 445 (1976) 579–597.

[100] Andreasson, L.-E., Malmström, B. G., Stromberg, Ch. and Vänngard, T., Eur. J. Biochem., 34 (1973) 434–439.

[101] O'Neill, P., Fielden, E. M., Morpurgo, L. and Agostinelli, E., Biochem. J., 222 (1984) 71–76.

[102] O'Neill, P., Fielden, E. M., Finazzi-Agro, A. and Avigliano, L., Biochem. J., 209 (1983) 167–174.

[103] Meyer, T. E., Marchesini, A., Cusanovich, M. A. and Tollin, G., Biochemistry, 30 (1991) 4619–4623.

[104] Kyritsis, P., Messerschmidt, A., Huber, R., Salmon, G. A. and Sykes, A. G., J. Chem. Soc., Dalton Trans., (1992) 731–735.

[105] Farver, O. and Pecht, I., Proc. Natl. Acad. Sci. USA, 89 (1992) 8283–8287.

[106] Yamazaki, I. and Piette, L. H., Biochim. Biophys. Acta, 50 (1961) 62–69.

[107] Farver, O. and Pecht, I., Mol. Cryst. Liq. Cryst., 194 (1991) 215–224.

[108] Andreasson, L.-E. and Reinhammar, B., Biochim. Biophys. Acta, 568 (1979) 145–156.

[109] Andreasson, L.-E., Bränden, R. and Reinhammar, B., Biochim. Biophys. Acta, 438 (1976) 370–379.

[110] Nakamura, T. and Ogawa, Y., J. Biochem. (Tokyo), 64 (1968) 267–270.

[111] Farver, O., Goldberg, M. and Pecht, I., FEBS Lett., 94 (1978) 383–386.

[112] Strothkamp, R. E. and Dawson, C. R., Biochem. Biophys. Res. Comm., 85 (1978) 655–661.

[113] Cole, J. L., Ballou, D. P. and Solomon, E. I., J. Amer. Chem. Soc., 113 (1991) 8544–8546.

[114] Clark, P. A. and Solomon, E. I., J. Amer. Chem. Soc., 114 (1992) 1108–1110.

[115] Curzon, G., Biochem. J., 100 (1966) 295–302.

[116] Curson, G. and Speyer, B. E., Biochem. J., 105 (1967) 243–250.

[117] Gunnarsson, P.-O., Nylen, U. and Pettersson, G., Eur. J. Biochem., 27 (1972) 572–577.

[118] Gunnarsson, P.-O. and Petersson, G., Eur. J. Biochem., 27 (1972) 564–572.

[119] Malkin, R., Malmström, B. G. and Vänngard, T., FEBS Lett., 1 (1968) 50–54.

[120] Bränden, R., Malmström, B. G. and Vänngard, T., Eur. J. Biochem., 36 (1973) 195–200.

[121] Strothkamp, R. E. and Dawson, C. R., Biochemistry, 16 (1977) 1926–1929.

[122] van Leeuwen, F. X. R., Wever, R., van Gelder, B. F., Avigliano, L. and Mondovi, B., Biochim. Biophys. Acta, 403 (1975) 285–291.

[123] Morie-Bebel, M. M., Morris, M. C., Menzie, J. L. and McMillin, D. R., J. Amer. Chem. Soc., 106 (1984) 3677–3678.

[124] Messerschmidt, A., Steigemann, W., Huber, R., Lang, G. and Kroneck, P. M. H., Eur. J. Biochem., 209 (1992) 597–602.

[125] Messerschmidt, A., Luecke, H. and Huber, R., J. Mol. Biol., 230 (1993) 997–1014.

[126] Malkin, R., Malmström, B. G. and Vänngard, T., Eur. J. Biochem., 7 (1969) 253–259.

[127] Hanna, P. H., McMillin, D. R., Pasenkiewicz-Gierula, M., Antholine, W. E. and Reinhammar, B., Biochem. J., 253 (1988) 561–568.

[128] Graziani, M. T., Morpurgo, L., Rotilio, G. and Mondovi, B., FEBS Lett., 70 (1976) 87–91.

[129] Reinhammar, B. and Oda, Y., J. Inorg. Biochem., 11 (1979) 115–120.

[130] Schmidt-Klemens, A. and McMillin, D. R., J. Inorg. Biochem., 38 (1990) 107–115.

[131] Graziani, M. T., Loreti, P., Morpurgo, L., Savini, I. and Avigliano, L., Inorg. Chim. Acta, 173 (1990) 261–264.

[132] Avigliano, L., Desideri, A., Urbanelli, S., Mondovi, B. and Marchesini, A., FEBS Lett., 100 (1979) 318–320.

[133] Morpurgo, L., Savini, I., Mondovi, B., and Avigliano, L., J. Inorg. Biochem., 29 (1987) 25–31.

[134] Merli, A., Rossi, G. L., Bolognesi, M., Gatti, G., Morpurgo, L. and Finazzi-Agro, A., FEBS Lett., 231 (1988) 89–94.

[135] Guss, J. M., Harrowell, P. R., Murata, M., Norris, V. A. and Freeman, H. C., J. Mol. Biol., 192 (1986) 361–387.

[136] Shepard, W. E. B., Anderson, B. F., Lewandowski, D. H., Norris, G. E. and Baker, E. N., J. Amer. Chem Soc., 112 (1990) 7817–7819.

[137] Nar, H., Ph.D. Thesis, Technische Universität München, 1992.

[138] Schilstra, M. J., Birker, P. J. M. W., Verschoor, G. C. and Reedijk, J., J. Inorg. Chem., 21 (1982) 2637–2644.

[139] Sorell, T. N. and Malachowski, M. R., Inorg. Chem., 22 (1983) 1883–1887.

[140] Kitajima, N., Fujisawa, R. and Moro-oka, Y., J. Amer. Chem. Soc., 111 (1989) 8975–8976.

[141] Magnus, K. and Ton-That, H., J. Inorg. Biochem., 47 (1992) 20–26.

[142] Andreasson, L.-E., Bränden, R., Malmström, B. G. and Vänngard, T., FEBS Lett., 32 (1973) 187–189.

[143] Aasa, R., Bränden, R., Deinum, J., Malmström, B. G., Reimhammar, B. and Vänngard, T., FEBS Lett., 61 (1976) 115–119.

[144] Aasa, R., Bränden, R., Deinum, J., Malmström, B. G., Reimhammar, B. and Vänngard, T., Biochem. Biophys. Res. Comm., 70 (1976) 1204–1209.

[145] Bränden, R., and Deinum, J., Biochim. Biophys. Acta, 524 (1978) 297–304.

[146] Cole, J. L., Avigliano, L., Morpurgo, L. and Solomon, E. I., J. Amer. Chem. Soc., 113 (1991) 9080–9089.

[147] Casella, L., Gullotti, M., Pallanza, G., Pintar, A. and Marchesini, A., Biochem. J., 251 (1988) 441–446.

[148] Casella, L., Gullotti, M., Pintar, A., Pallanza, G. and Marchesini, A., J. Inorg. Biochem., 37 (1989) 105–109.

[149] Hathaway, B. J., Compr. Coord. Chem., 5 (1987).

[150] Marcus, R. A. and Sutin, N., Biochim. Biophys. Acta, 811 (1985) 265–322.

[151] Farver, O., Goldberg, M., Wherland, S. and Pecht, I., Proc. Natl. Acad. Sci. USA, 75 (1978) 5245–5249.

[152] Lieber, C. M., Karas, J. L., Mayo, S. L., Albin, M. and Gray, H. B., in Proceedings of the Robert A. Welsh Conference on Chemical Research. Design of Enzymes and Enzyme Model., Robert A. Welsh Foundation, Houston, 1987, pp. 9–33.

[153] Avigliano, L., Rotilio, G., Urbanelli, S., Mondovi, B. and Finazzi-Agro, A., Arch. Biochem. Biophys., 185 (1978) 419–422.
[154] Gray, H. B. and Malmström, B. G., Biochemistry, 28 (1989) 7499–7505.
[155] McLendon, G., Acc. Chem. Res., 21 (1988) 160–167.
[156] Mayo, S. L., Ellis, W. R., Jr., Crutchley, R. J. and Gray, H. B., Science, 233 (1986) 948–952.
[157] Onuchic, J. N. and Beratan, D. N., J. Chem. Phys., 92 (1990) 722–733.
[158] Beratan, D. N., Onuchic, J. N. and Gray, H. B., in Sigel, H. and Sigel, A. (eds.), *Metal Ions in Biological Systems*, Dekker, New York, 1991, pp. 97–127.
[159] Beratan, D. N., Onuchic, J. N., Betts, J. N., Bowler, B. and Gray, H. B., J. Amer. Chem. Soc., 112 (1990) 7915–7921.
[160] Sykes, A. G., Struct. Bonding, 75 (1991) 175–224.

THE BIOINORGANIC CHEMISTRY OF ALUMINUM

Tamas Kiss and Etelka Farkas

OUTLINE

1. **Introduction** **200**
2. **Basic Methodology** **202**
   - 2.1 Analytical Methods 202
   - 2.2 Speciation Methods 204
3. **Coordination Chemistry of Al(III) with Relevant Biomolecules** **206**
   - 3.1 Inorganic Ligands 207
   - 3.2 Carboxylic Acids 211
   - 3.3 Hydroxycarboxylic Acids 212
   - 3.4 Amino Acids, Peptides, Proteins 214
   - 3.5 Organic Phosphates and Phosphonates 218
   - 3.6 Nucleotides 220
   - 3.7 Salicylates and Catecholates 223
   - 3.8 Hydroxamates 224
   - 3.9 Pyrone and Pyridinone Derivatives 225
   - 3.10 Porphyrins 228
4. **Ternary Complexes** **229**

Perspectives on Bioinorganic Chemistry
Volume 3, pages 199–250

ISBN: 1-55938-642-8

5. Aluminum Absorption and Metabolism **233**
6. The Toxic Effects of Aluminum **237**
6.1 Neurological Toxicity 237
6.2 Aluminum-Induced Bone Disease 238
6.3 Anemia 238
7. Speciation and Experimental Toxicology of Aluminum **239**
8. Aluminum and Medicines **240**
8.1 Pharmaceutical Uses of Aluminum Compounds 240
8.2 Chelation Therapy in Aluminum Detoxification 241
8.3 Aluminum and Silicon 241
Acknowledgments **242**
References **243**

1. INTRODUCTION

Aluminum is the most abundant metal and the third most abundant element after oxygen and silicon. It is locked in soils and minerals as oxides, and more commonly in the form of complex aluminum silicates. In contrast with its abundance in the Earth's crust, the ocean concentration is below 1 μg Al per liter, due to the low solubility of these minerals. Most natural waters contain insignificant amounts of Al(III), except for some volcanic regions and alum springs. When Al is disrupted from primary minerals, it commonly precipitates as a secondary mineral $Al(OH)_3$. Its bioavailability is therefore rather limited under normal conditions. Acid rain, however, can increase the mobility of this metal [1] and hence, the soluble Al(III) content of natural fresh waters and soils, mainly in the form of various inorganic and organic complexes.

In the early part of this century the phytotoxicity of Al in agroecosystems was already well known. Elaborate measures were taken to ameliorate Al toxicity through the application of lime and the selection of Al-tolerant species. While some plant species are highly sensitive, others are capable of thriving in soils of high Al content. The tea plant is able to grow in very acidic soils, where Al(III) is readily available for uptake by the roots. The tea leaves serve as a sink, accumulating over 10,000 ppm of Al(III) in some instances.

Aluminum research in the past 20 to 30 years focused in part on the effects of the increased mobilization of Al by acidic precipitation. Increased Al(III) concentrations and corresponding reductions in fish populations have been attributed to the mobilization of Al from the edaphic to the aquatic environment. It has been proved that an increased Al concentration is more damaging to fish than an increased acidity; fish can tolerate only nanomolar Al(III) levels.

For a long time, Al was considered to be an innocuous element for humans. It was not until the mid-1970s that aluminum toxicity was definitively proved. The advances in dialysis and the spread of hemodialysis as a routine and regular

Table 1. Estimated Daily Adult Intake of Aluminum from Various Sources

Category	*Source*	*mg Al/Day*
Diet	Total diet:	20–40
	Natural content	2–10
	Additives	20–50
	Cookware	3
Water	Natural content	1–2
Air	Dust, smoke, sprays	<1
Drugs	Antacids	50–1000
	Buffered aspirin	10–100

treatment for chronic renal failure played roles in the clarification of Al toxicity in humans [2]. Aluminum has already been identified as a causative or associative agent in various kinds of human disorders. Its toxic effects were first characterized on patients receiving long-term hemodialysis, primarily manifesting in dialysis encephalopathy, also known as dialysis dementia, with which vitamin D-resistant osteomalacia and non-iron-deficient microcytic anemia were later associated [2, 3]. Initial studies devoted to this syndrome suggested that dialysis encephalopathy and the accompanying bone lesions were due to the high levels of Al present in the tap water used for the dialysates. Evidence of Al intoxication in uremic patients ingesting Al(III)-containing phosphate binders has also been reported [3]. Aluminum has likewise been implicated in the etiology of neurological disorders such as Alzheimer dementia [4, 5]. Increasing concern has therefore been expressed about its widespread use in food and drugs. Table 1 lists sources of Al which might be ingested under everyday circumstances. Estimates are given for the daily amounts of Al typically expected from those sources where data are available [6].

To understand the effects of Al in the environment and also in an organism we need to know the gross amount present and also the multitude forms into which this element enters. Speciation or identification of species of Al(III) is also essential. It is now well understood that the toxicity of Al in aquatic and terrestrial systems does not correlate well with the total Al(III) concentration, but is a function of the concentration of the biologically active fraction [7, 8]. In terms of acute toxicity, the inorganic forms are believed to be the most toxic. However, organically bound species may be capable of crossing biological membranes and contribute to chronic bioaccumulation. In order to decrease the availability of Al from natural and surface waters and soils, as well as from medicines and medical treatments, the factors that exert the primary influence on Al bioavailability must be identified. Again, speciation studies are necessary to determine the chemical forms in which Al(III) is present in waters, soil-waters, and in the main biofluids. Analytical chemistry and coordination chemistry are the two main disciplines which can provide the necessary chemical information to help biologists, biochemists, toxicologists, and clinicists to understand the harmful roles of Al(III) in organisms.

This chapter is intended to provide a brief, fundamental account of the aqueous chemistry of Al(III) that is relevant or may be relevant to its environmental, biological, toxicological, or physiological contexts. We cannot give a complete review of this rapidly developing research area, but we refer the reader to some excellent books or chapters published on this topic during the past decade [2–12]. We shall set the scene by giving a simple outline of the analytical, coordination chemical, and kinetic methods used to determine and characterize the distribution and speciation of Al(III) in the environment and organisms. We shall then discuss the complex-forming properties of the relevant inorganic and organic ligands with Al(III) including binary and ternary complexes. This will be followed by a number of examples the biological applications and implications of the aqueous coordination chemistry of Al(III).

2. BASIC METHODOLOGY

Samples or materials in which Al has to be determined are extremely diverse. Aluminum is found in steel and related materials, in minerals, in biological or clinical samples, and in natural waters. Onishi [13] recently reviewed the classical and instrumental methods of separation and analysis which can be used in the quantitative analytical chemistry of Al. Here, we focus only on some general aspects of the trace analysis of aluminum. The key to the success of Al detection in the ppm–ppb range lies in the availability of accurate and precise analytical methods and detailed guidelines for contamination-free sample collection and preparation. Savory and Wills [14] gave a comprehensive review of the analytical methods for the determination of Al(III) in biological samples, discussed the problems associated with Al measurements, and provided recommendations for the preanalytical sample preparation and the control of contamination. A similar review was published by Bloom and Erich [15] on the determination of Al(III) in natural waters.

2.1 Analytical Methods

Several methods are available for the determination of total Al in biological samples.

Chemical and Physicochemical Methods

Aluminum can be measured by means of classical gravimetric, titrimetric, spectrophotometric, and fluorometric methods, although these are generally neither sensitive nor accurate enough for trace analysis. However, it is worthwhile to mention a recent publication in which a sensitive spectrofluorometric method with 0.1 ppb detection limit has been reported [16].

X-Ray Fluorescence

This method is very specific. The K_α line of Al at 0.8340 nm can be used to measure Al as a minor or a major constituent of solid complexes. The method, however, is not sensitive enough to detect the trace levels of Al(III) in biological samples.

Neutron Activation Analysis

The (n,γ) reaction of ^{27}Al can be used to determine trace amounts of Al. The technique has an excellent sensitivity and is relatively free from contamination since the sample is analyzed directly or with minimal pretreatment. One major disadvantage is the need for correction for the fast neutron reaction on phosphorus, which also produces the radioactive ^{28}Al nuclide. For this reason, neutron activation analysis (NAA) cannot be classified as a reference method [14].

Atomic Spectroscopy

Atomic spectroscopy, in either the absorption or the emission form, is very useful for the determination of small amounts or traces of Al in various matrices. Inductively coupled plasma atomic emission spectrometry (ICP AES) is very sensitive and offers a detection limit of as low as 0.3 ppb in pure solution, with a large range of linearity and excellent precision. Because of the relatively high cost of the instrumentation it is used mostly as a reference method for Al(III) measurements. Atomic absorption spectrometry (AAS), especially with a graphite furnace, is the most widely used technique today and produces reliable results. The sensitivity of the method extends down to approximately 1 ppb in biological materials, provided that the matrix effects are properly taken into account. Detailed working procedures are given and recommended for the determination of Al(III) in plasma/serum or urine [17]; in soft tissues, including brain and bone [14]; and in natural waters in [15].

Besides a knowledge of the gross amount of Al, monitoring of its cellular and subcellular localizations in tissues and cell cultures has become increasingly important. Procedures of this kind may be divided into two classes: (i) methods which permit a simple visual localization of sites of accumulation within a tissue, i.e. histochemical staining, and (ii) instrumental analyses which locate the metal and also measure its concentrations. Techniques which have been employed for the investigation of Al(III) distribution in tissues are electronprobe X-ray microanalysis (EPXMA) with energy dispersive or wavelength dispersive detection, laser microprobe mass spectrometry (LMMS), and nuclear microprobes (e.g. particle induced X-ray emission, PIXE). Excellent up-to-date reviews discuss the capabilities (sensitivity and resolution) and uses of these and also some newer, even more expensive, sophisticated techniques in comparison with the previously mentioned more general analytical methods [18].

Table 2. Brain Aluminum Levels in Alzheimer's Disease

Techniques	*Indication*	*References*
AAS, NAA	+	[20]
AAS, EPXMA	–	[21]
NAA	–	[22]
LMMS	+	[23]
LMMS	–	[24]
ICP-MS	+	[25]
PIXE	–	[26]
PIXE	+	[27]

In spite of the extensive range of analytical techniques available for bulk analysis and *in situ* measurements of Al(III), serious contradictory analytical results have been reported. As an illustration, controversial results on the accumulation of Al(III) in the brain of victims of Alzheimer's disease (AD) are listed in Table 2 (a + sign indicates accumulation as compared to a control group; data are mostly taken from the paper of Zatta [19]).

These discrepancies could be due to different important factors. From the aspect of analytical chemistry, the mode of sample collection and the possibility of contamination of samples during collection, storage, and preparation for analysis are two such factors which have to be taken into consideration. The preanalytical procedures used for sample preparation, such as extraction with organic solvents after a complexation process, or heat treatment with oxidizing acids in order to destroy the organic matrix, or the impurities in the solvents and reagents, can all introduce numerous contamination factors—all of which are important but difficult to control. A knowledge of the specificity and sensitivity of the instrumental technique and the proper understanding of the capability of the method, i.e. an intelligent use of the analytical equipment, is also essential. To overcome these technical and methodological problems, precise protocols for sampling and sample treatment, and established basic criteria for selecting an appropriate analytical technique including the detection limit and the standardization method, have to be prepared, accepted, and adopted by all laboratories involved in such investigations [18, 19].

2.2 Speciation Methods

Besides the total amount of Al in a sample, a knowledge of the chemical form of the element is also important. The different species may have very different properties, e.g. their charge might be essential in their membrane transport behavior, and it is known that not all chemical forms of Al(III) are equally toxic. Speciation means a more or less exact knowledge of the compositions of the different species, including their bonding modes or structures and charge. In analytical chemistry, the target is to determine the amounts of all Al(III) species after complete separation [28, 29]. Of course, this can be achieved only with relatively inert complexes, where the ligand exchange reactions between the different Al(III) species are slower than the time of the separation process, and equilibrium conditions are therefore not disturbed during the separation.

In the analysis of natural waters, three fractions of aqueous Al(III) can be and usually are defined and determined by different analytical speciation and separation techniques: acid-soluble Al(III) (mostly oligomeric, partly mixed hydroxo complexes of high molecular weight organic materials), monomeric Al(III) (complexes formed with low molecular weight Al(III) binder organic molecules), and monomeric inorganic Al(III) [15].

In the case of the formation of labile Al(III) complexes, a speciation description can be given by using various physicochemical methods permitting a study of solution equilibria and/or solution structure. pH-potentiometry is widely used to determine the stoichiometries and stability constants of the complexes formed, while NMR is applied to obtain information on the most likely binding modes of the species.

Al(III) is generally regarded as a moderately labile metal ion. The ligand exchange rates of the various Al(III) complexes cover a wide reaction rate range. While some monomeric Al(III) complexes with simple inorganic ligands are formed in less than a millisecond, the formation of oligomeric binary or ternary hydroxo complexes (which may involve slow intramolecular rearrangement reactions), even with simple organic ligands such as citric acid, may take days or weeks. Al(III) has a high tendency to hydrolyze, and readily forms hydroxo-bridged oligomeric complexes, which are generally formed in very slow processes. Speciation models can differ considerably, depending on whether only the relatively fast formation of the mostly monomeric complexes is monitored, or whether the true equilibrium state of the system is studied. The situation was nicely demonstrated by Öhman [30] on the Al(III)–citric acid system when "instant" pH-metric titration data were evaluated mostly in terms of monomeric species, while in the real thermodynamic equilibrium (achieved after a week) trimeric species $[Al_3(H_{-1}L)_3(OH)]^{4-}$ and $[Al_3(H_{-1}L)_3(OH)_4]^{7-}$ predominated in a wide pH range. The different time scales for pH-meter readings are the main reasons for the different speciation models reported for a given Al(III)–ligand system (see examples in Section 3). In solution equilibrium studies, when true pH equilibrium is measured independently of the time necessary to achieve it, species distribution can be described in a wider pH range, including the very slow olation/oligomerization reactions. (In cases, when hours/days/weeks are necessary to achieve this, of course, only a "batch" technique can be used instead of the normal titration method.) In the case when the experimental conditions are chosen to suppress these slow processes and/or the slow reactions are omitted from the evaluation, the technique allows speciation under more limited conditions (e.g. in a narrower pH range) and focuses on monitoring of the formation of mostly monomeric species. There is no question that the former method gives a more complete, thermodynamically valid speciation description of the system, but, in biological systems, for instance, where potential Al(III) binder molecules are generally in large excess (at the mM level compared with a μM level of Al(III)) and the time is never enough for true equilibrium to be reached (and thus slow processes are considerably suppressed), results obtained with the latter technique can be equally good or even more informative.

NMR is a powerful tool in Al(III) chemistry and is capable of providing much new information on the solution speciation of the metal ion. ^{27}Al is a nucleus with high intrinsic NMR sensitivity and high resonance frequency, and thus is a favorable nucleus for study. Its drawback is its quadrupole moment, the lines generally being broad; the minimum linewidths ($w_{1/2}$) observed so far are 2.0–2.5 Hz [31]. Asymmetry in the ligand field produces a considerable increase in the NMR linewidth, which varies from 3 to 6000 Hz; the more symmetric the complex, the narrower the linewidth. When the symmetry is fairly high the ^{27}Al spectral peak is sharp; for octahedral complexes of high symmetry the peakwidth never exceeds 100 Hz. ^{27}Al chemical shifts (δ) cover a range of about 300 ppm. The chemical shifts depend on the ligand type (complexes with different donors appear at different δ values) and the coordination number. Peaks for octahedral complexes appear at around 0 ppm, as referred to $[Al(H_2O)_6]^{3+}$ (between –20 and +50 ppm), whereas those for tetrahedral complexes appear further downfield (between +40 and +140 ppm). Thus, a combination of the chemical shift and the linewidth serves as a reasonable diagnostic probe for studying the chemical environment around the Al(III) in its complexes. Many Al(III) complexes display low ligand exchange rates on the NMR time scale, and thus different species possess separate lines, which can be utilized in speciation studies. Of course, not only ^{27}Al NMR can be used in this way to monitor complexation reactions of Al(III); the NMR of different ligand nuclei—such as ^{1}H, ^{13}C, ^{31}P, ^{19}F, etc.—can be equally efficient tools. One further important aspect of using NMR to examine Al(III) speciation in solution is the time dependence of species distribution (see above). Thus, it is essential that this be borne in mind when samples are measured by NMR, and information on the time elapsed from the sample preparation until the recording of the spectra has to be reported among the experimental details.

3. COORDINATION CHEMISTRY OF AL(III) WITH RELEVANT BIOMOLECULES

Although extensive and sometimes contradictory information is available on the bonding interactions of Al(III) with relevant inorganic and organic compounds, all these data are generally thermodynamic in nature rather than structural. Moreover, the thermodynamic information mostly refers to acidic aqueous conditions, which are required to ensure the kinetic feasibility of the potentiometric methodology employed to obtain them. In this section, information on the interaction of the metal center with potentially ligating sites present in biologically relevant small molecules such as aliphatic and aromatic hydroxycarboxylates, nucleotides, amino acids, small peptides, catecholamines, and amphiphilic molecules, relevant to (or models for) biological membranes, is summarized according to ligand groups.

3.1 Inorganic Ligands

Among the inorganic ligands, the hydroxide ion, phosphates, and the fluoride ion can bind very strongly to Al(III). Around the physiological pH range, however, only the fluoro complexes are water-soluble. An interesting, but rather unexplored field is the equilibrium chemistry of Al(III)–silicate complexes. No data are known for soluble complex formation between Al(III) and silicic acid [32, 33]. Analyzing the problem under blood plasma conditions (at 1–3 μM Al(III) concentration), Martin [32] found that six insoluble aluminosilicates can permit a free Al^{3+} concentration comparable with those permitted by citrate or transferrin. A computer modeling calculation reported recently by Vobe and Williams [34] for intestinal and blood plasma fluids showed that the coexistence of Al(III) and silicate ions as low molecular weight complexes in the presence of citrate and phosphate is impossible. The interaction between sulfate [35–37] or (hydrogen)carbonate [38, 39] and Al(III) is much weaker than the above one, but it occurs beyond doubt. No measurable interaction was found with Cl^-, Br^-, I^-, or S^{2-} and very little with SCN^- [40]. After many contradictory results on Al(III)–borate complexes, Öhman and Sjöberg [41], using potentiometry, ^{27}Al NMR, and ^{11}B NMR, convincingly demonstrated that no, or at best, only very weak complexes can be formed in the system.

Accordingly, only hydroxo, phosphato, fluoro, hydrogencarbonato, and sulfato complexes are detailed in this section.

Al(III)–OH⁻ System

All equilibrium works performed on a ligand–Al(III) system in aqueous solution need a speciation model and adequate stability data for the Al(III)–OH^- system since the ligand has to compete with the OH^- ion for Al(III) to form complexes. This fact stimulated enormous work in this field. It is supposed that Al toxicity in fish and plants is mostly caused by soluble Al(III)–hydroxo complexes [42–45], although there is some uncertainty in the stoichiometry of the toxic species. Whether solely the polynuclear species $[Al_{13}(OH)_{32}]^{7+}$ is toxic, or whether other species, especially the mononuclear hydroxo complexes, $[Al(OH)]^{2+}$ and $[Al(OH)_2]^+$ are too, is not yet clear.

In spite of the abundant publications on the Al(III)–OH^- system [37, 38, 46–59], numerous unsolved questions remain. To illustrate this, Table 3 lists some representative models with the corresponding stability data determined in the past decade by potentiometry.

Table 3. Speciation Models and Stability Constants[a] for the Al(III)–OH^- System

Species	*–log β_{pq}*	*Conditions*	*Reference*
$[Al(OH)]^{2+}$	5.52	I = 0.6 (NaCl)	[30]
$[Al(OH)_2]^+$	11.3	t = 25 °C	
$[Al(OH)_3]$	17.3		
$[Al(OH)_4]^-$	23.46		
$[Al_3(OH)_4]^{5+}$	13.57		
$[Al_{13}(OH)_{32}]^{7+}$	109.2		
$[Al(OH)]^{2+}$	5.34	I = 0.1 (NaCl)	[39]
$[Al_3(OH)_4]^{5+}$	13.70	t = 25 °C	
$[Al_{13}(OH)_{32}]^{7+}$	107.50		
$[Al(OH)]^{2+}$	5.52	I = 3.0 (NaCl)	[39]
$[Al_3(OH)_4]^{5+}$	13.96	t = 25 °C	
$[Al_{13}(OH)_{32}]^{7+}$	113.35		
$[Al(OH)]^{2+}$	5.55	I = 0.15 ($NaNO_3$)	[58]
$[Al_8(OH)_{20}]^{4+}$	68.46	t = 25 °C	
$[Al(OH)]^{2+}$	4.67		
$[Al(OH)_3]$	13.61	I = 0.15 (NaCl)	[57][b]
$[Al(OH)_4]^-$	23.96	t = 37 °C	
$[Al_3(OH)_{11}]^{2-}$	54.73		
$[Al_6(OH)_{15}]^{3+}$	39.47		
$[Al_8(OH)_{22}]^{2+}$	76.52		
$[Al(OH)]^{2+}$	5.33	I = 0.1 ($NaNO_3$)	[54]
$[Al(OH)_2]^+$	10.91	t = 25 °C	
$[Al_3(OH)_4]^{5+}$	13.13		
$[Al_p(OH)_q]^{3p-q}$	5.73 – 3.6p + 4.64q		

Notes: [a] $\beta_{pq} = [Al_pH_{-q}][H^+]^q/[Al^{3+}]^p$.

[b] Converted into β_{pq} values using the reported value of pK_w = 13.25.

$[Al(OH)]^{2+}$ seems to be a well-defined species (its formation was confirmed by NMR), with pK ~ 5.5. This means that it starts to be formed in measurable concentration at pH ~ 4.5. Below this pH, only $[Al(H_2O)_6]^{3+}$ exists. In the basic pH range, the existence of $[Al(OH)_4]^-$ is also clear-cut. The differences in the models occur in connection with species supposed to form in neutral or slightly acidic solution, where besides the solid compound $Al(OH)_3$, soluble polynuclear complexes may also be formed. The greatest problem is the very slow conversion of polynuclear hydroxo species to still larger oligomeric complexes, which are necessary precursors of the macromolecular polymer $Al(OH)_3$. The attainment of thermodynamic equilibrium may take hours or up to a day [53]. This also means that soluble hydroxo complexes may be maintained almost indefinitely in a metastable state under conditions such that solid $Al(OH)_3$ is the thermodynamically stable species. The analytical concentration of Al(III) likewise affects speciation [47, 55–57]. Thus, technical differences in the experimental method can cause significant differences in the speciation model. Among the polynuclear species, the formation of $[Al_{13}(OH)_{32}]^{7+}$ was proved unambiguously in the solid state by X-ray [59], and also in solution by ^{27}Al and ^{17}O NMR [47, 49–51]. The species $[Al_2(OH)_2]^{4+}$ was characterized by Akitt and Elders [47] using high-field NMR, but it was found not to be a significant species in a solution containing Al(III) at the mM level. This is similarly true for the species $[Al_3(OH)_4]^{5+}$, but this improved the fitting of the experimental potentiometric data in a somewhat higher concentration range, and it is therefore included in most equilibrium models established by potentiometry [37, 38, 52–57].

The above-mentioned features demonstrate the difficulty of Al(III)–OH^- model selection when an Al(III)–ligand–OH^- system is studied. The model proposed in the well-known book by Baes and Mesmer [59], or that determined by Öhman et al. [53], is often chosen. A further question still remains open, which concerns the lowest analytical concentration of Al(III) for which the formation of polynuclear species has to taken into account at all. In biological fluids or tissues, where the concentration of Al(III) can reach only the μM level, their formation is certainly suppressed. In toxicological studies, however, when for example, various soluble Al(III) toxins are administered in mM concentrations, they are not negligible.

Al(III)–Phosphate Systems

Al(III) readily forms a poorly soluble compound with PO_4^{3-}, often designated as $AlPO_4$ but probably of a more complex composition [60, 61]. In a recent paper, Öhman and Martin [61] reported that the poorly soluble species which may precipitate from body fluids is a mixed phosphato–hydroxo complex, with continuously variable proportions of phosphate and hydroxide. Under serum conditions, for instance, its composition may be described as $Al(PO_4)_{0.2}(OH)_{2.4}$.

The solution chemistry of this system can be directly studied only in the fairly acidic pH range, below pH 4, and this may be the main reason why it has been studied so rarely. A recent review [62] summarizes the present state, pointing out the main problems and difficulties. Potentiometric and various spectroscopic techniques (IR, Raman, ^{27}Al NMR, ^{31}P NMR) have been used to describe the Al(III)–PO_4^{3-} system [63–68]. Interaction was proved in strongly acidic solution [64], mostly between Al(III) and H_3PO_4 and to a smaller extent between Al(III) and $H_2PO_4^-$. The fully protonated molecule (H_3PO_4) may bind via an oxygen of the phosphoryl group [64]. The partly protonated $H_2PO_4^-$ and HPO_4^{2-} can coordinate in either a monodentate or a bidentate way, forming binding isomers. With Al^{3+}, chelating coordination is more probable than with dipositive alkaline earth metal and transition metal cations [69, 70]. The speciation models determined by potentiometry [65–68] include protonated complexes such as $[Al(H_2PO_4)]^{2+}$ and $[Al(HPO_4)]^+$ and some dinuclear complexes such as $[Al_2PO_4]^{3+}$, $[Al_2(OH)_2(PO_4)]^+$, and $[Al_2(OH)_3(PO_4)]$, either phosphato and/or hydroxo bridges being assumed in these latter species. Soluble $[Al(PO_4)]$ is not involved in these models and a bis complex, $[AlH(PO_4)_2]^{2-}$, is supposed in one case only [65]. The contradictions between the models and the considerable differences between the reported stability constants originate mostly from the difficulties involved in slow complexation processes, the formation of metastable oversaturated solutions, and precipitation reactions, as already discussed above. To overcome these difficulties, the approach of linear free energy relationships (LFERs) was used in two studies [68, 69] in order to characterize the interaction between Al^{3+} and PO_4^{3-} in solution. Harris [69] constructed LFERs between the overall basicity of the coordinating donor groups of a series of O-donor ligands and the corresponding Al binding constant. He found a well-defined linear relationship between the overall basicity and the Al binding affinity for bidentate ligands forming five-membered chelates. With the assumption that inorganic phosphate also coordinates as a bidentate ligand, a phosphate binding constant $\log \beta_{AlPO_4} = 14.10$ and, in a similar way, another stability constant for the hydrolyzed species $[Al(OH)(PO_4)]^-$, $\log \beta_{AlPO_4(OH)^-} = 8.37$ were determined. Atkári [68] constructed LFERs for a series of organic monophosphates (including mononucleotides) and phosphonates and obtained reliable stability constants for the monoprotonated soluble complex $[Al(HPO_4)]^+$ ($\log \beta_{AlPO_4H^+} = 17.60$), and the mixed hydroxo species $[Al(HPO_4)(OH)]$ and $[Al(OH)_2(HPO_4)]^-$ or $[Al(OH)(PO_4)]^-$ ($\log \beta = 13.5$ and 7.2, respectively). Unfortunately, no exact information is available on the actual stoichiometry of the species existing in the physiological pH range. This is why Al(III) speciation model calculations for biological fluids [61, 69–71] such as serum have led to contradictory results (see Section 5).

Di- and other oligophosphates can form stable six-membered mono and bis chelates with Al(III) via the coordination of adjacent phosphate groups. They can bind Al(III) about 4 orders of magnitude higher than can monophosphate, which can form only a four-membered chelate [35, 68]. It is worthy of mention that, in contrast with monophosphate, oligophosphates readily form bis chelated complexes, although the spatial requirements of the ligand molecules and also the electrostatic repulsion due to the coordination of a highly charged second ligand molecule are larger than in the case of the simple PO_4^{3-}. Chelate-type coordination of di- and triphosphates has been proved by ^{31}P NMR measurements too [68].

Cyclotriphosphate and cyclotetraphosphate have been shown to be capable of binding Al(III) only through one phosphate group, via the formation of four-membered chelate rings with the PO_4 tetrahedron [72–74]. Hence, their binding strength is more or less comparable with that of monophosphate ($\log K_{Al(P_3O_9)} \sim 3.1$, $\log K_{Al(P_4O_{12})} \sim 4.5$) [73]. For steric reasons, Al(III) can bind to two adjacent phosphate groups of the larger cyclohexaphosphate, which will result in a much higher binding strength [73].

Al–F⁻ System

The fluoride ion is a unique ligand in that it forms more stable complexes with Al(III) than with Fe(III). The stability constants for the Al(III)–fluoro complexes have been reviewed [33, 40]. The representative successive stability constants, i.e. log K values (6.4, 5.2, 3.9, 2.8, 1.1, and 0.4 [40]), show that the last two constants, relating to $[AlF_5]^{2-}$ and $[AlF_6]^{3-}$, are very small; this is due to both statistical reasons and, mostly, the large electrostatic repulsion. This may be the explanation why only four constants are published in many papers. The misinterpretation of some experimental results could also originate from this, as discussed by Martin [75]. Thus, in G-protein systems, $[AlF_4]^-$ was proposed as an activating species, binding as a tetrahedral phosphate analog to the protein. However, the presumed tetrahedral $[AlF_4]^-$ is actually a hexacordinated octahedral species, with the stoichiometry of $[AlF_4(H_2O)_2]^-$.

Al(III)–CO_2–OH^- System

Equilibria between Al^{3+}, CO_2 and OH^- have been studied in two publications [38,39], under different conditions (at different ionic strengths, at pH < 4.5, and with variation of the analytical concentration of Al(III) and p_{CO_2}). The hydrolysis of Al(III) was stated to dominate at low p_{CO_2}), but two ternary complexes were found at p_{CO_2} > 0.1 bar, with the compositions of $Al_2H_{-4}CO_2$ and $Al_3H_{-5}CO_2$ (minor species). These compounds were actually assigned as $[Al_2(OH)_2CO_3]^{2+}$ and $[Al_3(OH)_4(HCO_3)]^{4+}$ or $[Al_3(OH)_3CO_3]^{4+}$. Hedlund et al. [39] also performed model calculations concerning the equilibration of gibbsite and amorphous $Al(OH)_{3(s)}$ with water at different CO_2 pressures. A significant p_{CO_2}-dependent increase in the solubility of amorphous $Al(OH)_{3(s)}$ was found.

Al(III)–SO_4^{2-} System

It is rather difficult to study the weak complex formation between Al^{3+} and SO_4^{2-}, but some data are already known. ^{27}Al NMR was used by Akitt et al. [36] to monitor the formation of an inner sphere sulfato complex. They detected a decrease in the proportion of complexed Al(III) if the concentration of H^+ was increased by the addition HCl. This was explained in terms of protonation of the complexed sulfate anion. The formation of outer sphere sulfato complexes could also be detected [37].

3.2 Carboxylic Acids

In principle, monocarboxylic acids can bind to Al(III) in either a monodentate or a bidentate way, the latter yielding a four-membered chelate ring. Carboxylate can also behave as a bridging ligand and, similarly to carbonate [38, 39], can form dimeric species. This latter binding mode was assumed in the Al(III) complexes of simple aliphatic (acetic and propionic) acids, an alicyclic (cyclohexane-carboxylic) acid, and aromatic monocarboxylic (benzoic- and 3-hydroxybenzoic) acids—these complexes having the formula $[Al_2(OH)_2L]^{3+}$ [76, 77]. An approximately linear relationship was established between the equilibrium constants of the reaction $2Al^{3+} + L^- \rightleftharpoons [Al_2H_{-2}L]^{3+} + 2H^+$ and the acidity constant (pK_{HL}) of the carboxylic group. This plot can be used to estimate the stability of the complex $[Al_2(OH)_2L]^{3+}$ in other Al(III)–monocarboxylic acid systems. An even higher degree of oligomer formation (tetramer and hexamer $[Al_4L_{12}]$, $[Al_4(OH)_4L_6]^{2+}$, and $[Al_6(OH)_6L_{12}]$ was detected in nonpolar solvents such as octanol or benzene [78].

The dicarboxylic acids, oxalic and malonic, bind Al(III) much more strongly than do the monodentate monocarboxylic acids and they can prevent the precipitation of any Al(III) hydroxide or hydroxo mixed complex up to pH ~ 6. The dominating species are complexes of type $[AlL_n]$ ($n = 1, 2$ and 3) [79–82], each ligand forming five- or six-membered chelate rings. Close to the precipitation boundary (pH ~ 5–6), dinuclear $[Al_2(OH)_2L_2)]$ and trinuclear $[Al_3(OH)_3L_3]$ soluble hydroxo-bridged complexes were also found with oxalic acid. These species are then readily converted to the poorly soluble solid phases, $Al_3(OH)_7L_{(s)}$ and $NaAl(OH)_2L_{(s)}$, via the exchange of L^{2-} with OH^- [83].

The significantly weaker Al(III) binding ability of the simplest aromatic dicarboxylic acid, phthalic acid, as compared with that of oxalic acid or malonic acid, is due to the formation of two large seven-membered chelate rings [84]. For the same reason, the formation of oligonuclear mixed hydroxo complexes is more pronounced in the Al(III)–phthalic acid system [85].

3.3 Hydroxycarboxylic Acids

Hydroxycarboxylate ligands are of special interest in Al(III) biospeciation as many of them exist in the human body and can be low molecular weight binders of Al(III) in biological fluids.

Lactate forms weak complexes $[AlL]^{2+}$, $[AlL_2]^+$, and $[AlL_3]$ in acidic solutions where the hydroxy group retains its proton [86]. Protonated alcoholic-OH probably participates in the coordination since the stability of the complex $[AlL]^{2+}$ is about 1 log unit larger than that of the monocarboxylates [86, 87]. The crystal structure of the tris complex verified that the Al(III) in this complex is surrounded by a distorted octahedral coordination sphere of six oxygen atoms of the carboxylate and the hydroxy groups of three lactate molecules [87, 88]. The formation of (COO^-,OH)-type bis and tris complexes is negligible, indicating that the solid compound AlL_3 does not persist in solution. Toward neutral solutions, $[AlL]^{2+}$ and $[AlL_2]^+$ undergo loss of a lactate hydroxy group proton with $pK \sim 4$ to give the mononuclear $[Al(H_{-1}L)]^+$ and $[Al(H_{-1}L)L]$. NMR (^1H, ^{13}C, and ^{27}Al) spectral results [88,89] suggest the formation of alcoholate-O^- coordinated complexes at around neutral pH; however, proton dissociation from a coordinated water molecule and the parallel formation of mixed hydroxo complexes $[Al(OH)L]^+$ and $[Al(OH)L_2]$ cannot be completely ruled out. The pK values are much lower than the pK = 5.5 of a water molecule in $[Al(H_2O)_6]^{3+}$, indicating either proton liberation from the ligand molecule or a reduction in coordination number during ionization of a water molecule [90]. In these and similar Al(III)—hydroxy-carboxylate systems, it is likely that the formation of similar bonding isomers occurs much more frequently in solution than mentioned in the literature. The metal ion promoted deprotonation of alcoholic-OH and coordination of the alcoholate-O^- is more pronounced with malic acid, where the presence of another carboxylate group in the β position allows the tridentate coordination of malate via the formation of a (5+6)-membered joint chelate system [91]. Slow reactions produce a variety of polynuclear species containing OH^-, alcoholate-O^-, and -COO^- as bridging donors [92]. Such oligomeric species might also be important in the high metastability of an aqueous solution of Al(lactate)$_3$ [88]. In solution, glycolate shows basically similar coordination behavior toward Al(III) as does lactate [93]. In the solid state, however, it tends to form a five-coordinated complex containing a highly distorted AlO_5 trigonal bipyramid [94, 95].

Citrate is certainly the most extensively investigated and controversial ligand for Al(III) in aqueous solution [30, 96–109]. The speciation and binding modes of the complexes are still not unequivocally defined in solution in part because of the numerous coordination possibilities and also the protonation possibilities of the three carboxyl groups and one hydroxy group, and in part because of the rather slow olation/oligomerization reactions. In freshly prepared solution and at a high ligand excess, mononuclear 1:1 and 1:2 complexes are the predominant species [96]. The tridentate (COO^-, OH, COO^-)-coordinated chelate $[AlHL]^+$ (with one protonated terminal COOH function) loses three protons in a consecutive manner with pK ~ 2.7, 4, and 6, in overlapping processes, from the alcoholic-OH of citric acid, the terminal carboxylic group and metal ion bound water [30, 103, 110]. Among the monomeric complexes, the neutral [AlL] is a significant species at pH 2–5 [102–105]. Similar binding modes can be assumed in the bis complexes too. As nicely demonstrated by Öhman [30], in equimolar slightly acidic solutions monomeric species are transformed into a trimeric complex $[Al_3(H_{-1}L)_3(OH)]^{4-}$ in a very slow process (the half-time is about 10 hours), which becomes the only species at pH equilibrium, as proved by NMR [110]. The same trimeric complex $[NH_4]_4[Al_3(H_{-1}L)_3(OH)(H_2O)]$ was crystallized from neutral solution and was identified unambiguously by ^1H, ^{13}C, and ^{27}Al NMR spectroscopy and X-ray crystallography [107]. The complex trimeric anion consists of a trimeric Al_3O_4 core, with each citrate ligand coordinated to two or more Al(III) centers via the carboxylate- and deprotonated alcoholate-O atoms.

The more symmetrical arrangement of the two carboxylic and two alcoholic-OH groups in tartaric acid makes its coordination somewhat simpler. Potentiometric [111] and multinuclear NMR studies [112, 113] revealed mononuclear (1:1, 1:2 and 1:3) and binuclear (2:2) complexes with both octahedrally and tetrahedrally coordinated Al(III). ^{13}C and ^{27}Al chemical shifts showed that deprotonation of the Al(III)-coordinated hydroxy groups of tartaric acid starts at pH < 2. In the dimeric species the Al(III) centers are bridged by two quadridentate(4–) ligand molecules [111, 112]. It is worthwhile to note that the mononuclear tris complex $[Al(H_{-1}L)_3]^{6-}$, formed in basic solution displays remarkable stability and symmetry; this is ascribed to the formation of successive interligand hydrogen bonds between coordinated and deprotonated alcoholate groups and noncoordinated alcoholic hydroxy groups [112, 113].

The very few studied carboxylic acid derivatives of simple mono-saccharides, such as gluconic acid or saccharic acid, prove to be extremely efficient Al(III) binders [100,109]. The speciation description of these systems is not entirely certain as only monomeric complexes were reported in the early paper by Motekaitis and Martell [100]. The proposed bonding mode, with only alcoholate-O coordination of gluconic acid (with noncoordinated carboxylate function) and only α-ω terminal carboxylate coordination of saccharic acid, appears rather unlikely. Similarly as for the other hydroxycarboxylates discussed above, the primary coordination of the carboxylate function, as an anchoring donor which binds to Al(III) strongly enough to be able to promote dissociation and concomitant coordination of the weakly acidic alcoholic function(s), seems to be essential in the Al(III) binding properties of this ligand class. Although the actual binding modes in these Al(III) complexes are not clear, it was recently shown that gluconate can prevent the precipitation of $Al(OH)_3$ and keep Al(III) in solution even in the basic pH range without any detectable release of gluconate [109]. It is interesting that the mono-saccharide fructose displays a good binding ability to the solid $Al(OH)_3$ matrix via its alcoholic hydroxy groups, although this interaction is too weak to result in the dissolution of solid $Al(OH)_3$ [114].

3.4 Amino Acids, Peptides, Proteins

The α-carboxylate group of amino acids is weakly basic (pK ~ 2.4), which suggests a rather weak Al(III) binding capability. The usual sample composition in the pH-metric titration method (a metal ion concentration of a few mM with a metal ion to ligand ratio of from 1:1 to 1:5) is not suitable for the detection of any complex formation [115]. However, at much higher excesses of ligand (up to 1:40), simple α-amino acids such as glycine, alanine, serine, and threonine were found unambiguously to influence the speciation of $[Al(H_2O)_6]^{3+}$ [116, 117]. In these systems, complex formation can be safely detected even at high ligand excess since the buffer range of the ligand and the pH range of metal complex formation are well enough separated, and thus the pH effect of the proton displacement reaction due to metal complexation can be measured "clearly". The formation of monomeric species $[AlL]^{2+}$, mixed hydroxo species $[Al(OH)L]^{-}$, and hydroxo bridged dimeric complex $[Al_2(OH)_2L_2]^{2+}$ has been assumed. Although the exact speciation description of Al(III) simple amino acid systems has not yet been established, the stability constant characteristic for this interaction, log β = 5.8–5.9, obtained from direct potentiometric measurements [117], is in good agreement with the value derived from LFER calculations [118]. In the acidic pH range, ^{27}Al [119] and ^{1}H NMR [120] studies unambiguously indicate an Al(III)–carboxylate interaction (probably with the protonated amino group) for these simple amino acids. With a less basic amino group ($pK_{NH_3^+}$ = 5.2), complexation becomes conveniently observable with α-picolinate (2-pyridinecarboxylate), which yields mono and bis complexes $[AlL]^{2+}$, $[AlL_2]^+$ and $[Al(OH)L_2]$ [119–121].

The tridentate aspartic acid (Asp), containing two carboxylate-O and one central amino-N binding donors, is a much stronger Al(III) binder; the stability of the 1:1 complex is about 2 orders of magnitude higher than that of any simple amino acid, indicating the involvement of both carboxylate groups in metal binding [117, 122]. It is also interesting that no such strong complexation has been detected with either succinic acid or *N*-acetyl-Asp (both lack the central amino binding site), which would also indicate the involvement of the $-NH_2$ group in the bonding mode of Al(III)–Asp complexes. This strong interaction can be detected by means of ^{27}Al NMR: a relatively sharp signal at ~10 ppm (shown in Figure 1) suggests octahedral Al(III) in a fairly symmetrical chemical environment. This spectral behavior is reminiscent of that of the trimeric Al(III)–citrate complex [107]. The results may suggest that, besides negatively charged O donors—such as carboxylate, alcoholate, phenolate, and catecholate—in the event of a favorable steric arrangement, the amino group can also participate in binding to Al(III). More surprisingly, NMR measurements indicate some involvement even of the thiolate-S^- in the coordination in Al(III)–3-mercaptosuccinic acid complexes [117].

As the number of potential binding donor atoms increases in the aminopolycarboxylates [such as iminodiacetic acid (ida), nitrilotriacetic acid (nta)] and polyaminopolycarboxylates [such as ethylenediaminetetraacetic acid (edta), propanediaminetetraacetic acid (pdta), cyclohexanediamino-tetraacetic acid (cdta), and diethylenetriaminepentaacetic acid (dtpa)] they become more and more efficient Al(III) binders [123–129]. In consequence of the large number of coordinating donors, these ligands, with the exception of the tridentate ida [126, 127], form only 1:1 complexes. Coordination of the N-donor group(s) besides the carboxylate functions was demonstrated by X-ray [130] and NMR methods [127]. For a series of aminopoly-carboxylates, a linear relationship was observed between the ^{27}Al NMR chemical shift of the 1:1 Al(III) chelate and the denticity (**n**) of the ligand [127]. The additivity rule was applicable up to **n** = 6. When **n** = 6, all the water molecules in the coordination sphere are replaced, and hence a further increase in **n** would not be expected to result in a further increase in the chemical shift.

Similar binding properties, with somewhat less effective binding strength, were reported for the phosphonic acid analogues (the carboxylate groups are replaced by phosphonate groups) in this ligand class [131–138]. This behavior is explained by electrostatic and steric reasons, due to the larger negative charge repulsion and

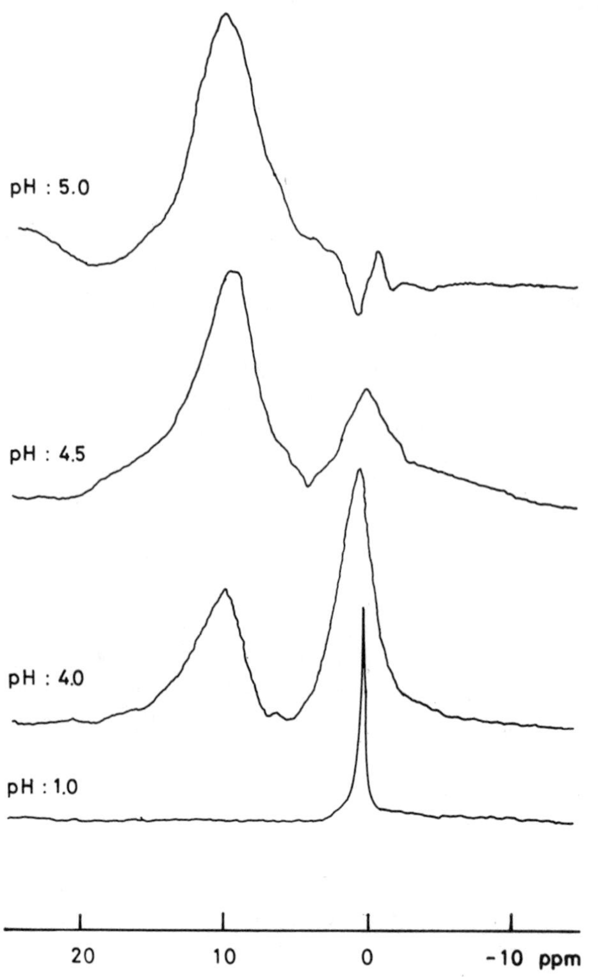

Figure 1. ^{27}Al NMR spectra of Al(III)–aspartic acid at 1:5 metal ion-to-ligand ratio and at different pH values.

greater spatial requirements of the tetrahedral 2– charged phosphonate group as compared with the flat 1– charged carboxylate group [134].

Al(III) complexes of Schiff bases formed from pyridoxal and several phosphonic acid analogues of tridentate amino acids were investigated in connection with mechanistic studies of the transamination and dephosphonylation of α-amino acids and analogues pertinent to the chemistry of vitamin B_6 catalysis [134–137]. It is thought that Al(III) can serve as a template for the synthesis of Schiff bases, and these Al(III) complexes may be intermediate in vitamin B_6 catalysis [137]. The Schiff bases formed are easily hydrolyzed in water in the absence of Al(III). We refer here to a more detailed review of the Schiff base complexes of Al(III) [139].

With oligopeptides, and especially with proteins, the side-chain donors of the skeleton amino acids are the primary metal binding sites. In the case of oligopeptides, the short peptide chain cannot provide the necessary closed arrangement of the potential binding donors to be able to bind Al(III) strongly enough. Thus, only weak binding of Al(III) was revealed by ^{1}H and ^{13}C NMR to a nonapeptide (Gly–Lys–Hyp–Gly–Glu–Hyp–Gly–Pro–Lys) via carboxylate donors [140] but no binding could be detected to the pentapeptide enkephalin (Tyr–Gly–Gly–Phe–Leu) in aqueous solution [118]. The interaction with Al^{3+} of calmodulin, an important Ca^{2+}-dependent regulating protein, was studied in detail (in connection with this being one possible manifestation of general Al(III) neurotoxicity). Calmodulin is a relatively small protein, consisting of 150 amino acids, and one-quarter of its amino acids residues bear carboxylate side chains. In this way, it should be an optimal Al(III)-binder protein. A series of papers by Haug et al. seem to indicate that Al(III) binds tightly to calmodulin (up to three Al(III) per calmodulin molecule with stability constant values greater than 10^{6}) and, in doing so, induces a significant conformational change (i.e. a 30% reduction in α-helical content) in the protein [141, 142]. In contrast, no evidence of a significant interaction between calmodulin and Al(III) was found in the pH range 5.5 to 6.5 in other investigations [143]. According to Martin [118], calmodulin bearing negatively charged carboxylate residues can bind multiply charged metal ions nonspecifically, as a polyelectrolyte. It may be mentioned, however, that in a very recent publication fairly strong interactions of the β-amyloid polypeptide fragments AβP(1-40) and AβP(25-35) with Al(III) were reported [144]. Circular dichroism spectroscopy was used to confirm that in a membrane-mimicking solvent (60% 2,2,2-trifluoroethanol/40% Tris buffer at pH 7.0), AβP(1-40), which adopts a partially helical conformation, lost this structure in the presence of physiologically relevant concentrations of Al(III).

Albumin, a water-soluble protein, having a molecular mass of ~70 kDa, is able to bind Al(III) *in vitro* at physiological pH. The ^{27}Al NMR spectra of Al(III) added to albumin revealed that at least two Al^{3+} ions bind to one albumin molecule [145]. The chemical shift value indicated that Al(III) is octahedrally coordinated by O donors. The most likely binding site is one consisting of carboxylates of albumin (probably at least three) to give a reasonably strong binding strength. The remaining ligands to the bound Al(III) are probably water molecules. The common albumin and globulin proteins of plasma are much weaker metal ion binders than another important transport protein of the plasma, transferrin. Recent quantitative spectroscopic determinations provided reliable successive stability constants characterizing the Al(III) binding strengths of the two sites of transferrin: $\log K_1 = 12.9$ and $\log K_2 = 12.3$ at pH = 7.4 and 27 mM HCO_3^-, $t = 25$ °C [146]; and $\log K_1 = 13.5$ and $\log K_2 = 12.5$ at pH = 7.4 and 100 mM NaCl, 25 mM Tris and 10 mM HCO_3^-, $t = 37$ °C [147]. X-ray data show that at both sites the metal ion is coordinated by two tyrosines, one histidine, and one aspartate, plus a bidentate carbonate ion in a pseudooctahedral geometry [148]. As far as the binding strengths of the two sites are concerned, the results are rather contradictory. The difference of 0.6 log units between the two constants determined by Martin et al. [146] suggests that, at pH = 7.4, Al^{3+} binds to the two transferrin sites with approximately equal intrinsic site binding constants. In contrast, Harris and Sheldon [147a] propose preferential binding of Al^{3+} to the C-terminal site, while Kubal et al. [147b] to the N-terminal site of human serum transferrin. From ^{27}Al and ^{13}C NMR measurements Aramini and Vogel [148] concluded that ovotransferrin exhibits an N-terminal site selectivity. Although this apparent difference between serum transferrin and ovotransferrin is not surprising in itself since they are only about 50% homologous, a recent criticism [149] postulates that the different experimental conditions in these studies prevent any conclusive statement about possible intrinsic differences in site specificity.

3.5 Organic Phosphates and Phosphonates

Probably it is the phosphate group to which Al^{3+} is most frequently and significantly complexed in biological systems. For the purpose of metal ion binding capabilities, soluble phosphates may be divided into two classes: basic phosphates and weakly basic phosphates [60]. Basic phosphates with pK = 6–7 are monosubstituted with the general formula $R–OPO_3^{2-}$; have a charge of 2–; and occur as the terminal phosphate in nucleoside mono-, di-, and triphosphates, phosphorylated proteins, and also in many other biophosphates. Inorganic mono- and polyphosphates all contain at least one basic phosphate group (cf. Section 3.1). Weakly basic phosphates with a single p$K < 2$ are disubstituted with the general formula $R–O(R'–O)PO_2^-$, have a charge of 1– and occur as internal phosphates in nucleoside di- and triphosphates, in the nucleic acids DNA and RNA, and also in phospholipids, phosphoglycolipids, etc.

Simple organic monophosphates, such as phenylphosphate or *p*-nitrophenylphosphate, and monophosphonates, such as methylphosphonate and ethylphosphonate, were found to bind Al(III) with a similar strength as for inorganic monophosphate (see Section 3.1), via the formation of mainly 1:1 parent and mixed hydroxo complexes [68]. None of these ligands is able to keep Al(III) in solution at physiological pH. 2,3-Diphosphoglycerate (dpg), which is an important constituent of red blood cells, is a much stronger Al(III) binder, as it forms mono and bis chelates via the carboxylate–2-phosphate donor pair. The stability constant of the bis chelate, $\log K_{AlL_2} = 12.5$, is comparable to that of ATP [150, 151]. Inositol-phosphates (IP), contain basic phosphate groups, and are therefore also potential Al(III) binders. D-Myo-inositol 1,4,5-triphosphate acts in a wide variety of cells by triggering the release of sequestered Ca(II) from the endoplasmic reticulum after stimulation of a specific intracellular receptor [152]. Another IP, D-myo-inositol 1,2,6-triphosphate, displays promising pharmacological properties, resulting from its extracellular action, which has still to be clarified [153]. Their chemical structure, which consists of three basic phosphate groups bound around the inositol ring,

Table 4. Proton (log *K*) and Aluminum (log β) Stability Constants of Myo-Inositol-Triphosphates[a] and Triamino-Inositol Derivatives[b]

	I(1,4,5)P3 [c]	*I(1,2,6)P3* [c]	*taci* [c]	*tdci* [c]
HL	8.74	9.48	9.7	9.68
H_2L	7.02	7.22	9.1	7.62
H_3L	5.80	5.70	8.9	5.89
H_4L	2.61	2.40	8.1	—
AlHL	18.98(6)	19.18(3)	—	—
AlL	13.37(7)	13.72(3)	11.8(1)	14.31(5)
AlL_2	—	19.72(7)	18.8(1)	26.4(1)
$AlH_{-1}L$	5.84(8)	6.10(4)	—	—
$AlH_{-2}L$	1.82(10)	—	—	—
AlL_2H	—	—	25.3(3)	30.61(5)
$AlH_{-1}L_2$	—	—	10.53(1)	—
$AlH_{-2}L_2$	—	—	1.63(1)	—
$AlH_{-3}L_2$	—	—	–7.45(1)	—
$AlH_{-4}L_2$	—	—	–17.16(1)	—

Notes: [a] Ref. 153, at I = 0.1 M tetra-*n*-butylammonium bromide, *t* = 25 °C.
[b] Refs. 154, 155 at I = 0.1 M KCl, *t* = 25 °C.
[c] I(1,4,5)P3: DL-myo-inositol 1,4,5-triphosphate,, I(1,2,6)P3: D-myo-inositol 1,2,6-triphosphate,, taci: 1,3,5-triamino-1,3,3-trideoxy-*cis*-inositol, tdci: 1,3,5-trideoxy-1,3,5-tris(dimethylamino)-*cis*-inositol.

suggest that they are capable of binding most of the metallic cations present in biological fluids. A comparative speciation study revealed that the two inositol-triphosphate isomers form similar complexes (see Table 4) and under physiological conditions they bind Al(III) more strongly than does ATP [153].

Synthetic inositol derivatives, such as 1,3,5-triamino-1,3,5-trideoxy-*cis*-inositol (taci) [154] and especially 1,3,5-trideoxy-1,3,5-tris(dimethylamino)-*cis*-inositol (tdci) [155], are also strong Al(III) binders; they are sterically constrained to facial coordination via three phosphates in the 1:1 complexes or six phosphates in the 1:2 complexes. Their stability constants are included in Table 4. Model calculations using the reported stability constants of these ligands reveal that I(1,4,5)P, I(1,2,6)P, and taci can barely compete with the hydrolysis of Al(III). However, tdci has a much higher binding affinity towards Al(III) and is even able to dissolve solid $Al(OH)_3$. The larger space filling of the Me_2N groups of tdci probably provides a more favorable constrained steric arrangement of the phosphates as compared with that of taci. This strong Al(III) sequestering ability of tdci may be utilized in solubilization of hydroxy-Al(III)-silicate deposits. Phytic acid (myo-inisitol hexaphosphate) readily forms insoluble complexes with essential metal ions, thereby affecting their bioavailability. It has been found that Al(III) forms both soluble and insoluble phytate complexes, presumably with 1:1 and 4:1 Al–phytate stoichiometries, respectively [156]. The reaction heats were endothermic, and the interaction of one Al(III) with one phosphate group could be characterized by an enthalpy change of ~29 kJ mol^{-1} in both complexes.

Phosphorylated proteins are important constituents of different body tissues. Phosphorylation and dephosphorylation reactions normally accompany cellular processes. The basic phosphate groups of any phosphorylated protein provide the necessary basicity and, in conjunction with juxtaposed carboxylate or other phosphate groups, become strong Al^{3+} binding sites [60]. Although there are no quantitative stability data characterizing the Al(III) binding strengths of these biomolecules, we can estimate a value of log K ~ 6 for the interaction of a single phosphate group with Al^{3+}. If other binding donors, such as a carboxylate or another phosphate, are also present in the proximity of this binding site, the stability constant can reach, as an upper limit, those of 2,3-dpg or diphosphate complexes [150]. *In vitro* circular dichroism studies have revealed that Al^{3+} is able to induce significant conformational changes in phosphorylated or over-phosphorylated (9–100 mol phosphate per mol protein) neurofilament proteins, yielding a high content of β-pleated structure [157]. This unambiguously suggests a direct interaction between Al^{3+} and the strongest binding site of the phosphorylated protein, the phosphate group [158]. In a recent enzymatic study, phosvitin, the major phosphoglycoprotein of egg yolk, was used as a model to study the interaction of Al(III) with multi-phosphorylated proteins [159]. A competition method revealed that the number of atoms of Al(III) bound per mole of phosvitin was 69 ± 3, and the dissociation constant was less than 10^{-9} M [160]. The results demonstrated that the action of Al^{3+} on phosvitin displayed significant similarities with the action of this metal ion on neurofilament proteins.

Phospholipids are the most abundant constituents of all biological membranes. They are derived from either glycerol or sphingosine, both polyalcohols. Phosphoglycerides consist of a glycerol backbone, two fatty acid chains, and a phosphorylated alcohol. Thus, in contrast with phosphorylated proteins, they contain only weakly acidic 1– charged phosphate groups; their metal binding ability is much weaker. Biophysical studies [161–163] have demonstrated that Al(III) in μM concentration promotes aggregation, fusion, and membrane rigidification, which is not limited to membranes composed exclusively of negatively charged phospholipids. Similar changes were induced in phosphatidylserine, phosphatidylethanolamine, and phosphatidylcholine containing lipid vesicles. This strongly suggests a role of the direct interaction of Al(III) with the phosphate binding site.

3.6 Nucleotides

Nucleotides contain three different metal binding sites: phosphate groups in the mono-, di- or triphosphate moieties, alcoholic hydroxyl(s) at the ribose or deoxyribose unit; and carbonyl-O and/or ring N donors in the nucleic base functions. The phosphate binding site can be weakly basic or nonbasic, as in the nucleic acids DNA and RNA, or it can be basic, as the terminal phosphate in nucleoside phosphates

Table 5. Proton (log K) and Aluminum (log β) Stability Constants of Adenosine 5′-phosphates (AMP, ADP and ATP)[a]

	AMP	*ADP*	*ATP*
HL	6.04(1)	6.19(1)	6.31(1)
H_2L	9.78(2)	9.98(2)	10.20(2)
AlHL	—	10.98(4)	11.30(4)
AlL	6.17(1)	7.82(3)	7.92(4)
$AlH_{-1}L$	2.02(9)	2.94(8)	2.46(7)
AlL_2	10.35(11)	12.16(4)	12.47(4)
$AlH_{-1}L_2$	b	5.01(7)	4.84(5)

Notes: [a] Ref. 151 at I = 0.2 M KCl, t = 25 °C.
[b] Precipitation.

and in many other biophosphates. Some stability constants recently determined for Al(III) binding to adenosine nucleotides [151] are featured in Table 5.

The basic terminal phosphate group is the primary binding site for Al(III) [151, 164–168]. No significant interaction with the other two binding sites has been detected so far by any structural investigation method. The stability constant for the formation of the equimolar complex $[AlL]^-$, the predominant species below pH ~ 4.5, is greater, than that for most metal ions, including Cu(II). Additionally, Al(III) forms strong bis complexes with a second ligand molecule. Although the free nucleotides are minimally stacked in dilute solutions at mM concentrations, the charge neutralization provided by the binding of Al(III) should promote base stacking, which may promote the formation of bis complexes. As may be seen in the species distribution for the Al(III)–ATP system presented in Figure 2, the bis complexes $[AlL_2]^{5-}$ and $[AlL_2(OH)]^{6-}$ predominate in the physiological pH range, 7.0–7.4, along with the mixed hydroxo mono complex $[AlL(OH)]^{2-}$, but this latter is present in higher concentration than $[AlL_2(OH)]^{6-}$ at lower ligand excess.

In the adenosine-5′-phosphate series, log K_{AlL} exhibits the sequence AMP < ADP ≤ ATP, suggesting Al(III) chelation to the two terminal phosphate groups in ADP and ATP, and coordination to the single phosphate in AMP. ^{1}H, ^{13}C, ^{31}P, and ^{27}Al NMR measurements established the formation of two main species in slow exchange with the free ligand [166, 169, 170]. In the pH range 3.8–7.3, at an Al to ATP ratio of 1:1, a dimeric (2:2) species was detected, while another species, a bis complex was formed in the pH range 4.2–8.1 in the case of an excess of the ligand. ^{31}P NMR data revealed that Al(III) binding occurs at P_β and P_γ in the phosphate chain, resulting in complexes containing octahedral Al(III) coordinated to six O atoms, as suggested by ^{27}Al NMR measurements [166]. On the basis of ^{1}H NMR spectroscopic results, it was proposed that Al(III) binds to N7 of the adenine base [165]. As discussed later [171, 172], the observed upfield H8 shift is due to nucleic base deprotonation at N1 or base stacking, rather than to Al(III)-N7 coordination, which should result in a downfield shift. The contradiction between

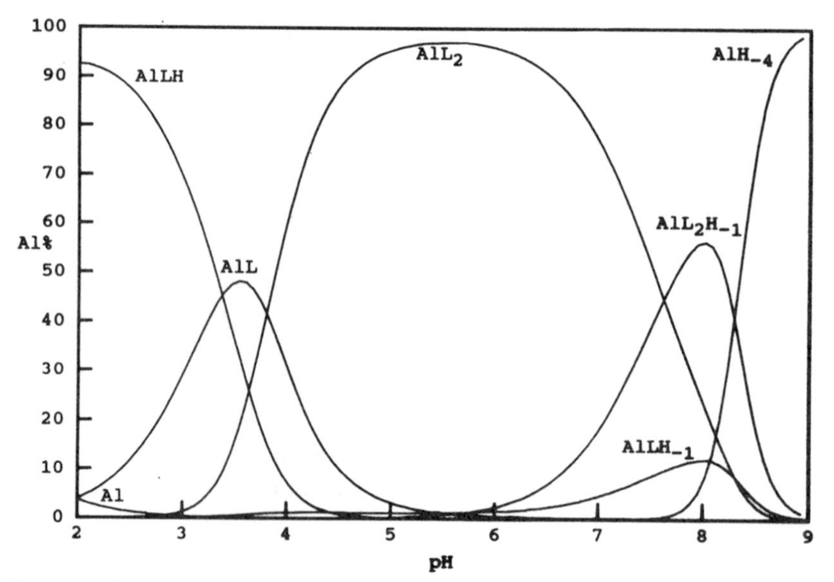

Figure 2. Speciation curves of the complexes formed in the Al(III)–ATP system as a function of pH. Calculated from stability constants listed in Table 5. C_{Al} = 2 μM, C_{ATP} = 20 mM.

the potentiometric and NMR results as concerns the monomeric or dimeric structure of the 1:1 complex appears to be only an apparent one since the NMR measurements were carried out on solutions one order of magnitude stronger, in the 10 mM Al^{3+} range (which would certainly favor dimer formation), while pH-metric measurements are not sufficiently sensitive to detect processes having no direct pH effect, such as the dimerization reaction 2 AlL $\rightleftharpoons$ Al_2L_2, and thus cannot differentiate between $[AlL]^-$ and $[Al_2L_2]^{2-}$. The formation of $[Al_2L_2]^{2-}$ via a stacking interaction between the adenine rings, however, is a very reasonable assumption.

Al(III) binds to all nucleoside phosphates, predominantly through the phosphate groups. The only basic phosphate group ($pK_a \geq 6$) is the terminal one. Since the nucleoside triphosphates exhibit a similar phosphate basicity to that of the nucleoside diphosphates and nucleoside monophosphates, the equilibrium constants in Table 5 can also be applied to other nucleoside phosphates. A great similarity in the Al(III) binding abilities of various adenine and guanine nucleotides was revealed by ^{27}Al NMR spectral measurements [173].

In contrast with nucleotides, nucleic acids contain only weakly basic phosphates ($pK_a < 2$), whose metal binding ability is rather weak. The interaction of Al(III) with DNA was found to be so weak that a quantitative study was limited to pH < 5.5 because of metal ion hydrolysis and precipitation [174]. A stability constant of log K = 5.6 was suggested as the upper limit to characterize Al(III) binding to DNA under intracellular conditions. In early work Karlik et al. [175, 176] detected several mono- and dinuclear Al(III) complexes with DNA at μM Al(III) concentrations by means of thermal denaturation and circular dichroism methods. The detailed structures of the Al(III)–DNA complexes remain to be specified. However, with very weak basic phosphates, DNA serves merely as a polyelectrolyte interacting with Al(III) weakly and nonspecifically. The value log $K = 5.6$ is more than 3 log units less than the conditional stability constants for nucleotides containing basic phosphate binding sites. Therefore, it is obvious that DNA cannot compete with ATP, other nucleotides, or biophosphates for Al(III). Hence, it can be concluded that the location of Al(III) in the chromatin of brain neuronal nuclei is due not to its coordination to DNA, but rather to some other ligand, most probably phosphorylated proteins [118] as discussed in the previous section.

3.7 Salicylates and Catecholates

Because of the high basicity of the donor groups, salicylates ($\Sigma pK \sim 17$) and especially catecholates ($\Sigma pK \sim 22$) chelate Al(III) through two negatively charged oxygens with high stabilities. This binding strength, however, is reduced considerably by proton competition in neutral aqueous solution. These ligands and their derivatives, as possible degradation products of the high molecular weight organic materials such as humic and fulvic acids and lignin, can be good model compounds to assess Al(III) binding ability and $Al(OH)_3$ solubilizing ability of organic soil materials.

Salicylic acid and its derivatives form mono and bis chelates; the latter loses two protons above pH~6, resulting in the mixed hydroxo complexes $[Al(OH)L_2]^{2-}$ and $[Al(OH)_2L_2]^{3-}$ [177–181]. In spite of some indications [182, 183], it does not readily form an octahedral tris complex since the binding strength of a third salicylate chelate is not competitive enough to suppress metal ion and complex hydrolysis, or even the precipitation of $Al(OH)_3$ above pH ~ 7. In contrast, catechol derivatives have a much higher affinity for Al(III) in the basic pH range, where the precipitation of $Al(OH)_3$ is prevented via formation of the octahedral tris complex $[AlL_3]^{3-}$ [181, 184–187]. The stability of this complex is so high that it can efficiently hinder formation of the very stable tetrahedral hydroxo complex $[Al(OH)_4]^-$, even above pH ~ 12. Oligomeric hydroxo-bridged species are also assumed [181, 184] to be present in low concentration in the weakly acidic pH range between 5 and 7. Not surprisingly, 1,2-dihydroxy-naphthalenes exhibit a fully similar complexation pattern [188]. However, when the number of available coordinating sites in the molecule is higher, as in gallic acid (3,4,5-trihydroxybenzoic acid) [189, 190], 2,3,4-trihydroxybenzoic acid or 2,3-dihydroxy-terephthalic acid [181], the tendency to oligomeric complex formation is also higher, while the tendency to metal ion hydrolysis is lower. Polyphenols of natural origin, such as epicatechin, its dimer, and the polymer tannin, which contains on average 10 catecholate units per molecule, can bind Al(III) with stability constants not greater than those of catechol. The dimeric epicatechin ligand, in which the chelating sites in the molecule are separated, can function only as a mono-catecholate [191], but in the polymeric tannin a small chelate effect arises from the coordination of remote 1,2-diphenolato sites [192].

Catecholamines that attain significant concentrations in various body fluids, e.g. in the brain cerebrospinal fluid can be important Al(III) binders in the human. Al^{3+}, as a typical hard metal ion, prefers O donors and chelates to the catecholate locus rather than to the side-chain amino group of catecholamines, and thus catechol-like binding predominates in a wide pH range (from acidic to basic). Even for L-dopa (3,4-dihydroxybenzoic acid), with its chelating glycinate locus, only catecholate coordination occurs [187]. At physiological pH, the main species is an (O^-,O^-)-coordinated tris complex with ammonium groups remaining protonated. *In vitro* enzymatic studies support the proposed binding mode, as Al^{3+} was found to inhibit enzymatic O-methylation, but not N-methylation of noradrenaline [193].

In consequence of their analytical chemical importance as complexing agents, the chelation of various alizarin ligands (alizarin-3-sulfonic acid and alizarin-4-sulfonic acid) has also been studied; they were found to form 1:1 and 1:2 complexes with stabilities similar to that of the simple catechol complex [194–196].

3.8 Hydroxamates

Interest in the Al(III)–hydroxamic acid interaction is based mainly on the fact that the microbial trishydroxamate siderophore, desferrioxamine B (DFO), has long been used in clinical therapy (Desferal^R) for the treatment of aluminum overload [186, 197] since it can bind Al(III) stronger than can either citrate or transferrin [146, 193] (and stronger than, for example, the well-known metal ion chelator, edta [186]). The formula of DFO in totally protonated form is shown in Scheme 1. Much work has been performed on different aspects and problems arising during treatment with DFO, but only a few publications deal with speciation in the Al(III)–DFO system [186, 198, 199]. In spite of this, mainly the work of Martell et al. [186, 197, 199, 200] that led to this system is fairly well characterized. Although the successive pK_a values for the three hydroxamic acid groups are quite high (8.32, 8.96 and 9.55 [199]), the coordination sphere of Al(III) is completed by the three hydroxamate chelates, and the complex $[Al(HDFO)]^+$ is formed in fairly acidic solution. This is the only species that exists up to pH ~ 8. Above this pH, the non-coordinated

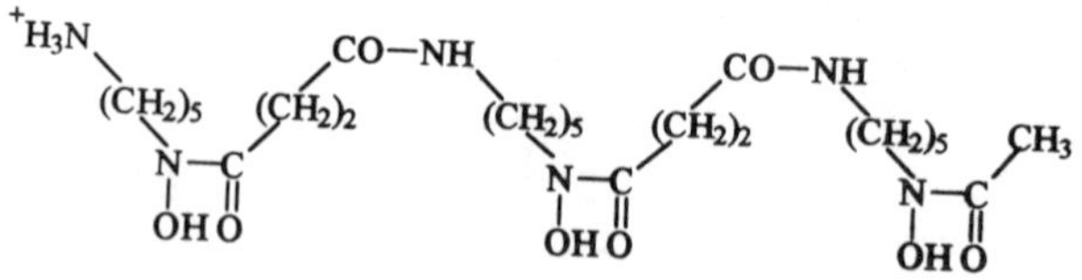

H_4DFO^+

Scheme 1.

NH_3^+ group releases its dissociable proton. OH^- can compete with DFO only at pH > 10. Nothing is mentioned in the relevant papers [186, 198–200] about mixed hydroxo complexes, but we can expect their formation to some extent since [Al(DFO)] is eventually converted to $[Al(OH)_4]^-$ [186]. One more fact should be noted relating to DFO as a metal ion chelator. It coordinates to Fe(III) much more strongly than to Al(III). The corresponding logarithmic stability constants are 30.60 and 24.50, respectively [186]. The problem of ligand design for the selective complexation of Al(III) was analyzed by Hancock and Martell [200]. They found that the stability sequence is determined by the affinity sequence of the metal ions towards OH^- and cannot be significantly altered by, for instance, variation of the length of the hydrocarbon chain connecting the hydroxamate functions. However, it may be altered by steric means and by changing the donor atoms.

Among the simple monohydroxamic acids, only the acetohydroxamic acid ($CH_3CONHOH$) has been included in equilibrium studies with the Al(III) ion [201, 202]. This ligand was found to be an effective chelator, suppressing the hydrolysis of Al(III) to an extent determined by the metal-to-ligand ratio, the pH, and the analytical concentrations [202] (see Figure 4 in Section 4). At any rate, the Al(III) remains dissolved in the whole pH region if at least an equimolar amount of acetohydroxamic acid is present in the solution. The monohydroxamic acid derivative of α-alanine [$(CH_3(NH_3^+)CHCONHOH)$] was found to coordinate to Al(III) in the same way as does acetohydroxamic acid, with the amino group remaining protonated up to slightly basic pH [202,203]. However, the complexes formed with α-alaninehydroxamic acid are much less stable than those formed with acetohydroxamic acid [203, 204]. The corresponding stability constants relating to the stability of the first chelate are 8.05 and 5.19 for Al(III)–acetohydroxamate and Al(III)–α-alaninehydroxamate, respectively. This significant decrease is explained by two effects: an electron-withdrawing effect of the NH_3^+ group, which makes the hydroxamic acid group more acidic, and an electrostatic repulsion between the NH_3^+ group and the coordinating Al(III) ion. The explanation is supported by the higher stability of the complex formed with β-alaninehydroxamic acid [$(NH_3^+)CH_2CH_2CONHOH$] where the NH_3^+ group is farther from the metal ion [203]. In case of aspartic acid-β-hydroxamic acid [$HOOCCH(NH_3^+)CH_2CON$-HOH] and glutamic acid-γ-hydroxamic acid [$HOOCCH(NH_3^+)CH_2CH_2CONHOH$] tridentate coordination of the ligands is assumed, with participation of the carboxylates in the Al(III) binding [204].

3.9 Pyrone and Pyridinone Derivatives

Maltol is a commonly used flavoring additive in various bakery products, malted beverages, and chocolate milk drinks, especially in the U.S. and Canada. It forms a thermodynamically very stable, water-soluble, tris complex $Al(ma)_3$ at physiological pH [205]. This low molecular weight, neutral complex can readily cross cell membranes via a passive transport mechanism, and is therefore easily absorbed

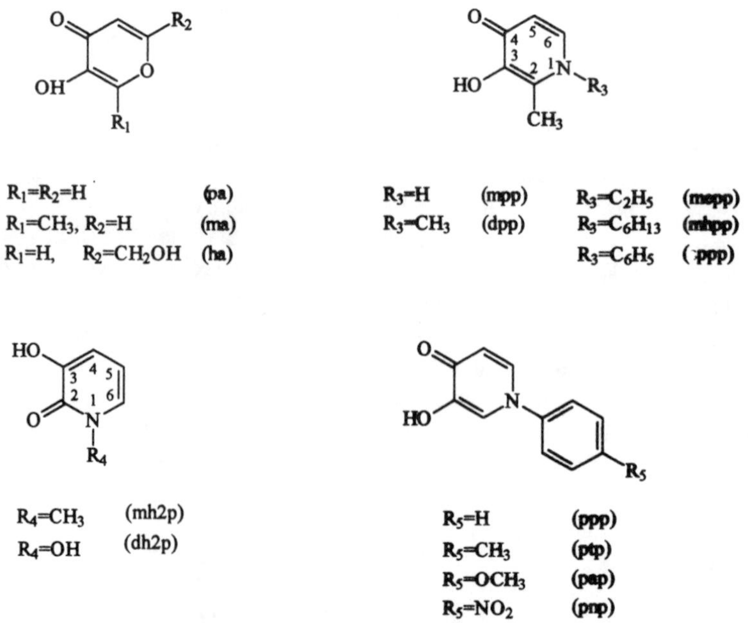

$R_1=R_2=H$ (pa)
$R_1=CH_3$, $R_2=H$ (ma)
$R_1=H$, $R_2=CH_2OH$ (ha)

$R_3=H$ (mpp)
$R_3=CH_3$ (dpp)
$R_3=C_2H_5$ (mepp)
$R_3=C_6H_{13}$ (mhpp)
$R_3=C_6H_5$ (ppp)

$R_4=CH_3$ (mh2p)
$R_4=OH$ (dh2p)

$R_5=H$ (ppp)
$R_5=CH_3$ (ptp)
$R_5=OCH_3$ (pap)
$R_5=NO_2$ (pnp)

Scheme 2.

by different test animals [206]. This fortunate combination of properties, i.e. water solubility, hydrolytic stability, and lipophilicity, has led to a more detailed research of Al(III) complexation with several 3-hydroxy-4-pyrones [207–209], 3-hydroxy-4-pyridinones [210–214] and 3-hydroxy-2-pyridinones [215], with the aim of their potential use in the study of Al neurotoxicity. The structures of the ligands studied are depicted in Scheme 2, while the stability data for the tris complexes are listed in Table 6.

Table 6. Stability Constants ($\log \beta_3$)[a] of Several Pyrone and Pyridinone Complexes of Al(III)

Ligand[b]	*log* β_3	*Ref.*	*Ligand*[b]	*log* β_3	*Ref.*
ma	22.47	118	mpp	32.05	123
ka	19.26	118	dpp	32.25	123
ima	14.45	127	mepp	32.17	123
mh2p	25.10	126	mhpp	31.71	123
dh2p	25.16	126	ppp	30.17	123

Notes: [a] $\beta_3 = [AlL_3]/[Al^{3+}][L^-]^3$.
[b] Abbreviations of the ligands are given in Scheme 2.

3-Hydroxy-4-pyridinone complexes proved to be the best candidates since their thermodynamic stability is extremely high (at least 10 orders of magnitude higher than that of the pyrone derivatives) (see Table 6), they have an easily functionalizable site on the ring, and the lipophilicity of the ligands (and of their Al(III) complexes too) can be varied by changes in the substituents on the ring N atom (without altering the high formation constants for the Al(III) complexes) [139]. The Al(III) complexes of 1-aryl-3-hydroxy-2-methyl-4-pyridinones proved to be the most lipophilic, as assessed by their partition coefficients between *n*-octanol and water, and this leads to their high lipid solubility and membrane permeability [213]. The higher stabilities of the Al(pyridinonate)$_3$ complexes as compared with those of the pyronates can be explained by the poorer ability of the N-containing ring to delocalize negative charge in the formation of the complex. In this way, because of the more polarized C(δ^+)–O(δ^-) bond, the hydroxy O is harder than in the hydroxypyrones [212]. Variable pH ^{27}Al NMR spectral measurements revealed that tris complexes are hydrolytically stable in the pH range from 4 to 9. Only a single ^{27}Al NMR peak is observed at around 38–39 ppm, which can be ascribed to a slightly distorted octahedral AlO_6 geometry [209, 210]. At higher pH, all complexes undergo hydrolysis to form the tetrahedral aluminate $[Al(OH)_4]^-$, but, they have a higher tendency to resist hydrolysis than most of the bidentate or tridentate ligands. It has been found that, under physiological conditions, dpp is a more effective Al(III) binder than edta, maltol, or catechol [129, 216].

Acetylacetone and other β-diketones form the corresponding mono-, bis-, and tris complexes with Al(III) as do the above pyrone or pyridinone derivatives. Taking into account the difference in basicity of the coordinating donors, their Al(III) binding ability (log $\beta = 21.4$) is about the same as that for the pyrones [217–220]; their water solubility is less, however, and this is also true for the Al(III) complexes [221,222]. For example, the *n*-octanol/water partition coefficient used to characterize the lipophilic–hydrophilic character of these compounds is almost 2 orders of magnitude larger for Al(acac)$_3$ than for Al(ma)$_3$—Al(acac)$_3$ being the more lipophilic [223]. The metal binding ability of this ligand group was studied in detail in view of their widespread application in analytical chemistry, mostly as solvent extraction reagents [218, 224].

Similar to the other ligands discussed in this group, ascorbic acid contains a "2-hydroxy-1-one" binding site, and thus it might be expected to bind Al(III) as strongly as does maltol. This similarity is only apparent, however, as both the acidity and the spatial arrangement of the donor groups are rather different. The acidity of the enolic-OH is pK_a ~ 4.0, while that of the 2-OH group of the pyrones or pyridinones is pK ~ 8–9. As regards its acid–base properties, ascorbic acid is much more similar to hydroxycarboxylic acids such as lactic acid. Further, the O–O distance in maltol, kojic acid, and tropolone is uniformly ca. 2.5 Å, while in ascorbic acid it is 2.98 Å. This probably means that the former ligands can all replace two water molecules from the inner coordination sphere of $[Al(H_2O)_6]^{3+}$ (O–O distance 2.66 Å), whereas the distance in ascorbic acid is too long to favor this substitution

[225]. Ascorbic acid forms a weak mononuclear species $[AlL]^{2+}$, together with some oligonuclear, probably trinuclear species [225]. The stability constant of AlL ($\log K = 1.38$) is about one order of magnitude lower than that of the corresponding lactate complex ($\log K = 2.36$) [86], but agrees well with the stability of Al(III)–monocarboxylates [76]. The oligonuclear complexes are formed close to the pH range where Al^{3+} starts to be hydrolyzed, and thus they are assumed to be hydroxo-bridged ternary species in which two hydroxy groups of ascorbic acid are deprotonated and coordinated to Al(III). Solution IR spectral results suggest metal chelation through O3 and O2 of the ascorbate ion. In the solid complex Al(ascorbate)$_3$, isolated from aqueous solutions at pH 6–7, the bonding mode is different: besides the two adjacent alcoholic-O atoms (the enolic one being deprotonated), the C1=0 carbonyl group is also involved in the coordination [226].

3.10 Porphyrins

Numerous Al(III)–porphyrins have been synthesized and characterized spectroscopically, electrochemically, or photochemically [227–231]. The aims of the studies were (i) to acquire a deeper insight into the structural and spectral properties of the biologically important Fe(III)–porphyrins, (ii) to obtain a better understanding of the role of chlorophyll in photosynthesis, and (iii) to study their possible catalytic activity [229]. Al(III)–porphyrins differ significantly in stereochemistry from the transition metal porphyrins. The main difference is the metal out-of-plane distance (ΔN), which is much larger for Al(III) complexes ($\Delta N \sim 0.4$–0.5 Å) than that for the transition metals ($\Delta N \sim 0.05$–0.15 Å) [180,181]. Al(III) porphyrins with a metal–carbon σ-bond in the axial direction were used to study the insertion of small molecules (such as carbon dioxide) between the metal ion and the carbon atom; such compounds may be photoactivated or sensitized by visible light [232].

We have given above a brief summary of the biologically relevant Al(III) complexes, with in some cases a characterization of the Al(III) binding ability of the bioligand in terms of the overall formation constants ($\log \beta$) of the complexes formed. The reader must bear in mind, however, that a direct comparison of these constants for a given set of ligands does not demonstrate the differences in binding strength at a particular pH. The competition between H^+ and Al^{3+} for the ligand must also be considered. When this proton competition is important the stability constants overstate the effective binding strengths of the ligands. The most practical method of allowing for proton–metal ion competition for a ligand is to calculate conditional stability constants, K_c, applicable at a given pH [193]. Conditional stability constant is one that corrects K or β for the fraction of unbound, deprotonated ligand (α) at a particular pH,

$$K'_c = K\alpha$$

where the value of α is between 0 and 1 and is governed by the acidity constant(s) of the ligand. Another commonly used parameter for a comparison of the Al(III)

Table 7. Conditional Stability Constants[a] and pM Values[b] for Al(III) Complexes at pH 7.4 and 25 °C

Ligand	*log* β_{cond}	*pM*	*Ligand*	*log* β_{cond}	*pM*
catechol(3)	15.5	12.3	citric acid (2)	12.7	12.9
dopa(3)	16.9	12.5	2,3-dpg(2)	12.2	12.1
dopamine(3)	17.1	12.7	ATP(2)	9.8	12.2
adrenaline(3)	18.2	13.2	nta(1)	11.6	13.3
noradrenaline(3)	18.4	13.4	edta(1)	14.7	16.4
tiron(3)	24.9	18.0	DFO(1)	19.2	20.9
kojic acid(3)	18.1	12.2	transferrin(1)	13.6	15.3
maltol(3)	18.4	12.6	—	—	—
isomaltol(3)	14.5	12.1	amorphous $Al(OH)_3$	—	10.7
hpp(3)	24.9	17.9	Al-phosphate[c]	—	11.4
salicylic acid(2)	11.7	12.1	$Al_2(OH)_4Si_2O_5$[d]	—	12.6

Notes: [a] β_{cond} values are calculated for the complex with maximum coordination number; it is given in parentheses after name of the ligands.

[b] Calculated for 1 μM total Al(III) and 50 μM total ligand concentration.

[c] 2 mM total phosphate, typical of plasma.

[d] 5 μM $Si(OH)_4$, typical of plasma.

binding abilities of the various ligands is the negative logarithm of the free Al^{3+} concentration, $-\log[Al^{3+}] = pAl$. This value is computed from conditional stability constants at known total metal ion and ligand concentrations. For biologically relevant Al(III) complexes, the value of pAl is commonly calculated for a solution containing 1 μM total metal and 50 μM total ligand concentrations [118]. Table 7 lists conditional stability constants and pAl values at the physiological pH 7.4, obtained from ref. 118 or recalculated from data reported in refs. 118,129. The higher the pAl value, the stronger the binding strength of the biomolecule.

4. TERNARY COMPLEXES

With water as solvent, a solution containing only the Al(III) ion is a real binary system. Because of the high affinity of Al(III) towards OH^-, a solution which contains this metal ion plus a single ligand besides OH^- is already a ternary system. Depending on the conditions (pH, metal-to-ligand ratio, and analytical concentrations) and on the character of the ligand, the formation of some mixed hydroxo species can be detected in almost all Al(III)–ligand systems. The effect of the pH is unambiguous. The formation of mixed hydroxo species can be expected only at pH higher than that is needed for the formation of the binary species $[Al(OH)]^{2+}$. On the other hand, in basic solutions, at pH > 9, the tetrahedral $[Al(OH)_4]^-$ generally

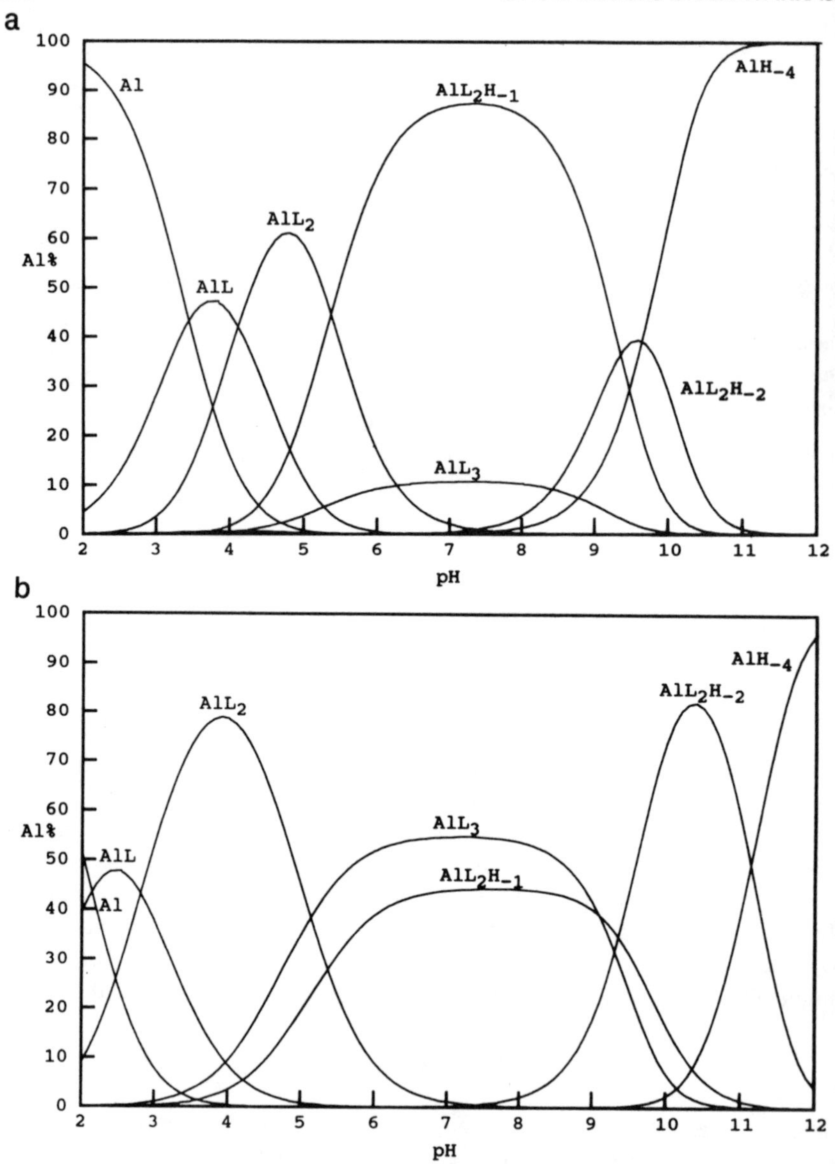

Figure 3. Speciation curves of the complexes formed in the Al(III)–acetohydroxamate system as a function of pH at 1:3 metal-to-ligand ratio. (**a**) C_{Al} = 2 mM, (**b**) C_{Al} = 40 mM.

predominates. Thus, the appearance of mixed hydroxo species is most likely in neutral and/or slightly basic solutions. The effect of the metal-to-ligand ratio is also well known. Increase of the ligand excess decreases the probability of formation of mixed hydroxo complexes, and vice versa. To analyze this problem, however, one has to take into account the Al(III)–ligand–OH^- ratios, which means that the analytical concentrations of Al(III) and the ligand are also very important. If the solution is diluted, the formation of mixed hydroxo species (as for hydrolysis in general) becomes more favored. This must be remembered when results determined at very different analytical concentrations are compared. The problem is illustrated in Figure 3, which depicts concentration distribution curves as representative examples for the Al(III)–acetohydroxamate system at 1:3 metal-to-ligand ratio. The analytical concentration of Al(III) was 2 mM for ^1H NMR (Figure 3a) and 40 mM (Figure 3b) for ^{27}Al NMR measurements [202].

The importance of the character of the ligand is also beyond doubt. A polydentate ligand completing the coordination sphere in a very stable 1:1 species (e.g. aminopolycarboxylates) hinders hydrolysis considerably. However, even in these cases mixed hydroxo species, e.g. [Al(OH)L], can be formed. It is an interesting result that the logarithmic stability constant for $[Al(OH)edta]^{2-}$, log $\beta \sim -5.8$ [232, 233], is quite close to the value for $[Al(OH)]^{2+}$, log $\beta = 5.5$ [52], showing that OH^- can displace one carboxylate in the above complex very easily. In contrast, the formation of $[Al(OH)_2edta]^{3-}$ is not observed at all. If the coordination sphere is completed by ligands in stepwise reactions, the hydrolysis becomes more significant. The details of mixed hydroxo complex formation with the different inorganic and organic ligands have been discussed in the appropriate sections of this chapter. One common phenomenon of these systems should be emphasized here. For multidentate ligands the hydrolysis constant characterizing the loss of the first proton from a complex [AlL] is usually smaller than that for the free ion $[Al(H_2O)_6]^{3+}$ and displays a nearly linear dependence on the stability of the [AlL] complex. This is in accordance with expectations since fewer (if any) water molecules are available for ionization in the Al complex than in the aquo ion. With mono-, bi-, or tridentate ligands, the complex [AlL] seems to contain a more acidic proton than does the $[Al(H_2O)_6]^{3+}$ ion. This suggests that the coordination sphere is reduced from hexacoordinate to a lesser number. Instead of a typical gap of at least 1 log unit between successive deprotonations for aqueous Al^{3+}, four deprotonations from $[Al(H_2O)_6]^{3+}$ to tetrahedral $[Al(OH)_4]^-$ occur over less than one pH unit. This cooperative process has been attributed to the reduction in coordination number [90]. Hence, the unusually acidic pK values may in all cases be ascribed to a concomitant reduction in coordination number [193].

Real biological systems always contain many potential metal ion binding biomolecules, and the formation of ternary complexes may therefore be of much greater importance than that of simple binary species. In spite of this obvious fact, little attention has been paid to ternary Al(III) systems. This may be explained by the relatively complicated nature of the binary Al(III) systems (as discussed above), the technical and methodological problems connected with the slow formation reactions, and the strong tendency of Al(III) to hydrolyze. These factors make equilibrium systems involving at least two different bioligands even more complicated and more difficult to monitor. For instance, 30 different complexes had to be taken into account in the speciation model for the quaternary Al(III)–nucleoside diphosphate–F^-–OH^- quaternary system [234]. Some studies on ternary Al(III) systems have been motivated by analytical problems [233, 235], e.g. to mask the metal ion for the determination of fluoride with an ion-selective electrode in the presence of Al(III) ion. The systems containing F^- plus one of the well-known masking reagents (such as nta, hedta, edta, and cdta) were found [233, 235] not to be applicable to solve the above problem, because of the formation of very stable ternary complexes, and further, strong competition was observed between $[Aledta]^-$ and F^- [233]. It is not surprising that no interaction was detected between $[Aledta]^-$ and Cl^-, Br^-, I^-, and SCN^-, but it is surprising that a very stable species, $[Al(edta)S]^{3-}$, was detected with the rather soft S^{2-} [233].

The interactions of Al(III)–nucleotides with F^- served as a model to study whether AlF_4^-, a tetrahedral pseudo-phosphate, can bind to guanosine diphosphate (or other nucleoside diphosphates (NDP)) in G-protein systems. In a recent paper, various ternary complexes $[(NDP)AlF_x]$ (x = 1 to 3) were identified by using ^{19}F and ^1H NMR techniques, but no $[(NDP)AlF_4]$ was found [234]. Ternary complexes were formed with a frequency predicted statistically on the basis of binary complex stabilities.

The possibilities of ternary complex formation have been studied in the Al(III)–adenosine-5′-phosphate(AMP, ADP, and ATP)–ligand B (oxalic acid, lactic acid, and malic acid) systems by pH-potentiometric and ^{31}P NMR methods [91]. Formation of the ternary complexes [AlHLB] and [AlLB] was favored in all systems in the acidic pH range. Under physiological conditions, however, Al(III) was bound mainly to the nucleotides: almost exclusively in the presence of the relatively weak bidentate Al(III) binders, oxalic acid and lactic acid, and to an extent of about 30% in the presence of the much stronger tridentate coordinating molecule, malic acid (see Figure 4).

Ternary Al(III)–phosphate–citrate complexes are assumed as potential small molecular weight Al(III) binders in the plasma. Although there are indications for ternary complex formation in the Al(III)–PO_4^{3-}–citrate system at acidic pH [236, 237], all our efforts to detect ternary complexes in this system and the Al(III)–ATP–citrate system at physiological pH have so far failed. The relatively unfavored formation of ternary Al(III) complexes at neutral and weakly basic pH might be connected with changes in the geometry of the complexes from octahedral to tetrahedral when OH^- can also be involved in the coordination, or with the different geometries of the binary species.

A computer model of the solution equilibria involving Al(III), Ga(III), and In(III) in the blood plasma has been constructed by Jackson [238] to investigate the *in vivo* fates of these metal ions. Besides the reported relevant binary stability constants and the very few ternary data on Al(III)–phosphate with cysteinate, malonate, and salicylate [236], and with oxalate and citrate [237], the extensive stability data were estimated from the stability constants of the respective binary species with a ternary complex stabilization factor of 1.0. This method is a built-in facility of the computer

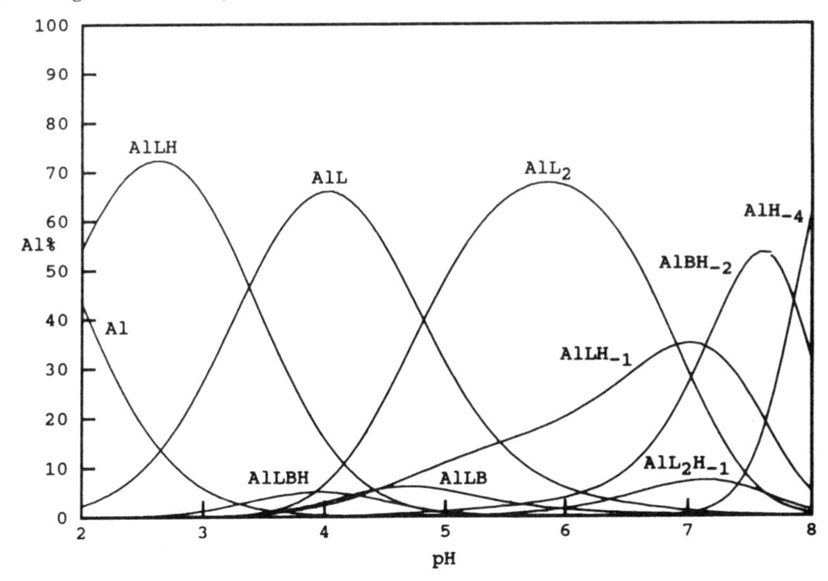

Figure 4. Species distribution curves for 1 μM total Al(III), 1 mM ATP (L), and 0.1 mM malic acid (B) using log β values given in Table 5 and HB, 4.57; H_2B, 7.73; AlHB, 7.34; AlB, 3.8; $AlH_{-1}B$, 1.32; $AlH_{-2}B$, −3.74; $AlH_{-1}B_2$, 4.74; AlHLB, 15.52; AlLB, 11.29.

program ECCLES [239], which is widely used, for example, in modeling the metal binding properties of the plasma or other biological fluids. However, it may be concluded from the above-mentioned results that more reliable data on the relevant model system would be necessary to draw any general conclusion on the actual importance of ternary complex formation in Al(III) binding processes in biological fluids. The above automatic procedure to estimate unknown ternary stability constants does not seem to be well established and accurate enough for model calculations, including those on Al(III) complexes.

5. ALUMINUM ABSORPTION AND METABOLISM

Because of the high degree of insolubility of its minerals, Al was unavailable for transport into primitive organisms during the early stage of biological evolution. As a result, most living systems have not adapted to elevated concentrations of Al(III) and cannot survive its large intracellular accumulation. The enhancement of environmental acidification has led to the Al content of natural waters increasing enormously in recent years [1]. As the bioavailability of Al(III) depends greatly on the chemical form in which it occurs, different methods have been elaborated to assess the distribution of Al in the different forms (see Section 2.2) by taking into account both homogeneous and heterogeneous reactions (see, for example, the relevant chapters in refs. 7 and 8). Chemical modeling based on equilibrium calculations can be a useful approach, although real systems are rarely in thermodynamic equilibrium. Dissolution and precipitation reactions are generally slower than reactions involving dissolved species. Furthermore, the precipitates initially formed are often amorphous and therefore metastable with respect to thermodynamically more stable solid phases. An additional complication with naturally occurring phases is that they seldom consists of pure solid phases [240]. For example, the solid phase suggested to control the aqueous concentration of Al(III) in natural waters is gibbsite [$Al(OH)_3$], whose equilibration time with water is in the range of days to years, depending on the chemical conditions. It has also been found that the actual solubility product of naturally occurring gibbsite is not identical with that of a synthetic, high-purity gibbsite [90]. Another difficulty with model calculations is the lack of a complete database with correct and reliable data for all possible equilibria. Although stability constant compilations [241–243] include large numbers of data on metal complexation, thermodynamic data on the surface complexation of aqueous Al species to pertinent solid phases are at present very rare. There are numerous computer programs capable of treating multicomponent, multiphase equilibrium systems which can be used to model the equilibrium speciation of Al(III) in various natural systems ranging from fresh waters and soil waters to various biological fluids [239, 240].

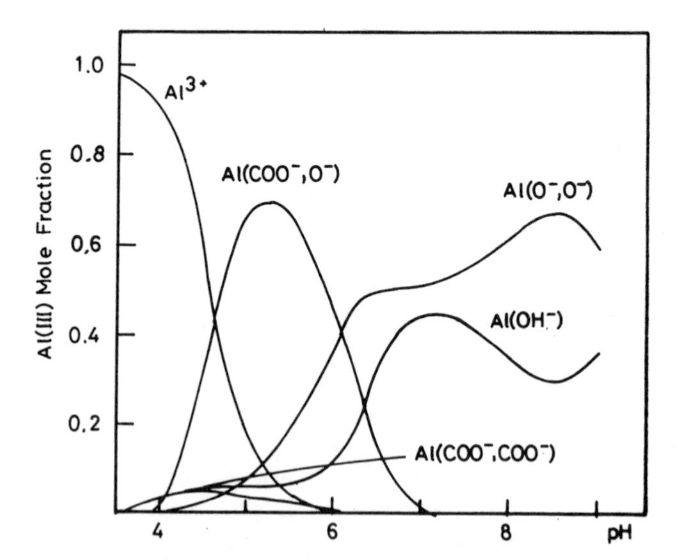

Figure 5. The sum of Al(III) mole fractions for Al(III)–phthalate [Al(COO⁻,COO⁻)], Al(III)–salicylate [Al(COO⁻,O⁻)], Al(III)–catecholate [Al(O⁻,O⁻)], and Al(III)–hydroxo [Al(OH⁻)] complexes as a function of pH. $C_{ligands}$ = 0.1 mM and crystalline gibbsite is the regulating solid phase (based on Ref. 240).

In applications to natural water, the main question to be answered is to what extent naturally occurring organic and inorganic ligands can influence the aqueous metal ion speciation in systems where a solid phase is also present and regulating the aqueous phase. Öhman and Sjöberg presented an illustrative example [240] on how simple aromatic and aliphatic binding sites occurring in the natural organic material humic and fulvic acids can influence the solubilization of gibbsite. Phthalic acid, salicylic acid, catechol, and citric acid were used as model compounds. The relative importance of these substances is indicated in Figure 5, which gives the fraction of Al(III) bound to each site when equal amounts (0.1 mM) of the ligands are allowed to equilibrate with gibbsite.

It is seen from the fractions of bound Al(III), that *o*-dicarboxylate is never an efficient binding site; the salicylic site exerts its maximum influence at pH ~ 5, while the catecholate site predominates at pH > 6. The calculations also demonstrate the importance of hydroxocomplex formation. Aliphatic citrate is much more effective (not shown in Figure 5) in the solubilization of gibbsite than any of the aromatic binding sites. Speciation studies with natural, chemically not fully characterized, humic and fulvic substances provided conditional metal binding constants which can be used in modeling calculations, and focused on the time-dependent nonequilibrium processes [244–246] that are of great importance in natural waters.

In human biological fluids, the metal ions present may be classified into different fractions: (i) those which are incorporated rigidly into the various metalloproteins and are nonexchangeable, (ii) those which are relatively loosely bound by other types of proteins and are exchangeable, (iii) those which are complexed by low molecular weight biomolecules including amino acids, carboxylates, phosphate, ascorbate, citrate, etc., and (iv) the free metal aquo ions. Concerning the occurrence of Al(III) in body fluids, the last three fractions represent the more or less mobile forms which can be transported to different organs, can transfer across membranes, and can interfere with normal metalloprotein actions. In general, very little Al(III) is absorbed in healthy humans since the very efficient defending system causes most of the ingested Al(III) to be eliminated via the feces in the form of poorly soluble $Al(OH)_3$ and $AlPO_4$. There is considerable evidence, however, that Al(III) can be absorbed when appropriately complexed in the stomach. There is convincing proof that citrate facilitates the absorption and incorporation of Al(III) into tissues [247–249]. Citrate has been shown in Section 3.3 to form the neutral [AlL] species at pH > 6; this can readily cross membranes by passive transport and thus is assumed to be mainly responsible for the absorption of Al(III) [118, 193].

As the absorbed metal is transported via the blood stream, speciation of Al(III) in blood plasma is an essential question. Potential plasma Al(III) binders in class (ii) are albumin and transferrin, while in class (iii) are citrate, phosphate, and hydroxide. Other plasma components—such as the various amino acids, lactate, oxalate, ascorbate, sulfate, and carbonate—bind Al(III) much weaker and are thus

Table 8. Suggested Species Distribution of Aluminum(III) in Serum[a]

	Ref. [238][b]	*Ref. [69]*[b]	*Ref. [61]*[c]	*Ref. [68]*[b]	*Ref. [250]*[d]
Al(III)–albumin	–	–	–	–	+
Al(III)–transferrin	63	81	89(5)	90	+
Al(III)–phosphate	23	16	–	2	?
Al(III)–citrate	8	2	11(5)	8	–
Al(III)–phosphate–citrate	5	+	?		

Notes: [a] Notation for qualitative assessment: +: positive indication, –: negative indication, ?: likely occurrence.
[b] Computer modeling.
[c] Solubility and stability data comparison.
[d] Chromatographic separation and NMR.

not efficient competitors. The possible species distribution results suggested by different authors using somewhat different approaches are listed in Table 8.

It can be seen from the results that transferrin is the most significant plasma Al(III) binder. *In vitro* studies have also demonstrated the weak nonspecific binding of Al to albumin [250]. *In vivo*, however, albumin is a much weaker metal binder, and is not assumed to be able to compete for significant amounts of Al(III) [69, 118, 193]. Transferrin, the main iron transport protein, binds Fe(III) ~ 10 orders of magnitude more strongly than Al(III) [146], but it is only 30% saturated with Fe(III) under normal conditions, thus the vacant sites can be filled by Al(III). The picture as to the most important low molecular weight Al(III) binders is much more controversial. On the basis of computer modeling calculations, Jackson [238] and Harris [69] propose inorganic phosphate as the main Al(III) binder. As discussed in Section 3.1, the species distribution in the Al(III)–monophosphate system at physiological pH cannot be obtained directly because of the precipitation of $AlPO_4$ at pH > 4. Extrapolation to pH ~ 7.4 seems to be the crucial point of serum model calculations. The stability data used by Harris for $[Al(PO_4)]$ and $[Al(OH)(PO_4)]^-$ are based on LFER calculations [69]. There is no evidence, however, of the existence of a soluble $[Al(PO_4)]$ species. At the same time, soluble species $[AlHPO_4]^+$ and $[Al(OH)(HPO_4)]$ are certainly formed [68]. The latter species has the overall stoichiometry of $[Al(PO_4)]$, the same as was assumed by Harris, but with an estimated stability constant about 1 order of magnitude lower: log $K_{Al(OH)(HPO_4)}$ =13.50 [68]. Estimation of reliable stability data for further possibly existing monomeric mixed hydroxo species, such as $[Al(OH)_2(HPO_4)_2]^{2-} \equiv [Al(OH)(PO_4)]^{2-}$, and dimeric complexes $[Al_2PO_4]^{3+}$, $[Al_2(OH)_2(PO_4)]^+$, or $[Al_2(OH)_3PO_4]$ [66, 67] (although they might be of minor importance at the μM Al^{3+} level), is even more difficult. Hence, the binding ability of phosphate under serum conditions seems somewhat overestimated by Harris [69] and Jackson [238]. Martin is strongly convinced that citrate is the only important low molecular weight Al(III) binder in the plasma [61], while Fatemi et al. deny its significance [250].

There is a similar uncertainty concerning citrate, although not so serious regarding phosphate speciation. Monomeric 1:1 and 1:2 complexes—$[Al(H_{-1}L)]^-$, $[Al(OH)(H_{-1}L)]^{2-}$, $[Al(H_{-1}L)L]^{4-}$, and $[Al(H_{-1}L)_2]^{5-}$—formed in relatively fast reactions, are likely to control the species distribution under physiological conditions. Oligomeric complexes predominating under equimolar equilibrium conditions [30] may be of much less importance. Formation of a variety of ternary complexes, such as Al(III)–citrate–phosphate, and with other ligands besides citrate or phosphate, is a reasonable assumption. However, the results discussed in Section 4 do not greatly support their significance in Al(III) binding.

Although modeling calculations and *in vitro* model studies have led to contradictory results, the binding of Al(III) to citrate has recently been detected directly by ^1H NMR in intact blood plasma samples [251]. In other body fluids, such as human milk, saliva, stomach, and intestinal fluid rich in citrate and phosphate, these two biomolecules are the most likely low molecular weight Al(III) binders [67]. In brain tissues and fluids low in citrate and transferrin, catecholamines may become important Al(III) binders [187]. A comparison of the conditional stability constants and pM values listed in Table 7 indicates that large excesses of catecholamines would be required to chelate Al(III) in the presence of citrate or nucleotides (ATP). In red blood cells, 2,3-dpg (see Section 3.5) present in mM concentration can be the main binder of Al(III) besides ATP [150].

6. THE TOXIC EFFECTS OF ALUMINUM

Absorbed Al(III) is mostly eliminated by the kidneys, and it is therefore not surprising that toxic effects of Al(III) were first observed in patients with advanced renal failure. It is now well established that Al(III) can exert toxicity on the neurological, skeletal, and hematological systems in humans with an impaired renal function. It is less well established under what circumstances Al(III) can be toxic in "healthy" individuals. The possible damaging effects of Al(III) may originate from its ability to replace Mg^{2+} and Ca^{2+} due to the stronger binding strength. For example, Al^{3+} binds organic phosphates including nucleotides 10^6–10^7 times more strongly than does Mg^{2+} [118]. Since Ca^{2+}, Mg^{2+}, and phosphate take part in numerous biochemical processes, Al^{3+} has many opportunities to disrupt functions in an organism. For details of Al(III) toxicity, and sometimes controversial views on the involvement of Al(III) in various neurological disorders, we refer the reader to some excellent reviews and recent papers [252–255].

6.1 Neurological Toxicity

In the 1970s, increased serum and brain Al(III) levels were observed in dialysis patients with a renal insufficiency. Alfrey et al. [256] first described the neurological syndromes of this dialysis encephalopathy or dialysis dementia. Increased Al(III) concentrations could also be detected in several brain regions of patients with

Alzheimer's disease (AD). Brain autopsies of AD victims reveal the deposition of fibrillar amyloid proteins as paired helical filaments that contain high levels of Al(III), and of senile plaques that may contain aluminosilicate. Whether Al(III) causes the formation of neurofibrillary tangles (NFT) with paired helical filaments, or only has an affinity to bind to abnormal tangles, is uncertain. It is known that AD neurons are rich in phosphorylated proteins. Abnormally and over-phosphorylated proteins have been found in brain cells, especially in the nuclear chromatin region [257]. It is demonstrated in Section 3.5 that the phosphate groups of these proteins are basic enough, and together with neighboring carboxylate or other phosphate groups, are strong Al(III) binding sites. Al^{3+} induces conformational changes in these neurofilament proteins, yielding a high content of β-pleated sheet structure [157]. The active involvement of Al(III) in NFT formation is also suggested by the fact that Al^{3+} was found to be a potent initiator (even at the nM level) of the normally Mg^{2+}-promoted tubulin polymerization into microtubules [258]. This extremely effective stimulatory role of Al^{3+} was explained by the fact that Al^{3+} binds $10^{7.5}$ times more strongly than Mg^{2+} to the tubulin–guanosine-5′–triphosphate complex, via the chelating site of the nucleotide molecule. Formation of this ternary complex is assumed to be essential for polymerization [258, 259].

6.2 Aluminum-Induced Bone Disease

Osteomalacia is a pathological condition of the adult skeleton, characterized by defective calcification of the bone matrix and a consequent reduction in bone strength [260]. Al(III) can induce osteomalacia in patients with chronic renal failure. It is associated with a reduced mineralization rate, with an excess of osteoid (nonmineralized bone matrix) formation, and with preferential $AlPO_4$ deposition at the interface between the osteoid and the calcified matrix. The action of Al(III) on bone cells is a result of the much lower solubility of $AlPO_4$ than that of hydroxy-apatite $Ca_5(PO_4)_3(OH)$, a principal constituent of bones and teeth. According to Martin's model calculation [193], the free Ca^{2+} concentrations permitted by hydroxyapatite under bone-cell conditions is 10^8 times greater than the free Al^{3+} concentrations allowed by insoluble $AlPO_4$. Thus, thermodynamically, Al^{3+} can easily displace Ca^{2+} from phosphate binding. It has been reported that Al(III) [and also Fe(III)] are extremely effective inhibitors of calcium phosphate precipitation [261], which is greatly potentiated by the addition of a small amount of citrate. It is known that Al(III) functions by adsorbing to the crystal surface and thereby prevents the incorporation of lattice ions.

6.3 Anemia

Beside dementias and osteomalacia, Al(III) overload is classically associated with anemia too. Its morphological appearance resembles the iron-deficient disease, the red blood cells being shrunken (microcytosis), but it does not show any response to iron therapy, indicating a normal iron level. The mechanisms by which

Al(III) leads to microcytic anemia are not yet fully understood and are probably complex. On the basis of *in vitro* studies, it has been postulated that Al(III) decreases heme synthesis since it has been shown to be capable of inhibiting heme-synthesizing enzymes such as ferrochelatase or uroporphyrin decarboxylase [262]. Al(III) is transported by transferrin in the blood plasma, thus it can enter cells in the same way as Fe(III) via the transferrin–transferrin receptor system. In a red blood cell, 2,3-dpg present at 2–4 mM concentration seems to be the predominant small ligand. Since the amount of 2,3-dpg considerably exceeds that of hemoglobin with which it interacts in the deoxygenated form, Al(III) levels would probably not rise high enough to withdraw significant quantities of 2,3-dpg from interacting with hemoglobin. However, a 2,3-dpg complex of Al(III) may be an active deleterious species. As to the Al(III) binding ability of 2,3-dpg (see Table 7), this is comparable with those of both transferrin and citrate, the two important plasma binders of Al(III); hence at equilibrium there should be comparable concentrations of Al(III) on both sides of the red cell membrane [150]. In this way, Al(III) can accumulate in the blood cells and interfere with hemoglobin synthesis.

7. SPECIATION AND EXPERIMENTAL TOXICOLOGY OF ALUMINUM

The observed phenomenological connection between Al accumulation and human pathologies has stimulated extensive experimental toxicological work. However, the results obtained with various Al toxins (mostly with an aqueous solution of some inorganic Al salt) are rather controversial. Understanding the solution chemistry of Al(III), it is clear that the chemical form of Al(III) administered is crucial to the expression of its biological effects. In a recent *in vitro* study, Exley et al. [263] have demonstrated that the enzyme inhibitory effect of Al(III) is strongly dependent upon the equilibrium state of the Al(III) stock solution used in the measurements. Reliable results can be obtained only if the administered toxin is precisely identified chemically and that it can reach the biological target organ [264, 265]. The solution state of an Al(III) salt at physiological pH is determined by the very low metal ion solubility and very slow ligand exchange reactions, generally resulting in undefined metastable conditions (see Section 2.1). Al(III) solutions intended to be used as toxins can be stabilized against precipitation with comparable concentrations of O-donor ligands, such as lactate, citrate, tartrate, gluconate, maltolate, and acetylacetonate. The thermodynamics of the Al(III) complexes discussed in Sections 3 and 4 can provide a description of the equilibrium solution state of the Al toxins, but this is severely affected by kinetic factors, and thus it can be far from the actual toxicologically relevant solution state.

Using IR and NMR techniques, Corain et al. studied the speciation of different Al(III)–hydroxy-carboxylate complexes (in part of commercial origin and in part lab-synthesized) after dissolution in water and adjustment of the pH to 7.5 [109, 266]. He found significant disagreement with the thermodynamic expectations based on speciation calculations, at least in fairly concentrated solutions (100 mM). Al(III)–lactate, and Al(III)–tartrate were shown to be in a metastable state in solution, containing binary and/or ternary hydroxo complexes and the free ligand. Al(III)–gluconate, in contradiction with the speciation model, did not release the ligand, probably as a consequence of the particular kinetic inertness of the complex. Al(III)–citrate was found to behave according to the thermodynamic prediction. This lends support to the suggestion of Martin that in toxicological studies Al(III) should be administered in the form of the soluble citrate complex, using a citrate-to-metal ion mole ratio of at least 2:1 [60]. It must also be kept in mind that both the speciation and the kinetic properties of the Al toxins are strongly influenced by the metal concentration and the metal ion-ligand ratio (cf. Section 5), and thus further investigations are necessary. Specific work in this respect would be of great help to experimental toxicologists.

At physiological pH, monoanionic bidentate ligands can form neutral tris complexes which are particularly suitable for toxicological studies. $Al(lact)_3$ was shown to be hydrolytically unstable [88], while $Al(acac)_3$, $Al(malt)_3$, and similar Al(III) complexes of various hydroxypyridinone derivatives are structurally well characterized and hydrolytically stable [139], thereby allowing precise dose-response relationships to be established with these latter compounds.

Both $Al(malt)_3$ [267] and $Al(acac)_3$ [268] have been reported to be about 2 orders of magnitude more neurotoxic than $Al(lact)_3$, which is obviously connected with the large difference in their hydrolytic stability. Further, the hydrophilic–lipophilic character of Al(III) complexes of the hydroxy-pyridinone derivatives can be easily varied by changes in the ring substituents [139], with a resulting alteration in their ability to cross cell membranes, which can be an important toxicological goal.

8. ALUMINUM AND MEDICINES

8.1 Pharmaceutical Uses of Aluminum Compounds

Various Al compounds are used in pharmaceuticals, e.g. hydroxides, hydroxycarbonates, some organic complexes, and simple inorganic salts. They may be employed as antacids, buffers, gastrointestinal protectives and adsorbents, pepsin inactivators, adjuvants in vaccines, phosphate binders, astringents, and antiperspirants [269].

Amorphous aluminum hydroxide gel is a very efficient antacid having high buffer capacity and a fairly low buffering pH. Dried aluminum hydroxide suitable for the formulation of tablets or capsules proved to be less effective than suspensions due to the crystallization of $Al(OH)_3$. Lyophilization and drying from a water-miscible, nonaqueous vehicle have successfully produced dried aluminum hydroxide that is highly amorphous and exhibits the antacid properties of the gel [269].

The urinary excretion of phosphate represents a major route of phosphate loss from the body, thus chronic renal failure patients retain phosphate and develop high serum phosphate levels. Amorphous aluminum hydrocarbonate (used also as an antacid) is frequently administered to lower the serum phosphate level. Two mechanisms account for this effect: (i) phosphate is adsorbed by aluminum at the surface of the particles [270], and (ii) soluble Al(III) species formed through acid neutralization in the stomach react with phosphate in the intestine to form insoluble $AlPO_4$ [271]. Phosphate-binding gels which produce soluble Al(III) species can cause a systemic aluminum accumulation. Crystalline forms of $Al(OH)_3$ are non-reactive with gastric acid, but having a high surface area could provide phosphate-binding properties without producing soluble Al(III) species, resulting in the gastrointestinal absorption of Al(III). Different Al-free agents, such as $CaCO_3$ [272]-, Ca^{2+}-, and Fe^{3+}-containing heteropolyuronic acid polymers [273] have also been elaborated and chemically tested as an alternative means of treating hyper-phosphatemia.

8.2 Chelation Therapy in Aluminum Detoxification

Aluminum removal is essential in treating patients with Al accumulation. Currently the most effective method is chelation of Al(III) with DFO. As it is discussed in Section 3.8, DFO forms an extremely stable complex with Al(III) (log $K = 22$); under physiological conditions DFO is the strongest Al(III) binder (see Table 7). Thermodynamically DFO should extract Al(III) even from plasma transferrin. However, owing to the tight binding and slow release of Al(III) by transferrin, Al(III) transfer to DFO is likely to be negligible during a typical treatment time [193]. Upon administration of the ligand, Al(III) is mobilized from the tissues and the plasma Al(III) level increases as a result of formation of the [AlDFO] chelate, which may be removed by dialysis [274]. One advantage of DFO treatment is that it forms much stronger complexes with tripositive metal ions than with dipositive essential metal ions, thus it does not disturb their metabolism. As a side effect in long-term treatment, there is a decrease in ferritin levels since it binds Fe(III) much more strongly (log $K = 31$) than Al(III). A normal iron level, however, can be restored relatively easy via a temporary withdrawal of DFO therapy and/or iron supplementation [275]. DFO is now widely and effectively used in the therapy of dialysis patients suffering from dialysis dementia, osteomalacia, and anemia. Further, it has been shown that the sustained use of low-dose DFO slows the progression of the dementia of AD patients [276].

8.3 Aluminum and Silicon

As discussed in Section 3.1, Al(III) and silicic acid readily form poorly soluble hydroxyaluminosilicates under physiological conditions [32]. There may be two major biological implications of hydroxyaluminosilicate formation: the first relates to the availability of Al to living organisms via the fish gill, plant root membranes, and the gastrointestinal tract. The second relates to the *in vivo* transport, binding, and cellular toxicity of Al(III) [277].

Epidemiological studies have indicated a relationship between the Al(III) content of domestic water and the incidence of AD [278], and an inverse relationship with the silicon content of the water [279]. The monomeric silicic acid in water is conducive to a lower bioavailability of Al(III) via the formation of insoluble aluminosilicates.

There are indications that silicic acid increases aluminum excretion in humans. Beside citric acid, silicic acid is also assumed to contribute to Al(III) binding in the plasma, keeping Al(III) in low molecular weight complexes which can be rapidly excreted, prompting the mobilization of Al(III) [277].

Because of the strong Al(III)–phosphate interaction, a disturbing effect of Al(III) on the inositol–phosphate second messenger system and hence on intracellular Ca^{2+} homeostasis is presumed. *In vitro* studies have shown that the injection of silicic acid leads to a protective effect against the inhibitory effect of Al(III) [280].

Finally, aluminosilicate deposition can occur in senile plaques in dead brain tissues. According to Birchall this is likely to be connected with the observed change in the binding of Al(III) between phosphate and silicate at around pH ~ 6.6 [277]:

Al(III) phosphate species + free silicic acid	← pH < 6.6	$Al(OH)_x^{(3-x)+}$ PO_4^{3-} $Si(OH)_4$	pH > 6.6 →	Hydroxy-aluminosilicate species + free phosphate

The intraneuronal Al(III) (at pH ~ 6.6) is assumed to be bound to phosphate sites of phosphorylated proteins (see Section 3.4). When neurons die (partly as a result of the toxicity of Al), Al(III) is exported into the extracellular environment (pH ~ 7.4), and the higher pH will allow its binding to silicic acid and its deposition in senile plaques.

These results suggest that the beneficial effect of silicic acid is to aid the exclusion of Al(III) from organisms, to "sequester" the metal, thereby reducing its effect on enzymic processes, and to reduce tissue retention, promoting the excretion of Al(III).

ACKNOWLEDGMENTS

The authors are grateful to Professor R.B. Martin (University of Virginia, Charlottesville) for helpful comments on the manuscript and Mrs. Á. Gönczy for her help in the technical work. This publication is sponsored by the U.S.–Hungarian Science and Technology Joint Fund in cooperation with Department of Health and Human Services, U.S. and Ministry of

Social Welfare, Hungary under Project No. 182/92b and by National Science Research Fund, Hungary under Project No. OTKA/T7458.

REFERENCES

[1] Tamm, S. C. and Williams, R. J. P., J. Inorg. Biochem., 26 (1986) 35.
[2] De Broe, M. E. and Van de Vyver, F. L. (eds.), *Aluminum a Clinical Problem in Nephrology,* Clinical Nephrology, 24, Supplement 1, 1985.
[3] Coburn, J. W. and Alfrey, A. C. (eds.), *Conference on Aluminum Related Disease,* 1984, Kidney International, 28, Supplement 18, 1986.
[4] Lord Walton of Detchant (ed.), *Alzheimer's Disease and the Environment,* Royal Soc. Med. Services, Round Table Series 26, London, 1991.
[5] *Aluminum in Biology and Medicine,* Ciba Foundation Symposium 169, Wiley, Chichester, 1992.
[6] Epstein, S. G. (ed.), *Aluminum and Health, The Current Issues,* The Aluminum Assoc., Washington, 1991.
[7] Sposito, G. (ed.), *The Environmental Chemistry of Aluminum,* CRC Press, Boca Raton, 1989.
[8] Lewis, T. E. (ed.), *Environmental Chemistry and Toxicology of Aluminum*, Lewis Publ., Chelsea, 1989.
[9] Massey, R. and Taylor, D. (eds.), *Aluminum in Food and the Environment,* Royal Soc. Chem. Spec. Publ. No.73., London, 1989.
[10] Gitelman, H. J. (ed.), *Aluminum and Health,* Marcel Dekker, New York, 1989.
[11] De Broe, M. E. and Coburn, J. W. (eds.), *Aluminum and Renal Failure,* Kluwer Acad. Publ., Dordrecht, 1990.
[12] Nicolini, M., Zatta, P. F. and Corain, B. (eds.), *Aluminum in Chemistry, Biology and Medicine,* Vol. 1, Cortina International, Verona, 1991; Vol. 2, in Life Chemistry Reports, 11, 1994.
[13] Onishi, H., in *Chemistry of Aluminum, Gallium, Indium and Thallium,* Blackie Academic, London, 1993, Ch. 10, pp. 491–510.
[14] Savory, J. and Wills, M. R., in Gitelman, H. J. (ed.), *Aluminum and Health,* Marcel Dekker, New York, 1989, pp. 1–26.
[15] Bloom, P. R. and Erich, M. S., in Sposito, G. (ed.), *Environmental Chemistry of Aluminum,* CRC Press, Boca Raton, 1989, pp. 1–26.
[16] Carrillo, F., Pérez, C. and Cámara, C., Anal. Chim. Acta, 243 (1991) 121.
[17] Scheller, K., Letzel, S. and Angerer, J., in Seiler, H. G., Sigel, A. and Sigel, H. (eds.), *Handbook on Metals in Chemical and Analytical Chemistry,* Marcell Dekker, New York, 1993, pp. 217–226.
[18] Nicolini, M., Zatta, P. F. and Corain, B. (eds.), *Aluminum in Chemistry, Biology and Medicine,* Vol. 2, in Life Chemistry Reports, 11, 1994, Section I, pp. 3–98.
[19] Zatta, P. F., Trace Elements and Medicines, 10 (1993) 120.
[20] Krishnan, S. S., McLachlan, D. R., Krishnan, B., Fenton, S. S. A. and Harrison, J. E., Sci. Total. Environ., 71 (1988) 59.
[21] Jacobs, R. W., Doung, T., Jones, R. E., Trapp, G. A. and Scheibel, A. B., Can. J. Neurol. Sci., 16 (1989) 498.
[22] Ehmann, W. D., Markesbery, W. R., Kasarksis, E. J., Vance, D. E., Khare, S. S., Hord, J. D. and Thompson, J. R., Biol. Trace Elem. Res., 13 (1987) 19.
[23] Perl, D. P. and Good, P. F. in International Conference on Aluminium and Health, The Aluminium Association, Orlando, 1990.
[24] Stern, A. J., Perl, D. P., Munoz-Garcia, D., Good, P. F., Abraham, C. and Selkoe, D. J., Neuropathol. Exp. Neurol. (Abstract), 45 (1986) 361.
[25] Corrigan, F. M., Reynolds, G. P. and Ward, N. I., BioMetals, 6 (1993) 149.
[26] Landsberg, J. P., McDonald, B. and Watt, F., Nature, 360 (1992) 65.

[27] Watt, F., Grime, G. W., Gadd, G. M., Candy, J. M., Oakley, A. E. and Edwardson, J., in Proceedings XI. International Congress X-ray Optics and Microanalysis, University of Western Ontario, 1986, pp. 127–136.
[28] Datta, A. K., Wedlund, P. J. and Yokel, R. A., J. Trace Elem., Electrolytes Health Dis., 4 (1990) 107.
[29] Van Landeghem, G. F., D'Haese, C. D., Lamberts, L. V. and De Broe, M. E., Anal. Chem., 66 (1994) 216.
[30] Öhman L.-O., Inorg. Chem., 27 (1988) 2565.
[31] Akitt, J. W., Progr. NMR Spectroscopy, 21 (1989) 1.
[32] Martin, R. B., Polyhedron, 9 (1990) 193.
[33] E. Högfeldt (ed.), *Stability Constants of Metal-Ion Complexes, Part A, Inorganic Ligands,* Pergamon Press, Oxford, 1982.
[34] Vobe, R. A. and Williams, D. R., Chem. Speciation Bioavability, 4 (1992) 85. (Chem. Abstr., 118 (1993) 57004p).
[35] Knoche, W. and Lopez-Quintela, M. A., Thermochim. Acta, 62 (1983) 295.
[36] Akitt, J. W., Farnsworth, J. A. and Lettellier, P., J. Chem. Soc., Faraday Trans. I, 81 (1985) 193.
[37] Matsushima, Y., Matsunaga, A., Sakai, K. and Okuwaki, A., Bull. Chem. Soc. Japan, 61 (1988) 4259.
[38] Öhman, L.-O. and Forsling, W., Acta Chem. Scand., A35 (1981) 795.
[39] Hedlund, T., Sjöberg, S. and Öhman, L.-O., Acta Chem. Scand., A41 (1987) 197.
[40] Smith, R. M. and Martell, A. E., *Critical Stability Constants,* Vol. 1, Plenum Press, New York, 1974.
[41] Öhman, L.-O. and Sjöberg, S., Marine Chem., 17 (1985) 91.
[42] Hunter, D. and Ross, D. S., Science, 251 (1991) 1056.
[43] Kinraide, T. B., Plant and Soil, 134 (1991) 167.
[44] Shann, J. R. and Bertsch, P. M., Soil. Sci. Soc. Amer. J., 57 (1993) 116.
[45] Flaten, T. P. and Gurrato, R. M., J. Theor. Biol., 156 (1992) 129.
[46] Akitt, J. W. and Elders, J. M., J. Chem. Soc., Faraday Trans. I, 81 (1985) 1923.
[47] Akitt, J. W. and Elders, J. M., J. Chem. Soc., Dalton Trans., (1988) 1347.
[48] Akitt, J. W., Elders, J. M., Fontaine, X. L. R. and Kundu, A. K., J. Chem. Soc. Dalton Trans., (1989) 1897.
[49] Thompson, A. R., Kunwar, A. C., Gutowski, H. S. and Oldfield, E., J. Chem. Soc., Dalton Trans., (1987) 2317.
[50] Bertsch, P. M., Layton, W. J. and Barnhisel, R. I., Soil. Sci. Soc. Amer. J., 50 (1986) 1449.
[51] Bertsch, P. M., Soil. Sci. Soc. Amer. J., 51 (1987) 825.
[52] Sjöberg, S., Öhman, L.-O. and Ingri, N., Acta Chem. Scand., A39 (1985) 93.
[53] Öhman, L.-O., Inorg. Chem., 27 (1988) 2565.
[54] Brown, P. L., Sylva, R. N. and Batley, G. E., J. Chem. Soc., Dalton Trans., (1985) 1967.
[55] Milic, N. B., Bugarčič, Z. D. and Djurdievic, P. T., Can. J. Chem., 69 (1991) 28.
[56] Milic, N. B., Bugarčič, Z. D. and Niketic, S. R., Gazetta Chim. Italiana, 121 (1991) 45.
[57] Venturini, M. and Berthon, G., J. Chem. Soc., Dalton Trans., (1987) 1145.
[58] Hayden, P. L. and Rubin, A. J., Separation Sci. and Techn., 21 (1986) 1009.
[59] Baes, C. F. and Mesmer, R. E., *The Hydrolysis of Cations*, Wiley, New York, 1976.
[60] Martin, R. B., Acc. Chem. Res., 27 (1994) 204.
[61] Öhman, L.-O. and Martin, R. B., Clin. Chem., 40 (1994) 598.
[62] Corain, B., Kiss, T. and Zatta, P. F., Coord. Chem. Rev., 149 (1996) 329.
[63] Teng, Q. and Waki, H., Polyhedron, 10 (1991) 659.
[64] Wolfram, R. and Steger, W. E., Z. Phys. Chem., 176 (1992) 185.
[65] Jackson, G. E. and Voyi, K. V. V., S. Afr. Chem., 41 (1988) 17.
[66] Daydé, M., Filella, M. and Berthon, G., J. Inorg. Biochem., 38 (1990) 241.

[67] Duffield, J. R., Edwards, K., Evans, D. A., Morrish, D. M., Vobe, R. A. and Williams, D. R., J. Coord. Chem., 23 (1991) 277.
[68] Atkári, K., Thesis, Kossuth University, Debrecen, Hungary, 1994.
[69] Harris, H. R., Clin. Chem., 38 (1992) 1809.
[70] Berthon, G. and Daydé, S., J. Amer. Coll. Nutrition, 11 (1992) 340.
[71] Berthon, G., in Anastassopoulou, I., Gollery, Ph., Etienne, J.C. and Theophanides, Th. (eds.), *Metal Ions in Biology and Medicine,* Vol. 2, John Libbey Eurotext, Paris, 1992, p. 253.
[72] Gushikem, Y., Giesse, R. and Volpe, P. L. O., Thermochimica Acta, 68 (1983) 83.
[73] Feng, Q. and Waki, H., Polyhedron, 10 (1991) 1527.
[74] Miyajima, T., Maki, H., Sakurai, M., Sato, S. and Waranabe, K., Phosphorus Res. Bull., 3 (1993) 31.
[75] Martin, R. B., Biochem. Biophys. Res. Commun., 155 (1988) 1194.
[76] Marklund, E., Öhman, L.-O. and Sjöberg, S., Acta Chem. Scand., A43 (1989) 641.
[77] Öhman, L.-O., Acta Chem. Scand., 45 (1991) 258.
[78] Yamada, H., Hayashi, H., Fujii, Y. and Mizuta, M., J. Chem. Soc. Japan, 59 (1986) 789.
[79] Sjöberg, S. and Öhman, L.-O., J. Chem. Soc. Dalton Trans., (1985) 2665.
[80] Thomas, F., Masion, A., Bottero, J. Y., Rouiller, J., Genèvrier, F. and Boudot, D., Environ. Sci. Technol., 25 (1991) 1553.
[81] Marklund, E. and Öhman, L.-O., Acta Chem. Scand., 44 (1990) 353.
[82] Powell, H. K. J. and Town, R. M., Aust. J. Chem., 46 (1993) 721.
[83] Bilinski, H., Horváth, L., Ingri, N. and Sjöberg, S., Geochim. Cosmochim. Acta, 50 (1986) 1911.
[84] Hedlund, T., Bilinski, H., Horváth, L., Ingri, N. and Sjöberg, S., Inorg. Chem., 27 (1988) 1370.
[85] Lövgren, L., Geochim. Cosmochim. Acta, 55 (1991) 3639.
[86] Marklund, E., Sjöberg, S. and Öhman, L.-O., Acta Chem. Scand., A40 (1986) 367 .
[87] Bombi, G. G., Corain, B. and Sheikh-Osman, A. A., Inorg. Chim. Acta, 171 (1990) 79.
[88] Corain, B., Longato, B., Sheikh-Osman, A. A., Bombi, G. G. and Macca, C., J. Chem. Soc., Dalton Trans., (1992) 169.
[89] Karlik, S. J., Tarien, E., Elgavish, G. A. and Eichorn, G. L., Inorg. Chem., 22 (1983) 525.
[90] Martin, R. B., J. Inorg. Biochem., 44 (1991) 141.
[91] Kiss, T., Sóvágó, I., Martin, R. B. and Pursiainen, J., J. Inorg. Biochem., 55 (1994) 53.
[92] Marklund, E., Öhman, L.-O., Acta Chem. Scand., 44 (1990) 228.
[93] Venema, F. R., Peters, J. A. and van Bekkum, H., J. Chem. Soc., Dalton Trans., (1990) 2137.
[94] Cruickshank, M. C. and Glasser, L. S. D., Acta Crystallogr., C41 (1985) 1014.
[95] Alemany, L. B. and Kirker, G. W., J. Am. Chem. Soc., 108 (1986) 6158.
[96] Pattnaik, R. K. and Pani, S., J. Ind. Chem. Soc., 38 (1961) 379.
[97] Toy, A. D., Smith, T. D. and Pilbrow, J. R., Aust. J. Chem., 26 (1973) 1889.
[98] Jackson, G. E., S.-Afr. Tydskr. Chem., 35 (1982) 89.
[99] Öhman, L.-O. and Sjöberg, S., J. Chem. Soc., Dalton Trans., (1983) 2513.
[100] Motekaitis, R. J. and Martell, A. E., Inorg. Chem., 23 (1984) 18.
[101] Greenway, F. T., Inorg. Chim. Acta, 116 (1986) L21.
[102] Martin, R. B., J. Inorg. Biochem., 28 (1986) 181.
[103] Venturini, M. and Berthon, G., J. Chem. Soc., Dalton Trans., (1987) 1145.
[104] Venturini, M. and Berthon, G., J. Inorg. Biochem., 37 (1989) 69.
[105] Gregor, J. E. and Powell, H. K. J., Aust. J. Chem., 39 (1986) 1851.
[106] Lopez-Quintela, M. A., Knoche, W. and Weith, J., J. Chem. Soc., Faraday Trans. I, 80 (1984) 2313.
[107] Feng, T. L., Gurian, P. L., Healy, M. D. and Barron, A. R., Inorg. Chem., 29 (1990) 408.
[108] Van Duin, M., Peters, J. A., Kieboom, A. P. G. and van Bekkum, H., Recl. Trav. Chim. Pays-Bas, 108 (1989) 57.
[109] Sheikh-Osman, A. A., Bertani, R., Tapparo, A., Bombi, G. G. and Corain, B., J. Chem. Soc., Dalton Trans., (1993) 3229.

[110] Kiss, T., Decock, P. and Barbry, D., in preparation.
[111] Marklund, E. and Öhman, L.-O., J. Chem. Soc., Dalton Trans, (1990) 755.
[112] Venema, F. R., Peters, J. A. and van Bekkum, H., Inorg. Chim. Acta, 191 (1992) 261.
[113] Venema, F. R., Peters, J. A. and van Bekkum, H., Inorg. Chim. Acta, 197 (1992) 1.
[114] Tonković, M., Bilinski, H. and Smith, M. E., Inorg. Chim. Acta, 197 (1992) 59.
[115] Marklund, E. and Öhman, L.-O., Acta Chem. Scand., 44 (1990) 353.
[116] Djurdjevic, P. T. and Jelic, R., Z. Anorg. Allg. Chem., 575 (1989) 217.
[117] Kiss, T., Sóvágó, I., Tóth, I. and Martin, R. B., in preparation.
[118] Martin R. B., in Nicolini, M., Zatta, P. F. and Corain, B. (eds.), *Aluminum in Chemistry, Biology and Medicine,* Cortina Intl., Raven Press, New York, 1992, pp. 3–20.
[119] Jons, O. and Johansen, E. S., Inorg. Chim. Acta, 151 (1988) 129.
[120] Findlow, J. A., Duffield, J. R. and Williams, D. R., Chem. Speciation Bioavail., 2 (1990) 3.
[121] Feng, T. L., Tsangaris, J. M. and Barron, A. R., Monat. Chem., 121 (1990) 113.
[122] Charlet, P., Deloume, J. P., Duc, G. and Thomas-David, G., Bull. Soc. Chim. France, (1984) L222.
[123] Macarovici, C. G. and Chis, E., Rev. Roum. Chim., 24 (1979) 1457.
[124] Macarovici, C. G. and Chis, E., Rev. Roum. Chim., 25 (1980) 95.
[125] Taqui Khan, M. M. and Hussain, A., Ind. J. Chem., 19A (1980) 50.
[126] Dubey, S. N., Singh, A. and Puri, D. M., J. Inorg. Nucl. Chem., 43 (1981) 407.
[127] Iyer, R. K., Karweer, S. B. and Jain, V. K., Magn. Res. Chem., 27 (1989) 328.
[128] Öhman, L.-O., Polyhedron, 9 (1990) 199.
[129] Clevette, D. J. and Orvig, C., Polyhedron, 9 (1990) 151.
[130] Valle, G. C., Boruli, G. G., Corain, B., Favarato, M. and Zatta, P., J. Chem. Soc., Dalton Trans., (1989) 1513.
[131] Motekaitis, R. J. and Martell, A. E., J. Coord. Chem., 14 (1985) 139.
[132] Dyatlova, N. M., Medved, T. Ya., Rudomino, M. V. and Kabachnik, M. I., Izv. Akad. Nauk SSSR, 4 (1970) 815.
[133] Larcsenko, V. E., Popov, K. I., Grigorjev, A. I. and Dyatlova, N. M., Koord. Khim., 10 (1984) 1187.
[134] Kiss, T., Lázár, I. and Kafarski, P., Metal Based Drugs, 1 (1994) 247.
[135] Szpoganicz, B. and Martell, A. E., Inorg. Chem., 24 (1985) 2414.
[136] Szpoganicz, B. and Martell, A. E., Inorg. Chem., 25 (1986) 327.
[137] Martell, A. E. and Szpoganicz, B., Inorg. Chem., 28 (1989) 4199.
[138] Taylor, P. A. and Martell, A. E., Inorg. Chim. Acta, 152 (1988) 181.
[139] Orvig, C., in Robinson, G. H. (ed.), *The Coordination Chemistry of Aluminum,* VCH, New York, 1993, Chapter 3, pp. 85–121.
[140] Gervais, M., Commenges, G. and Laussac, J.-P., Magn. Res. Chem., 25 (1987) 594.
[141] Siegel, N., Coughlin, R. and Haug, A., Biochem. Biophys. Res. Comm., 115 (1983) 512.
[142] Siegel, N. and Haug, A., Biochim. Biophys. Acta, 744 (1983) 36.
[143] You, G. and Nelson, D. J., J. Inorg. Biochem., 41 (1991) 283.
[144] Exley, C., Price, N. C., Kelly, S. M. and Birchall, J. D., FEBS Letters, 324 (1993) 293.
[145] Fatemi, S. J. A., Williamsom, D. J. and Moore, G. R., J. Inorg. Biochem., 46 (1992) 35.
[146] Martin, R. B., Savory, J., Brown, S., Bertholf, R. L. and Wills, M. R., Clin. Chem., 33 (1987) 405.
[147] (a) Harris, W. R. and Sheldon, J., Inorg. Chem., 29 (1990) 119; (b) Kubal, G., Mason, A.B., Sadler, P. J., Tucker, A. and Woodworth, R. C., Biochem. J., 285 (1992) 711.
[148] Aramini, J. M. and Vogel, H. J., J. Am. Chem. Soc., 115 (1993) 245.
[149] Sola, M., Chemtracts-Inorg. Chem., 5 (1993) 201.
[150] Sóvágó, I., Kiss, T. and Martin, R. B., Polyhedron, 9 (1990) 189.
[151] Kiss, T., Sóvágó, I. and Martin, R. B., Inorg. Chem., 30 (1991) 2130.
[152] Berridge, M. J. and Irvine, R. F., Nature, 341 (1989) 197.

[153] Mernissi-Arifi, K., Bieth, H., Schlewer, G. and Spiess, B., J. Inorg. Biochem., 57 (1995) 127.
[154] Hegetschweiler, K., Ghisletta, M., Fässler, T. F., Nesper, R., Schmalle, H. W. and Rihs, G., Inorg. Chem., 32 (1993) 2032.
[155] Kradolfer, T. and Hegetschweiler, K., Helv. Chim. Acta, 75 (1992) 2243.
[156] Evans, W. J. and Martin, C. J., J. Inorg. Biochem., 34 (1988) 11.
[157] Hollósi, M., Ürge, L., Perczel, A., Kajtár, J., Teplán, I., Ötvös, L. Jr. and Fasman, G.D., J. Mol. Biol., 223 (1992) 673.
[158] Laczkó, I. and Hollósi, M., Magyar Kém. Foly., 100 (1994) 112.
[159] Geladopoulos, T. P. and Sotiroudis, T. G., J. Inorg. Biochem., 54 (1994) 247.
[160] Rowatt, E. and Williams, R. J. P., J. Inorg. Biochem., 55 (1994) 249.
[161] Deleers, M., Servais, J-P. and Wuelfer, E., Biochim. Biophys. Acta, 813 (1985) 195.
[162] Deleers, M., Servais, J-P. and Wuelfer, E., Biochim. Biophys. Acta, 855 (1986) 271.
[163] Panchalingam, K., Sachedina, S., Pettegrew, J. W. and Glonej, T., Int. J. Biochem., 23 (1991) 1453.
[164] Jackson, G. E. and Voyi, K. V. V., S. Afr. Chem., 41 (1988) 17.
[165] Karlik, S. J., Elgavish, G. A. and Eichhorn, G. L., J. Am.. Chem. Soc., 105 (1983) 602.
[166] Laussac, J. and Commenges, G., Nouv. J. Chim., 7 (1983) 579.
[167] Bock, J. L. and Ash, D. E., J. Inorg. Biochem., 13 (1980) 105.
[168] Jackson, G. E. and Voyi, K. V., Polyhedron, 6 (1987) 2095.
[169] Feng, Q. and Waki, H., Inorg. Chem., submitted for publication.
[170] Lussac, J. and Laurent, J., C.R. Acad. Sci. Paris, 291 (1980) 157.
[171] Scheller, K. H., Scheller-Krattiger, V. and Martin, R. B., J. Am.. Chem. Soc., 103 (1981) 6833.
[172] Scheller, K. H., Hofstetter, F., Mitchell, P. R., Prijs, B. and Sigel, H., J. Am. Chem. Soc., 103 (1981) 247.
[173] Karlik, S. J., Elgavish, G. A., Pillai, R. P. and Eichhorn, G. L., J. Magn. Res., 49 (1982) 164.
[174] Dyrssen, D., Haraldsson, C., Nyberg, E. and Wedborg, M., J. Inorg. Biochem., 29 (1987) 67.
[175] Karlik, S. J., Eichhorn, G. L., Lewis, P. N. and Crapper, D. R., Biochemistry, 19 (1980) 5991.
[176] Karlik, S. J., Eichhorn, G. L. and Crapper McLachlan, D. R., Neurotoxicology, 1 (1988) 83.
[177] Öhman, L.-O. and Sjöberg, S., Acta Chem. Scand., A37 (1983) 875.
[178] Zonnevijlle, F. and Brunisholz, K., Inorg. Chim. Acta, 102 (1985) 205.
[179] Lajunen, L. H. J., Kokkonen, P. and Anttila, R., Finn. Chem. Lett., 15 (1988) 101.
[180] Charlet, P., Deloume, J. P., Duc, G. and Thomas-David, G., Bull. Soc. Chim. France, (1985) 683.
[181] Kiss, T., Atkári, K., Jezowska-Bojczuk, M. and Decock, P., J. Coord. Chem., 29 (1993) 81.
[182] Lajunen, L. H. J., J. Indian Chem. Soc., 59 (1982) 1238.
[183] Havelková, L. and Bartusek, M., Collection Czechoslov. Chem. Comm., 34 (1969) 3722.
[184] Öhman, L.-O. and Sjöberg, S., Polyhedron, 2 (1983) 1329.
[185] Kennedy, J. A. and Powell, H. K. J., Aust. Chem., 38 (1985) 659.
[186] Martell, A. E., Motekaitis, R. J. and Smith, R. M., Polyhedron, 9 (1990) 171.
[187] Kiss, T., Sóvágó, I. and Martin, R. B., J. Am. Chem. Soc., 111 (1989) 3611.
[188] Öhman, L.-O., Sjöberg, S. and Ingri, N., Acta Chem. Scand., A37 (1983) 561.
[189] Öhman, L.-O. and Sjöberg, S., Acta Chem. Scand., A35 (1981) 201.
[190] Öhman, L.-O. and Sjöberg, S., Acta Chem. Scand., A37 (1982) 47.
[191] Kennedy, J. A. and Powell, H. K. J., Aust. J. Chem., 38 (1985) 879.
[192] Powell, H. K. J. and Rate, A. W., Aust. J. Chem., 40 (1987) 2015.
[193] Martin, R. B., Metal Ions in Biol. Syst., 24 (1988) 1.
[194] Conturier, Y., Bull. Soc. Chim. France, (1987) 963.
[195] Conturier, Y., Bull. Soc. Chim. France, (1989) 756.
[196] Rowley, D. A. and Cooper, J. C., Inorg. Chim. Acta, 147 (1988) 257.
[197] Martell, A. E. and Motekaitis, R. J., in Lewis, T. E. (ed.), *The Environmental Chemistry and Toxicology of Aluminium,* Lewis Publ., Chelsea, 1989. Ch. 1, pp. 3–17.
[198] Schwarzenbach, G. and Schwarzenbach, K., Helv. Chim. Acta, 46 (1963) 1390.

[199] Evert, A., Hancock, R. D., Martell, A. E. and Motekaitis, R. J., Inorg. Chem., 28 (1989) 2189.
[200] Hancock, R. D. and Martell, A. E., Chem. Rev., 89 (1989) 1875.
[201] Anderegg, G., L'Eplattenier, F. and Schwarzenbach, G., Helv. Chim. Acta, 46 (1963) 1409.
[202] Farkas, E., Kozma, E., Kiss, T., Tóth, I. and Kurzak, B., J. Chem. Soc., Dalton Trans., in press.
[203] Farkas, E., Szõke, J., Kiss, T., Kozlowski, H. and Bal, W., J. Chem. Soc., Dalton Trans., (1989) 2247.
[204] Farkas, E., Kiss, T. and Kurzak, B., J. Chem. Soc., Perkin Trans., (1990) 1255.
[205] Finnegan, M., Lutz, T. G., Nelson, W. O., Smith, A. and Orvig, C., Inorg. Chem., 26 (1987) 2171.
[206] Crapper-McLachlan, D. R., Neurobiol. Aging, 7 (1986) 525.
[207] Hedlund, T. and Öhman, L.-O., Acta Chem. Scand., A42 (1988) 702.
[208] Finnegan, M. M., Rettig, S. J. and Orvig, C., J. Am. Chem. Soc., 108 (1986) 5033.
[209] Finnegan, M. M., Lutz, T. G., Nelson, W. O., Smith, A. and Orvig, C., Inorg. Chem., 26 (1987) 2171.
[210] Nelson, W. O., Karpishin, T. B., Rettig, S. J. and Orvig, C., Inorg. Chem., 27 (1988) 1045.
[211] Nelson, W. O., Rettig, S. J. and Orvig, C., Inorg. Chem., 28 (1989) 3153.
[212] Clevette, D. J., Nelson, W. O., Nordin, A., Orvig, C. and Sjöberg, S., Inorg. Chem., 28 (1989) 2079.
[213] Zhang, Z., Rettig, S. J. and Orvig, C., Inorg. Chem., 30 (1991) 509.
[214] Clarke, E. T. and Martell, A. E., Inorg. Chim. Acta, 191 (1992) 57.
[215] Clarke, E. T. and Martell, A. E., Inorg. Chim. Acta, 196 (1992) 185.
[216] Lutz, T. G., Clevette, D. J., Rettig, S. J. and Orvig, C., Inorg. Chem., 28 (1989) 715.
[217] Stary, J. and Hladky, E. Anal. Chim. Acta, 28 (1963) 227.
[218] Stary, J. and Liljenzin, J. O., Pure Appl. Chem., 54 (1982) 2557.
[219] Izatt, J., Fernelius, W. C., Haas, C. G. and Bloch, B. P., J. Phys. Chem., 59 (1955) 170.
[220] Krishen, A. and Freiser, H., Anal. Chem., 31 (1959) 923.
[221] Patel K. S. and Adimado, A. A., J. Inorg. Nucl. Chem., 42 (1980) 1241.
[222] Kolesov, B. A. and Igumenov, I. K., Spectrochim. Acta, 40A (1984) 233.
[223] Tapparo, A. and Perazzolo, M., Int. J. Environ. Anal. Chem., 36 (1989) 13.
[224] Starý, J. and Hladký, E., Anal. Chim. Acta, 28 (1963) 227.
[225] Öhman, L.-O. and Nordin, A., Acta Chem. Scand., 46 (1992) 515.
[226] Tajmir-Riahi, H. A., Inorg. Biochem., 44 (1991) 39.
[227] Kaizu, Y., Misu, N., Tsuji, K., Kaneko, Y. and Kobayashi, H., Bull. Chem. Soc. Japan, 58 (1985) 103.
[228] Hirai, Y., Murayama, H., Aida, T. and Inoue, S., J. Am. Chem. Soc., 110 (1988) 7387.
[229] Nam, W. and Valentine, J. S., J. Am. Chem. Soc., 112 (1990) 4977.
[230] Guilard, R., Zrineh, A., Tabard, A., Endo, A., Han, B. C., Lecomte, C., Sonhasson, M., Habbon, A., Ferhat, M. and Kadish, K. M., Inorg. Chem., 29 (1990) 4476.
[231] Guilard, R., Lecomte, C. and Kadish, K. M., Struct. Bonding, 64 (1987) 205.
[232] Anderegg, G., *IUPAC Chemical Data Series, No. 14, Critical Survey of Stability Constants of EDTA Complexes,* Pergamon Press, Oxford, 1977.
[233] Tóth, I., Brücher, E., Zékány, L. and Veksin, V., Polyhedron, 8 (1989) 2057.
[234] Nelson, D. J. and Martin, R. B., J. Inorg. Biochem., 43 (1991) 37.
[235] Yuchi, A., Hotta, H., Wada, H. and Nakagawa, G., Bull. Chem. Soc. Japan., 60 (1987) 1379.
[236] Ramamoorthy, S. and Manning, P., J. Inorg. Nucl. Chem., 37 (1975) 363.
[237] Arp, P. and Meyer, W., Can. J. Chem., 63 (1985) 3357.
[238] Jackson, G. E., Polyhedron, 9 (1990) 163.
[239] May, P. M., Linder, P. W. and Williams, D. R., J. Chem., Soc. Dalton Trans., (1977) 588.
[240] Öhman, L.-O. and Sjöberg, S. in Kramer, J. R. and Allen, H. E. (eds.), *Metal Speciation: Theory, Analysis and Application,* Lewis Publ., Chelsea, 1988.

[241] Martell, A. E. and Smith, R. M., *Critical Stability Constants,* Vol. 1–6, Plenum, New York, 1974–1989.
[242] L. D. Pettit and Powell, H. K. J. (eds.), *Stability Constants Database*, IUPAC/ Academic Software, Otley, 1993.
[243] Martell, A. E. (ed.), *NIST Critical Stability Constants and Related Thermo-dynamic Constants of Metal Complexes Database,* NIST, Gaithersburg, 1993.
[244] Tipping, E., Backes, C. A. and Hurley, M. A., Wat. Res., 5 (1988) 597.
[245] Shuman, M. S., Environ. Sci. Technol., 26 (1992) 593.
[246] Browne, B. A. and Driscoll, C. T., Environ. Sci. Technol., 27 (1993) 915.
[247] Slanina, P., Frech, W., Ekstrom, L.-G., Loof, L., Slorach, S. and Cedergren, A., Clin. Chem., 32 (1986) 539.
[248] Domingo, J. L., Gomez, M., Llobet, J. M. and Corbella, J., Kidney Int., 39 (1991) 598.
[249] Hewitt, C. D., Poole, C. L., Westerwelt, F. B. Jr., Savory, J. and Wills, M. R., Lancet 2 (1988) 849.
[250] Fatemi, S. J. A., Kadir, F. H. A., Williamson, D. J. and Moore, G. R., Advances Inorg. Chem., 36 (1990) 409.
[251] Bell, J. D., Kubal, G., Radulovic, S., Sadler, D. J. and Tucker, A., Analyst, 118 (1993) 241.
[252] Banles, W. A. and Kartin, A., Neurosci. Biobehavioral Rev., 13 (1989) 47.
[253] Deloncle, R. and Guillard, O., Neurochem. Res., 15 (1990) 1239.
[254] Eichhorn, G. L., Exp. Gerontology, 28 (1993) 493.
[255] Meiri, H., Banin, E., Roll, M. and Rousseau, A., Progr. Neurobiol., 40 (1993) 89.
[256] Alfrey, A. C., Ann. Rev. Med., 29 (1978) 93.
[257] Sternberger, N. H., Sternberger, L. A. and Ulrich, J., Proc. Nat. Acad. Sci. USA, 82 (1985) 4274.
[258] MacDonald, T. L., Humphreys, W. G. and Martin, R. B., Science, 236 (1987) 183.
[259] Zatta, P. F., Cervellin, D. and Zambenedette, P., Life Chem. Reports, 11 (1994) 111.
[260] Bonucci, E., Ballanti, P., Berni, S. and Della Rocca, C., Life Chem. Reports, 11 (1994) 225.
[261] Meyer, J. L. and Thomas, W. C. Jr., Kidney International, 29 (1986) S20.
[262] Drücke, T. B., Life Chem. Reports, 11 (1994) 231.
[263] Exley, C., Price, N. C. and Birchall, J. D., J. Inorg. Biochem., 54 (1994) 297.
[264] Corain, B., Tapparo, A., Sheikh-Osman, A. A. and Bombi, G. G., Coord. Chem. Rev., 112 (1992) 19.
[265] Corain, B., Nicolini, M., and Zatta, P. F., Coord. Chem. Rev., 112 (1992) 33.
[266] Corain, B., Sheik Osman, A. A., Bertani, R., Tapparo, A., Zatta, P. F. and Bombi, G. G., Life Chem. Reports, 11 (1994) 103.
[267] Crapper McLachlan, D. R., Neurobiol. Aging, 7 (1986) 525.
[268] Zatta, P. F., Giordano, R., Corain, B., Favarato, M. and Bombi, G. G., Toxicol. Lett., 39 (1987) 185.
[269] Hem, S. L. and White, J. L., in Gitelman, H. J. (ed.), *Aluminum and Health,* Marcel Dekker, New York, 1989, pp. 257–282.
[270] Lin, J.-C., Feldkamp, J. R., White, J. L. and Hem, S. L., J. Pharm. Sci., 73 (1984) 1355.
[271] Van Riemsdijk, W. H. and Lyklema, J., Colloid Surf., 1 (1980) 33.
[272] Fournier, A., Moriniere, P., Boudaillier, B., Lalan, J. D., Renaud, H., Hocine, C., Bellrik, S. and Westel, P.-F., in De Broe, M. E. and Coburn, J. W. (eds.), *Aluminum and Renal Failure,* Kluwer Publ., Dordrecht, 1990, pp. 325–343.
[273] Schneider, H. W., Kulbe, K. D., Weber, H. and Streicher, E., Kidney International, 29 (1986) S120.
[274] Milliner, D. S., Hercz, G., Miller, J. H., Shinaberger, J. H., Nissenson, A. R. and Coburn, J. W., Kidney International, 29 (1986) S100.
[275] McLachlan, D. R., Fraser, P. E. and Dalton, A. J., in Chadwick, D. J. and Whelan, J. (eds.), *Aluminum in Biology and Medicine,* Wiley, Chichester, 1992, pp. 92–94.

[276] Charhon, S. A., in De Broe, M. E. and Coburn, J. W. (eds.), *Aluminum and Renal Failure,* Kluwer Publ., Dordrecht, 1990, pp. 309–323.
[277] Birchall, J. D., in Gitelman H. J. (ed.), *Proceedings Second International Conference on Aluminum and Health,* Tampa, 1992, pp. 47–49.
[278] Martyn, C. N., in Chadwick, D. J. and Whelan, J. (eds.), *Aluminum in Biology and Medicine,* Wiley, Chichester, (Ciba Found. Symp. 169), 1992, pp. 69–86.
[279] Birchall, J. D. and Chappel, J. S., in Thornton, I. (ed.), *Geochemistry and Health,* Science Reviews Ltd, Northwood, 1988, pp. 231–342.
[280] Petersen, O. H., Wakui, M. and Petersen, C. C. H., in Chadwick, D.J. and Whelan, J. (eds.), *Aluminum in Biology and Medicine,* Wiley, Chichester (Ciba Found. Symp. 169), 1992, pp. 237–253.

THE ROLE OF NITRIC OXIDE IN ANIMAL PHYSIOLOGY

Anthony R. Butler, Frederick W. Flitney, and Peter Rhodes

OUTLINE

1. Introduction 252
2. Solution Chemistry of NO 252
3. Vascular Smooth Muscle Relaxation 253
4. Identification of the EDRF 255
5. Nitrovasodilators 256
6. Activation of Guanylate Cyclase 258
7. Platelet Aggregation 261
8. NO and the Immune System 262
9. NO in the Nervous System 264
10. NO, NO^+, and NO^- 267
11. Nitric Oxide Synthase 268
12. Medical Uses of NO Gas 271

References 273

Perspectives on Bioinorganic Chemistry
Volume 3, pages 251–277

ISBN: 1-55938-642-8

1. INTRODUCTION

There are several oxides of nitrogen, the most important of which are nitrous oxide (N_2O), nitric oxide (NO), dinitrogen trioxide (N_2O_3), and dinitrogen tetroxide (N_2O_4). Nitrous oxide (or "laughing gas") has, of course, been used medically for many years as a general anesthetic but it is its cousin, nitric oxide (hereafter called NO), which has caused much excitement in biological and medical circles in recent years.

The electronic configuration of the NO molecule is $(\sigma_1)^2(\sigma_1^*)^2(\sigma_2\pi)^6(\pi^*)$ and the unpaired π^* electron makes it a radical, although a stable one. This electron also renders the molecule paramagnetic and partly cancels the effect of the π-bonding electrons to give a bond order of 2.5. This bond order is confirmed by measurement of the interatomic distance of 1.15 Å, intermediate between a double and triple bond. The π^* electron is relatively easily lost to give the nitrosonium ion NO^+ and this may, or may not, be of biological significance. The matter will be discussed in more detail later.

That a molecule as simple as NO should have a major role or, more correctly, many roles in animal physiology was an unexpected discovery. For many biological chemists it was a disturbing discovery in that the molecule contains no carbon atom, but inorganic chemists have long known that non-carbon containing materials, particularly metals, have a vital role in animal physiology. Some biologists also found NO a difficult molecule to comprehend as it is a radical and radicals are generally regarded as highly reactive and destructive, which NO is not. The number of papers now appearing in the scientific literature concerning the role of NO in animal physiology is very large and hardly a month goes by without an unexpected development in the saga of NO in biology or medicine.

The topic has been reviewed a number of times but any review is somewhat out-of-date before the ink is dry, let alone before it appears in print. A short review, intended for chemists, is that by Butler and Williams [1]. A more detailed one by Moncada et al. [2] concentrates more on the pharmacological aspects. The present review makes no claim to be up-to-date on all aspects of NO biochemistry or to be comprehensive. Instead we will try to emphasize the inorganic aspect of NO's biochemistry. The range of journals in which relevant articles appear is great and, even with modern abstracting systems, it is easy to miss a publication. Comprehensiveness would make the review too long and we have made what we hope is a judicious selection of topics. Many workers have contributed to the unfolding of the NO story and to reference everyone would render the list of references unwieldy. With this in mind, we hope that colleagues will not be offended if their work is not given the prominence it deserves.

2. SOLUTION CHEMISTRY OF NO

The gas phase chemistry of NO, particularly its reaction with oxygen, has been much investigated. The gas phase reaction with oxygen is so fast that, when a physiological role for NO was first suggested, some felt that its speed of oxidation made this unlikely. However, in both the gas phase and in solution the reaction is first order in oxygen and second order in NO and, at concentrations relevant to the physiological situation, the overall third order makes the reaction rather slow. The kinetics of the aqueous reaction have been investigated by Wink et al. [3] and by Kharitonov et al. [4] and the magnitude of the third-order rate constant is such that with both dioxygen and NO at 10^{-6} M the half-life for the reaction is 17.5 h. The physiological half-life of NO (less than 60 s), therefore, cannot be ascribed to its rapid oxidation within the cell. It is difficult to decide from the older literature if there is a direct reaction between NO and water but it appears that there is not. There is, however, a well authenticated reaction between NO_2 and water to give a mixture of nitrite and nitrate:

$$H_2O + 2NO_2 \rightarrow NO_2^- + NO_3^- + 2H^+$$

It might be supposed that these products would be obtained by the reaction of NO with oxygenated water but this turns out not to be the case. The only product obtained from this reaction over the pH range 1–13 is nitrite [5]. The reaction scheme is:

$$2NO + O_2 \rightarrow 2NO_2$$

$$NO_2 + NO \rightarrow N_2O_3$$

$$N_2O_3 + H_2O \rightarrow 2HNO_2$$

It appears that reaction of NO_2 with NO is faster than that of NO_2 with water. Once nitrite has formed it may well then be oxidized to nitrate and NO in plasma [6] to eventually give a mixture of nitrite and nitrate. In arterial blood, nitrite may be rapidly oxidized to nitrate by oxyhemoglobin and nitrite is found only at low levels [7]. However, in most biological studies analysis for nitrite by the Griess test [8] gives a reasonably good measure of NO formation. It is now possible to measure NO concentrations by more direct means, such as NO-sensors [9].

3. VASCULAR SMOOTH MUSCLE RELAXATION

Although a frequently recounted story, it is still of value to recall the manner in which the role of NO in vascular smooth muscle (VSM) relaxation was discovered. It is first necessary to describe the biochemistry of the relaxation process.

The resistance to flow in different regions of the cardiovascular system varies as the reciprocal of the fourth power of the radius of the vessels. The smallest arteries and arterioles constitute the major site of increased resistance to blood flow and, for this reason, they are called resistance vessels. Arterioles are crucially important in regulating pressure and flow into local capillary beds. Their resistance is not fixed but can be varied by a layer of muscle cells which surround the lumen.

Contraction of VSM cells of an arteriole causes the lumen to constrict and more kinetic energy is dissipated as friction. The pressure gradient across the capillary bed served by the vessel decreases and blood flow is reduced. Conversely, relaxation of VSM cells dilates the lumen and increases the pressure gradient which facilitates flow through the capillary bed. The contractile state of VSM cells throughout the vascular system is thus a major determinant of systemic blood pressure overall and of local tissue perfusion.

The question we must address now is: how is VSM contractility regulated?

Contraction of muscles which are responsible for voluntary movements of the body (skeletal muscle) is brought about by the interaction of two proteins—actin and myosin—which form macromolecular arrays, or myofilaments, inside muscle cells. Surface projections on the myosin filaments are able to form temporary "cross-bridges" which undergo repeated cycles of attachment to actin, tilting, and detachment. This provides a motive force which causes the two sets of filaments to slide relative to one another in such a way that the muscle cell shortens [10]. The "trigger" for the process is an increase in the intracellular calcium concentration (= $[Ca_i^{2+}]$): this "switches on" contraction, whereas a reduction in $[Ca_i^{2+}]$ causes muscles to relax [11]. The free $[Ca_i^{2+}]$ is extremely low in a muscle at rest (ca. 0.1–0.2 μM). This is because it is sequestered inside a closed-off membrane system, the sarcoplasmic reticulum (SR), and hence unavailable to the actin and myosin filaments. When a nerve impulse reaches a muscle fiber it stimulates the release of Ca^{2+} from the SR [12], initiating contraction. Relaxation is brought about by resequestration of Ca^{2+} into the SR. Hence, the contractile state of a skeletal muscle is "finely tuned" by cellular mechanisms which regulate the shuttling of Ca^{2+} to and fro between the SR and myofilaments.

The contractile state of VSM cells is likewise determined by the interaction of actin and myosin and the process is also regulated by $[Ca_i^{2+}]$. However, the source of the Ca^{2+} required and the cellular processes that control free $[Ca_i^{2+}]$, as well as the nature of its involvement in initiating contraction, are all somewhat different. First, in VSM cells some of the Ca^{2+} is released from storage sites within the cells [13] but, additionally, some enters from the extracellular fluid through ionic channels in the surface membrane [14]. Second, a rise in $[Ca_i^{2+}]$ "triggers" contraction indirectly by stimulating the enzymic phosphorylation of myosin, an obligatory prerequisite which permits actin and myosin to interact [15]. And third, relaxation results from several processes with resequester Ca^{2+} into internal membrane systems [16] and extrude it to the extracellular fluid [17], both energy (ATP)-consuming since calcium is being moved against a concentration gradient, and also by a mechanism located at the cell membrane which exchanges Na^{+} from outside in return for Ca^{2+} [18]. Together, these processes lower $[Ca_i^{2+}]$ and cause dephosphorylation of myosin with the consequent loss of its ability to interact with actin.

The mechanisms involved in reducing $[Ca_i^{2+}]$ in VSM cells, outlined above, are in turn regulated by a "second messenger" molecule, guanosine 3′,5′ cyclic monophosphate (cGMP), formed in VSM cells from guanosine triphosphate (GTP) by the action of an enzyme called guanylate cyclase (GC). Increased [cGMP]i results in activation of cGMP-dependent protein kinase, an enzyme which then phosphorylates proteins involved in resequestering calcium [19] and/or extruding it to the exterior [20], increasing their activity and thus facilitating calcium removal.

The contractile state of VSM depends ultimately upon the control of GC and thus of [cGMP]i. It is here that NO plays a crucial role. To trace our understanding of the role of NO in regulating VSM contraction we turn first to what is now a classic experiment performed by Furchgott and Zawadzki [21] some 15 years ago. It was known at the time that injections of acetylcholine (ACh) into animals invariably lowers blood pressure. This is because ACh invariably relaxes VSM *in vivo* and is therefore a powerful *vasodilator*. However, *ex vivo* experiments with ACh using isolated, precontracted blood vessels gave confusing results. ACh sometimes caused their VSM to relax, "mimicking" the *in vivo* vasodilator action, but other times it either did nothing or even caused the vessel to contract further. Furchgott and Zawadzki cleared up the mystery by showing that it was essential to preserve the integrity of a layer of cells lining the vessels, called the endothelium, in order to obtain vasodilator-type responses. Their experiments demonstrated that the endothelium is an extraordinarily delicate structure which is easily damaged during dissection. Their important contribution was to realize that ACh must act first upon the endothelium to release a potent vasodilator substance which then diffuses to the outer VSM layer causing the cells to relax. They termed this substance *endothelium-derived relaxing factor* (EDRF).

4. IDENTIFICATION OF THE EDRF

During the 1980s various attempts were made to ascertain the chemical nature of the EDRF. A number of suggestions could not be sustained on further examination. The resolution of the problem came about in rather a surprising manner. It had been known for some time that it was possible to activate guanylate cyclase *in vitro* by means of a number of "NO-containing compounds" like glyceryl trinitrate and sodium nitroprusside and, indeed, by NO itself [22]. Furchgott [23] first made the suggestion that since NO is effective *in vitro* perhaps it is also the naturally occurring EDRF. With hindsight it appears so straightforward but, at the time, to suggest that cells generate NO was highly innovative. Two groups obtained experimental proof that this was the case. Palmer et al. [24] used a bioassay (a cascade of strips of rabbit aorta) and showed that the characteristics of the EDRF from cultured endothelial cells were identical with those of an aqueous solution of NO. Ignarro et al. [25] used a similar experimental procedure and came to the same conclusion. Of the early confirmatory experiments one of the most convincing was the demonstration by Palmer et al. [26] that NO could be detected in a cell culture of activated endothelial cells. The cells were activated by bradykinin or the calcium ionophore A23187 and the NO detected by bioassay, chemiluminescence, and mass spectrometry. In addition, mass spectrometry studies using ^{15}N-labeled L-arginine

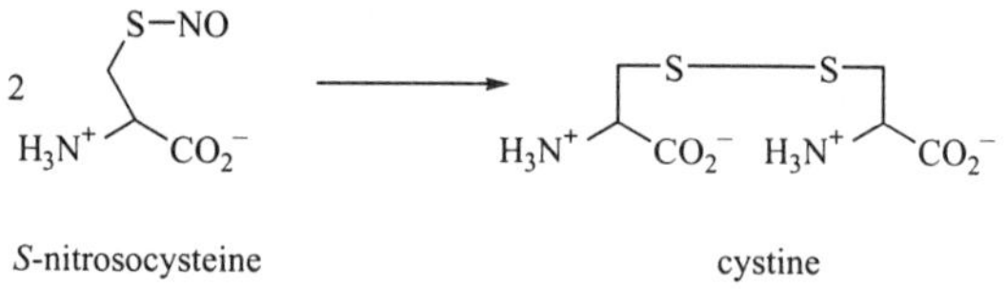

2 *S*-nitrosocysteine → cystine

Figure 1.

indicated that NO is formed from one of the terminal guanidino nitrogen atoms. This aspect will be described in more detail when the enzyme NO-synthase is discussed.

The publication of these results occasioned a response in which Harrison et al. [27] produced evidence that they claimed showed the EDRF resembles *S*-nitrosocysteine more than NO. This nitrosothiol is an unstable compound giving, on decomposition, cystine and NO (Figure 1) and so any attempt to distinguish between NO and *S*-nitrosocysteine is difficult. Convincing evidence that *S*-nitrosocysteine is not the EDRF has been provided by Feelisch et al. [28]. From a reading of the biological literature it might be supposed that NO reacts with cysteine to give *S*-nitrosocysteine. This is clearly not the case unless the reaction is preceded by an oxidative process to give NO^+. Although it appears unlikely that *S*-nitrosocysteine is the EDRF, nitrosothiols do appear to play some role in the biochemistry of NO. This matter is discussed in the next section.

5. NITROVASODILATORS

A number of compounds containing NO, NO_2, or NO_3 groups have been used over many years as vasodilators although the mechanism of action had not been fully understood. For example, glyceryl trinitrate is commonly used for the relief of the symptoms of angina pectoris, a condition caused by narrowing of the arteries in the heart itself. Amyl nitrite has also been used in the same way and its use was first reported as long ago as 1867 [29]. It now appears that alkyl nitrites and nitrates can be transformed into NO in the body by a series of enzymatic and nonenzymatic processes [30]. It has been proposed that thiols are necessary for the conversion of organic nitrate into NO but nitrosothiols are not formed [31]. The topic has been reviewed [32].

Sodium nitroprusside [SNP; $Na_2Fe(CN)_5NO$] is a particularly interesting vasodilator [33]. It is used in hospital practice as an infusion for the control of blood pressure and as a ready source of NO in physiological experiments. According to some reports it spontaneously decomposes *in vivo* to give iron ions, cyanide, and NO, and it is the latter which is responsible for its biological action. In fact, the formation constants of all cyanoferrates are very high and so spontaneous decomposition is unlikely. In both aqueous solution and in blood SNP is quite stable [34] but ready NO release does occur photochemically [35]:

$$[Fe(CN)_5NO]^{2-} \xrightarrow{h\nu} [Fe(CN)_5H_2O]^{2-} + NO$$

Whether this plays any part in its clinical use it is difficult to say since few physiological experiments involve the vigorous exclusion of light; however Flitney and Kennovin [36] reported that a potent effect of SNP on isolated frog heart is abolished all together if experiments are carried out in the dark. However, it does appear that SNP is a vasodilator even in the absence of light and this is most readily explained by a reaction between SNP and thiols to give nitrosothiols, which decompose to release NO. At high pH this is not the reaction between thiol and SNP [37] but, at pHs closer to physiological pH, nitrosothiols are probably produced in a very slow reaction.

An interesting inorganic NO-donor drug is Roussin's Black Salt [RBS; $NaFe_4S_3(NO)_7$]. This remarkable ionic compound has seven NO ligands in each anion and NO is released thermally, photochemically, and on oxidation. Although ionic, RBS has the unusual property of being more soluble in organic solvents than in water. Studies using isolated rat tail artery showed [38] that RBS is a potent vasodilator due to release of NO. However, the effect of a bolus injection of RBS is not a transient dilation, as is the case with nitroprusside. Instead there is a loss of muscular tone which persists for several hours because RBS, due to its lipid solubility, lodges in the endothelial cells and slowly releases NO. Unfortunately, RBS is not effective in whole animals since it binds very strongly to albumin in serum. RBS has proved to be an effective way of delivering NO to cell cultures in a number of physiological studies where its rapid release can be effected by illumination.

A class of organic compounds which have aroused considerable interest since the role of NO in human physiology was first proposed is nitrosothiol or thionitrite. These are colored compounds made by the nitrosation of thiols in acid solution:

$$RSH + NO^+ \rightarrow RSNO + H^+$$

Most of them are unstable and impossible to isolate but that obtained by the acidic nitrosation of *N*-acetylpenicillamine (*S*-nitroso-*N*-acetylpenicillamine; SNAP) can be obtained as stable green crystals [39]. It is a potent vasodilator in whole animals [40] as well as in *ex vivo* experiments [41]. It is easy to see why this is so as nitrosothiols readily decompose with release of NO:

$$RNSO \longrightarrow RS^{\bullet} + NO$$

$$RS^{\bullet} \downarrow 1/2\ RSSR$$

Kinetic studies showed erratic behavior for the *in vitro* decomposition of SNAP and this was eventually traced to the very powerful catalysis of decomposition by copper ions [42]. There are sufficient copper ions in ordinary distilled water to have a dramatic effect on the decomposition of SNAP. Although the exact reason for this effect has yet to be fully explained, copper must bind to the SNAP molecule so as to weaken the S–N bond of the nitrosothiol group. Whether the presence of metal ions is necessary for the *in vivo* release of NO from SNAP is difficult to ascertain. Metal ions are involved in so many essential processes within the living cell that removing them and then testing the effect of SNAP is experimentally impossible. The binding of copper ions to SNAP is relatively strong since the catalytic effect of this ion can be detected even when the Cu^{2+} is present in a complexed state.

The related nitrosothiol *S*-nitrosocysteine has come under considerable scrutiny. Its confusion with NO as the EDRF has already been described. It is not isolable and can only be used for physiological study as an unstable species in solution. We suggest that its instability relative to SNAP is due to stronger binding to metal ions. Nitrosothiols do occur naturally in bodily fluids [43] but, at present, their role is unknown. They could be stores or pools from which NO is released on demand. When the nitrosothiol group is part of a large molecule (a peptide, for example) decomposition does not occur readily, as it does with SNAP. This could be due to the poorer binding of copper ions to a peptide. There is considerable evidence that an enzymatic process is involved in the release of NO from an S-nitrosated peptide [44].

6. ACTIVATION OF GUANYLATE CYCLASE

Cell–cell communication by NO is most often mediated inside the target cell by activation of the enzyme guanylate cyclase (GC). GC catalyzes the hydrolysis and cyclization of GTP to form cGMP) and pyrophosphate (Figure 2). cGMP functions as an intracellular "second messenger".

It can alter cellular processes directly, for example, by acting on cGMP-gated ion channels or on cGMP-regulated cyclic nucleotide phosphodiesterases; or indirectly, by activating cGMP-dependent protein kinase (G-kinase). Activated G-kinase can then phosphorylate key protein substrates which modify cellular function.

GC is a polymorphic enzyme which exists in both soluble and membrane-bound (particulate) isoforms. The soluble isoenzyme (sGC) is a heme-containing heterodimer [45] with α and β subunits of M_r in the range 73–88 kDa and 70 kDa, respectively. Progress in understanding the molecular mechanisms which regulate the enzyme has been slow for two reasons: first, even relatively mild isolation procedures can cause dissociation of the loosely bound heme moiety, which turns out to be essential for its regulation; and second, because its activity is profoundly influenced by redox reactions. Early biochemical studies [22] showed that many sGC preparations could be activated by agents such as SNP, glyceryl trinitrate, and amyl nitite, all of which are potential sources of NO. Direct exposure of the enzyme

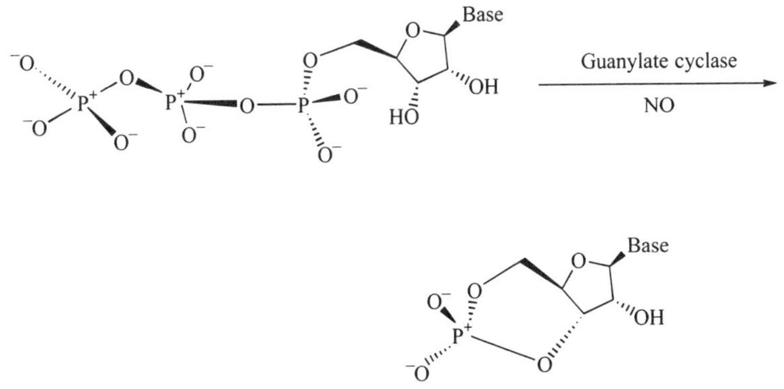

Figure 2.

to NO also caused activation [46]. These observations led to the idea that NO is the proximal activator of sGC by nitrovasodilators and that it acts by forming a nitrosyl–heme complex. In support of this hypothesis, preparations which were insensitive to NO were found to be deficient in heme and their susceptibility to NO or NO "donors" could be restored by the addition of hemoglobin, hematin, or catalase [47]. Later, the separation of heme-containing and heme-deficient fractions from partially purified enzyme preparations revealed that the former were sensitive to NO and NO "donors", whereas the latter were not [48]. Significantly, protoporphyrin IX, the de-metalated form of ferroprotoporphyrin IX (ferrous heme), maximally activated the enzyme even in the absence of NO, while the introduction of ferrous iron into the molecule abolished this type of activation. Furthermore, the addition of preformed nitrosyl–hemoprotein (e.g. nitrosyl–hemoglobin, nitrosyl–catalase, and nitrosyl–cytochrome P450) to partially purified sGC preparations which were insensitive to NO resulted in maximal activation. Enzyme activity could not be increased further by NO donors, nor was it potentiated by reducing agents [49]. The physiological significance of NO-dependent activation of sGC became apparent only after the discovery that EDRF and NO were chemically and pharmacologically indistinguishable (see Section 4).

The observations outlined above suggested that ferrous heme *suppresses* sGC activity and that binding of NO causes enzyme activation by "disinhibition" [50]. EPR studies indicated that binding of NO to sGC weakens the coordinate bonds which hold iron in the plane of the porphyrin ring [51]. It was postulated [52] that this would result in an "out-of-plane" movement of the central iron atom, effectively transforming the heme–NO complex into a protoporphyrin-like structure, resulting in activation of the enzyme.

Stone and Marletta [53] made a spectroscopic study of bovine lung sGC, purified to homogeneity by an improved isolation procedure which retained ~1 heme group/heterodimer. The electronic spectrum exhibited a Soret band at 431 nm and a single α/β peak centered on 555 nm, consistent with a 5-coordinate ferrous–heme structure with imidazole (histidine) as the proximal ligand. Binding of NO activated their enzyme 130-fold. The Soret band shifted from 431 to 398 nm and prominent α and β peaks appeared at 572 and 537 nm, respectively, together with a distinct "shoulder" at 485 nm. These observations show that a 5-coordinate structure is retained after binding NO, confirming that formation of the nitrosyl–heme complex causes the axial imidazole ligand to be released. Interestingly, the spectrum was unaltered by placing the enzyme in 100% dioxygen, demonstrating that the heme moiety does not have a high affinity for oxygen and is therefore quite unlike that of hemoglobin or myoglobin.

In addition to activation by endogenous NO, agents that covalently modify thiols profoundly influence sGC activity [22]. Thiols which are susceptible to alkylating agents and/or are able to participate in mixed disulfide formation are crucial for expression of basal and stimulated sGC activity. Both basal and stimulated activity are inhibited by alkylating agents (e.g. ethacrynate, maleimide) and by mixed disulfide formation (e.g with cystine, cystamine). In one study, the incorporation of ^{35}S-labeled cystamine into a purified rat lung sGC closely parallelled the loss of activity. The process was reversed on treatment with dithiothreitol, an agent which converts disulfides into free thiols [54].

Enzyme activity is also critically influenced by redox reagants. Oxidation of thiol groups by incubation under aerobic conditions increases activity of crude extracts or purified enzyme [55, 56]. Generally speaking, thiol reducing agents (dithiothreitol, ascorbate, glutathione, and cysteine) potentiate activation by NO or NO donors, whereas thiol-oxidizing agents (diamide, oxidized glutathione) inhibit the enzyme. However, Wu et al. [57] recently demonstrated reversible activation of sGC in intact platelets using the thiol-oxidizing agents, diamide and the reactive disulfide compound 4,4′-dithiopyridine. Moderate concentrations of either compound caused a time-dependent (i.e. transient) activation of the enzyme and also potentiated the stimulatory action of SNP, whereas high concentrations were inhibitory. Activation of the enzyme was parallelled by the formation of protein-bound glutathione and by diminished intracellular reduced glutathione (GSH) levels. Enzyme activity was inhibited when GSH levels dropped below the detection limit. These experiments suggest that oxidation of a thiol-containing domain, most likely located at the catalytic site, activates sGC, whereas "overoxidation" of this group results in loss of enzyme activity. They conclude that protein disulfide formation is involved in the activation process and that binding of NO to the heme group triggers an additional, synergistic activation.

sGC can be inhibited *in vivo* and *in vitro* by a variety of agents which have been used to investigate the role of the L-arginine–NO–cGMP pathway in mediating cellular responses. Methylene blue (MB) has been used extensively for this purpose. Hemoglobin also inhibits sGC indirectly, by scavenging NO. Additionally, both MB and Hb are able to generate O_2- [58], a reactive radical species that may alter the oxidation state of the enzyme directly and/or "inactivate" NO by forming peroxynitrite (see Section 8).

sGC activity *in vivo* can be up- or downregulated. Supersensitivity of blood vessels to nitrovasodilators following acute periods of NO deprivation (e.g. after blockade of nitric oxide synthase with inhibitors or removal of the endothelium) is well documented [59]. This phenomenon appears to be at the level of sGC though the underlying mechanism has not been studied fully. Conversely, sGC can become desensitized following prolonged treatment with nitrovasodilators [60]. This is known as "tolerance" and is especially important in the clinical context where patients are on long term nitrovasodilator therapy. Oxidized (but not un-oxidized) low-density lipoprotein (LDLox) attenuates the effect of NO donors on sGC from bovine lung without influencing the formation of endothelial NO [61]. This appears to be a direct effect of LDLox on sGC and may afford an explanation for the hyporesponsiveness of atherosclerotic vessels to endogenous NO and to NO donors.

Generally, NO is synthesized from L-arginine as required (see Section 11). However, evidence for a molecular "store" of NO emerged recently from studies of light-induced relaxation (photorelaxation) of vascular smooth muscle. Photorelaxation is associated with increased cGMP synthesis; both can be prevented by inhibitors of sGC and by oxyhemoglobin acting as a scavenger of NO [62]. The NO store is photolabile and can be depleted rapidly (< 5 min) by exposing precontracted arteries to visible laser light (514 nm; 6.3 mW cm^{-2}), after which it spontaneously reforms in the dark. The repriming process displays an absolute requirement for endothelium-derived NO and it can be prevented by the thiol alkylating agent ethacrynate [63]. Repeated short exposures (2 min) of endothelium-denuded arteries to polychromatic light results in a more gradual depletion of the store, with no little or no repriming between periods of irradiation [64]. Interestingly, in this type of preparation each interval between successive exposures is marked by a stepwise increase in vasoconstrictor tone. This result suggests that a component of "basal" vasodilator tone may be attributable to continuous activation of sGC by NO liberated from the store in a "dark" reaction.

7. PLATELET AGGREGATION

It has been said that NO can inhibit the clotting of blood, but this is true only if the word clotting is used in a very general sense of a process which prevents blood flow (hemostasis). The correct definition of clotting is the initiation of a cascade system within the blood which leads to the generation of thrombin and thus to the conversion of the soluble plasma protein fibrinogen into the insoluble fibrin polymer. Red blood cells become enmeshed by the strands of fibrin to form a clot. Thrombosis (or thrombus formation), on the other hand, involves blood platelets—small particles which occur normally in blood. Blood platelets adhere to sites of endothelial cell loss or areas where the endothelial cells are otherwise abnormal and, in the course of undergoing a shape change, release a number of potent factors which lead to further processes: the aggregation of further platelets at the site of the initial adhesion, the local formation of fibrin, changes in vessel permeability, and a stimulatory effect on the connective tissue cells in the underlying vessel wall. Strands of fibrin around the platelets serve to stabilize the platelet mass. The purpose of platelet aggregation is the prevention of bleeding from a damaged blood vessel, but a thrombus can also result in hemostasis. The consequences of thrombus formation are severe. For example, myocardial infarction is associated with formation of thrombi in places where cardiac muscle has been replaced by noncontractile scar tissue. Thrombus formation may also occur in a coronary vessel coated with atherosclerotic plaque (a coating on the inside of a blood vessel containing quantities of cholesterol).

Concurrently with the discovery of the role of NO in effecting vascular muscle relaxation came the discovery that NO inhibits both platelet aggregation and adhesion [65]. Prostacyclin and NO act synergistically to inhibit platelet aggregation and to disaggregate platelets, but there is no parallel synergism in platelet adhesion [66]. The role of NO in this matter seems to be as a feedback mechanism to counteract the effect of substances in the body, produced after injury, which promote aggregation and adhesion. The NO utilized by the platelets is mostly derived from endothelial cells with which the platelets come in contact, but there is also an enzyme system in platelets themselves which acts on arginine to produce NO [67].

NO-donor drugs, like SNAP and *S*-nitrosoglutathione, will both prevent platelet aggregation and disaggregate collections of platelets [68]. Thrombus formation is not an easy condition to treat and it may well be that the role of NO in effecting platelet disaggregation is of greater clinical significance than that of its role in the control of vascular muscle relaxation since there are already many highly successful drugs available for the treatment of high blood pressure.

8. NO AND THE IMMUNE SYSTEM

Production of NO is a normal, constitutive activity but its formation may also be induced in response to certain stimuli associated with disease. Such production, broadly referred to as inducible production, is effected by enzymes synthesized by particular cells, in locations clearly related to disease processes. NO production by inducible systems has been implicated in DNA damage and cell mutation [69], tumoricidal activity *in vivo* [70], and pathological features associated with sepesis [71].

The immune system provides a range of defense mechanisms that protect the individual against foreign proteins and invasion of the intracellular environment by microorganisms. The system, when activated, produces a whole range of reactive oxygen intermediates (ROIs) in response to the inflammatory stimuli. The first indication of a place for NO in the immune response came when high levels of nitrate were found in patients suffering from gastroenteritis. Treatment of rats with *E. coli* lipopolysaccharide also leads to high nitrate concentrations in these animals. Demonstrations then followed which showed that cultured macrophages, one of the main cell types involved in the immune response, also produce nitrate, as well as nitrite and nitrosothiols, a process that is dependent upon the presence of arginine in the cell culture medium [72]. These observations led to the conclusion that the arginine–NO pathway can be induced in macrophages. The ability of macrophages to kill abnormal cells, such as tumor cells, was then shown to be dependent upon arginine, and NO was proposed as a natural cytotoxic agent for the defense of the body against foreign agents and against abnormal cells within the body [73].

The important question now for many scientists is the nature of the cytotoxic action of NO. The effects of NO produced by inducible systems are certainly very different to those produced by constitutive systems. Are these differences simply related to the *quantity* of NO produced, or are there important *qualitative* events involved also? As NO is a radical, it was initially suggested that this may in itself be sufficient to explain its toxic nature, although the benign, constitutive production of NO posed a difficulty. However, unlike many other radicals, NO is not particularly reactive. When tested against *Clostridium*, an aqueous solution of NO is not a significant antibacterial agent (74). Since macrophage activity also produces ROIs, it has been proposed that NO produced by macrophages reacts with superoxide (O_2–) to give the peroxynitrite ion ($ONOO^-$) [75]. This exists as the protonated form under physiological conditions and may decompose to give hydroxyl radicals, which are certainly very highly reactive [76]. However, not all experimental evidence supports this view. When tested against *Leishmania major* *S*-nitroso-*N*-acetylpenicillamine (SNAP) was found to be an effective toxic agent, the action of which was enhanced by addition of superoxide dismutase (SOD), which removes any superoxide anion present. In contrast, peroxynitrite was not found to be particularly toxic [77]. It may be that the reaction of peroxynitrite with cellular targets, rather than formation of the hydroxyl radicals, is responsible for its biological activity and this will depend upon the environment in which it finds itself [78]. A similar conclusion was reached in a study of the effect of peroxynitrite on the aggregation of washed platelets [79].

Two studies of the antimicrobial action of NO-donor drugs suggest that, under the circumstances of these studies, it is neither the NO nor the hydroxyl radical which is responsible for their toxic action. Sodium nitroprusside ($Na_2[Fe(CN)_5NO]$; SNP) contains NO as a ligand where it is formally present as NO^+. SNP is an effective toxic agent against *Clostridium spirogenes* [74]. NO is also present in Roussin's Black Salt [$NaFe_4S_3(NO)_7$; RBS] as NO^+, and this salt appears to be the most effective of a range of NO donor drugs against *E. coli* [80]. The exact significance of this finding is at present unclear, but it may be that the crucial element in the toxic effect of SNP and RBS is the transfer of NO^+ to a physiologically important thiol.

The involvement of NO in inflammatory responses is by no means a blanket one. There may be many situations in which its role is carefully targeted, or in which it has no role at all. One of the most interesting contrasts to have emerged in this area is between the two main forms of inflammatory bowel disease—Crohn's disease and ulcerative colitis [81]. In both of these conditions, there is chronic inflammation of the intestinal wall, and the inflammatory reactions are often difficult to tell apart on histological grounds. In ulcerative colitis there is high activity of inducible NO-synthase, whereas in Crohn's disease the activity is normal, or even reduced. Such findings bring into question the factors that may regulate the inclusion of NO in the inflammatory response, and these factors have been little explored.

9. NO IN THE NERVOUS SYSTEM

It has been known for over two decades that excitation of nerve cells in the central nervous system (CNS) is accompanied by an increase in cGMP concentrations. This phenomenon is now known to be caused by NO, and a major trigger for its release in the CNS is the activation of glutamate receptors. The most direct demonstrations of NO production in the CNS to date have come from use of the Shibuki probe [82]. The clearest role for NO has been established in the peripheral autonomic nervous system where NO is released from non-adrenergic non-cholinergic (NANC) nerves and mediates relaxation of smooth muscle in many tissues. In the CNS, most interest has revolved around synaptic plasticity, regulation of blood flow, and neurodegeneration.

The autonomic nervous system regulates involuntary body functions and, as such, can be contrasted with the nerves that supply skeletal muscle used for voluntary actions. Inhibitory NANC neurones are important components of the autonomic innervation to many organs, and the chemical identity of their neurotransmitter has been much debated. There is now substantial evidence that it is NO. However, there are several important experimental findings that have led to proposals that NO-related compounds, rather than NO itself, may serve as NANC transmitters. For example, smooth muscle relaxation evoked by NANC nerve stimulation can be blocked by inhibitors of NO-synthase (the enzyme responsible for the biosynthesis of NO; see later), yet is unaffected by NO scavengers able to eliminate the relaxant effects of exogenously applied NO. Also, superoxide generators inactivate exogenous NO but not NANC-induced relaxation of rat gastric muscle. Unfortunately, this type of work has not given consideration to the diffusion properties of NO. Analysis of the way NO moves has reduced the acceptability of some of these studies [83]. If allowances are made for the likely kinetic and concentration profiles of NO when interpreting such studies, they can probably be seen to support the idea of NO as the NANC transmitter, and may not require the existence of alternative transmitters to be postulated [83, 84].

In the gastrointestinal tract NO seems to mediate many forms of relaxation, including adaptive dilation of the stomach after ingestion of food [85]. Histochemical studies of biopsies from infants with hypertrophic pyloric stenosis and adults with achalasia suggest that these conditions are associated with a lack of NO-synthase in pyloric and gastro-oesophageal tissue. In the intestine, muscle relaxation involved in peristalsis is also mediated by NO [86]. In the gastrointestinal tract, as in the cardiovascular system, there seems to be a constant NO-dependent dilator tone, crucial to the function of these organs.

NO is responsible for the relaxation of the blood vessels and smooth muscle of the male corpus cavernosum, and thus development of penile erection. Immunohistochemical evidence of NO-synthase in nerves has been found in penile tissue [87]. Other work has indicated dysfunction of the NO system in disorders associated with male impotence [88]. Functional studies with isolated human and animal penile tissue have shown that relaxation induced by electrical stimulation of autonomic nerves can be abolished by blockade of NO-synthase. The penile flaccidity found in diabetes may be a direct consequence of impaired NO synthesis and from a clinical perspective this provides a rationale for the treatment of certain types of impotence by intracavernosal administration of vasodilators. Studies of both rat and canine tissue have shown that NO-synthase is in fact widely distributed throughout the urogenital tract. High NO-synthase activity found in the urethra and its involvement in relaxation of the bladder neck suggest that NO may be important in the regulation of micturition and urine continence [87, 89]. The widespread system of nerves throughout the body that has been known as the NANC system thus appears to use NO as a transmitter. It has as its main theme the relaxation of smooth muscle. New insights into the nature of the transmitter are proving to be as important clinically as any gained from the study of nerves where the chemical nature of the neurotransmitter is well established.

The production and release of NO from the CNS has been demonstrated most directly with an electrochemical probe. Shibuki [82] has applied the sensor to measurement of NO in the CNS and claims to detect NO at 5×10^{-9} mol L^{-1} after endogenous production in the rat cerebellum in response to electrical stimulation of the white matter. NO may be able to affect cellular function over spherical volumes with an influence that is both predictable and constant. Close to a production source, yet over a volume sufficient to encompass many hundreds of neurones, this will depend mainly on the diffusion properties of NO since it is chemically unreactive enough to persist unchanged as it diffuses away from the production source to fill its sphere of influence. NO appears to influence a number of processes occurring in the brain and these will be described.

Synaptic plasticity is a process by which neuronal connections may be reinforced or altered and it is involved in cerebral functions such as memory formation and complex automated functions like playing a piano concerto fluently. Part of this phenomenon involves retrograde messengers which are able to return from a stimulated neurone to relay information back to the preceding neurone. NO is an attractive candidate as a retrograde messenger. It may mediate use-dependent changes in synaptic transmission by acting at presynaptic sites after it has been formed in a postsynaptic region [90]. The most studied example of synaptic plasticity is long-term potentiation in the hippocampus, a region of the brain involved in memory and learning, but the role of NO in this phenomenon remains controversial and uncertain [84].

Another example of synaptic plasticity is hyperalgesia, a state of enhanced sensitivity to painful stimuli. The mechanism of hyperalgesia shares several features with hippocampal long-term potentiation, particularly its endurance, dependence on afferent fiber activity, and its blockade by some antagonists. A range of evidence suggests a role for NO in the plastic changes giving rise to hyperalgesia, but further experimental work is required to understand the basic mechanisms involved [91, 92].

The possibility that neuronally derived NO may be important in the regulation of cerebral circulation has received much attention and it is believed to act within the cardiovascular centers in the CNS to regulate sympathetic outflow, vascular tone, and arterial pressure [93, 94]. Fibers that contain NO-synthase originate in one of the cranial autonomic ganglia and innervate major cerebral vessels in order to provide a neurogenic control over the cerebral circulation [95]. NO generated by neurones within the brain may also be a factor that links neuronal activity to local increases in cerebral blood flow. These mechanisms may all be fundamental to the participation of NO in regulation of cerebral vascular tone [96].

One of the most serious of cerebral pathologies is subarachnoid hemorrhage, a condition in which blood extravasates into the tissues that are in immediate proximity to the brain itself. Although many patients survive the initial events, the cerebral blood vessels frequently constrict in the period that follows the bleed and dramatically cut down blood flow to the brain, a condition known as cerebral vasospasm. For many years it has been considered that the problem could be related to the presence of free hemoglobin in the cerebral tissue, but the precise changes that this brings about have been unclear. It now seems that this condition could be elegantly explained if the free hemoglobin in the region of the bleed were to sequestrate NO (NO binds strongly to hemoglobin) and reduce the ambient vasodilator tone. Further research is required to identify the precise events that may occur in these circumstances, but there appears to be potential here again for a major stride forward in a previously troublesome pathology.

The possibility that NO can contribute to neurodegeneration in the CNS has been raised because it is a potentially toxic molecule and is able to inactivate key enzymes either directly or, possibly, by formation of peroxynitrite [76]. In cultured activated microglia, macrophage-like cells in the CNS, sufficient NO can be produced to kill cocultured cerebral neurones; it is therefore possible that NO production *in vivo* could lead to neuronal cell death [97]. In *in vivo* models of ischemia (oxygen starvation), NO-synthase inhibitors have been found to give marked protection, in some studies rescuing up to 70% of cortical neurones [98]. However, these inhibitors can also increase damage [99]. This is an exciting area where the chemical synthesis of selective inhibitors of neuronal NO-synthase is urgently needed. These would avoid the potentially detrimental action of blocking vital NO-dependent actions such as platelet disaggregation and cerebral vessel dilation but, at the same time, inhibit NO production where it causes damage. In view of the apparently conflicting roles of NO in the brain (neuroprotective and neurodestructive), it has been proposed that the redox forms of NO (NO^+ and NO^-) may play parts in the activity of NO. Although this may at first seem an attractive solution to the apparently paradoxical roles of NO, the chemistry of these species is unlikely to allow them a place among the biological mediators that are related to NO. The matter is discussed in the next section.

10. NO, NO^+, AND NO^-

To understand how NO can successfully fulfill so many roles in animal physiology is something of a problem. As described in the previous section, sometimes these roles appear to be in conflict. In endothelial cells NO acts as a benign messenger molecule for the activation of guanylate cyclase, but NO produced by macrophages appears to have a serious cytotoxic action. Why does NO not kill the smooth muscle cells it enters? How does it avoid all the sensitive targets in a muscle cell and alight selectively upon guanylate cyclase? In the brain the situation is even more puzzling as sometimes NO appears to be neuroprotective and at other times neurotoxic. Stamler et al. [100] have attempted to resolve this apparent conflict by suggesting that the nitrosonium ion (NO^+), as well as NO itself, is active in the brain. The nitrosonium ion is a well-known chemical species thought to be responsible for the nitrosation of a number of functional groups [101]. The difficulty of assigning NO^+ a role in animal physiology is that NO^+ is too reactive to exist in aqueous solution, where it reacts immediately to form nitrous acid:

$$H_2O + NO^+ \rightarrow HNO_2 + H^+$$

In strongly acid solution the reverse of this reaction occurs and it is NO^+ that is responsible for the conversion of a thiol (RSH) into a nitrosothiol (RSNO):

$$RSH + NO^+ \rightarrow RSNO + H^+$$

This reaction is the basis for the only possible role for NO^+ in biology in that nitrosothiol can transfer NO^+ to another nucleophilic site, probably another thiol [102]:

$$R^1SNO + R^2S^- \rightarrow R^2SNO + R^1S^-$$

This could be the route by which naturally occurring nitrosothiols, such as the nitrosothiol of human serum albumin, generate NO. There is likely to be transfer of NO^+ to any cysteine [103] present in the cell to give *S*-nitrosocysteine, which can then decompose homolytically to thiyl radicals (which dimerize to cystine) and NO:

$$R^2SNO \rightarrow R^2S + NO$$

Thus, in saying that NO^+ has a role in biology, we are implicating nitrosothiols. So far there appears to be no direct evidence for the transfer of NO^+ *in vivo* between nucleophiles, but this mechanism should be borne in mind.

The possibility that the anion NO^- (the reduced form of NO) has a role in animal physiology has been little explored, although it is possible. NO^- results from the ionization of nitroxyl (HNO), but it is rarely seen in the laboratory as HNO undergoes dimerization to give nitrous oxide:

$$2HNO \rightarrow H_2O + N_2O$$

However, as dimerization is a second-order process, the reaction rate is critically dependent on concentration and would be slow at the concentrations of HNO which might occur in cells. The pK_a of HNO is 4.7 and so ionization will occur to give NO^- at physiological pHs. Little has been published on the solution chemistry of NO^- and so it is difficult to comment on its suitability as a cell mediator. It does, however, react with reduced cytochrome *c* [104]. NO can readily be converted into NO^- by the enzyme superoxide dismutase [105] and this could be the biosynthetic path for its *in vivo* production. However, there is no direct evidence for a physiological role for NO^-.

11. NITRIC OXIDE SYNTHASE

The substrates for the endogenous production of NO are the amino acid arginine and dioxygen, and the enzyme responsible is known as nitric oxide synthase (NOS; EC1.14.13.39) [106]. The other product of reaction is citrulline (Figure 3). As the arginine–NO pathway occurs in a number of different tissues it is not surprising that there is a family of NOSs, each one having structural and functional similarities with other members of the family, but each especially suitable for the production of NO in the tissue where it is found. The concentration of arginine within cells is clearly important for NO production, and homeostasis of arginine metabolism depends upon regulating the rate of arginine degradation rather than its synthesis [107].

The first indication of the mechanism of NOS action came from experiments involving the use of ^{18}O-labeled dioxygen [108]. These studies showed that the oxygen in both NO and citrulline comes from molecular dioxygen and none comes from the solvent. The reaction occurs in two steps, both of which are monooxygenations. N^{ω}-Hydroxyarginine (NHA) is the intermediate, formed by a reaction utilizing one dioxygen and one NADPH and requiring the presence of tetrahydrobiopterin [108, 109]. The second step is oxidation of NHA to citrulline and NO, a process which also involves dioxygen and NADPH (Figure 4). Depending on the NOS, both processes may require the calcium-containing protein calmodulin.

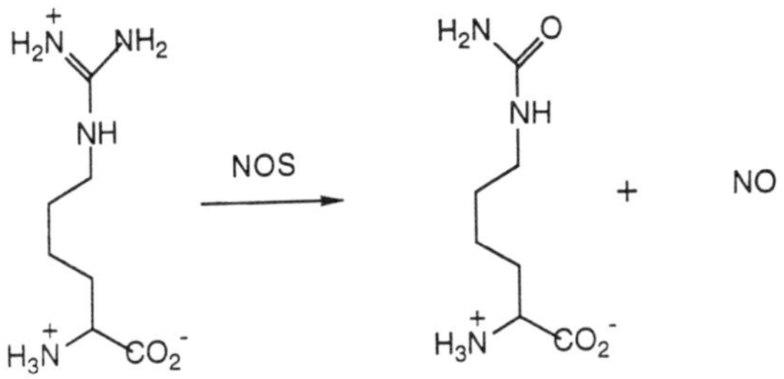

Figure 3.

Isoenzymes of the NOS family may be described either in terms of their origin or in terms of the manner in which they act. The principal distinction according to the latter criterion is whether the isoenzyme is constitutive or inducible. In the former case the synthase is present all the time and can respond rapidly to stimulation, but inducible NOS is formed only when required. The form present in the brain and peripheral nerves is neuronal, constitutive NO synthase, nNOS, or NOS type1. nNOS from the rat forebrain was the first isoenzyme to be purified [110] and cloned [111]. It is calcium-dependent and is present in fairly large quantities, not only in the rat forebrain but also in peripheral nerves [112]. Curiously the human nNOS appears also to occur in human skeletal muscle [113]. nNOS is stimulated in the brain by glutamate.

The isoenzyme present in endothelial cells (eNOS or NOS type III) was, of course, the first to be recognized. It is constitutive and calcium-dependent and may be associated with a membrane [114]. Its continuous presence in the walls of arteries and veins is partially responsible for the dynamic state of the cardiovascular system, according to the demands of the body. It is stimulated by a wide range of factors, including acetylcholine.

An inducible isoform of NOS (iNOS or NOS type II) was first detected in cultured macrophages [115] but is obtained on exposure of several cell types and tissues to cytokines [116] or bacterial products [117]. iNOS appears to be calcium-

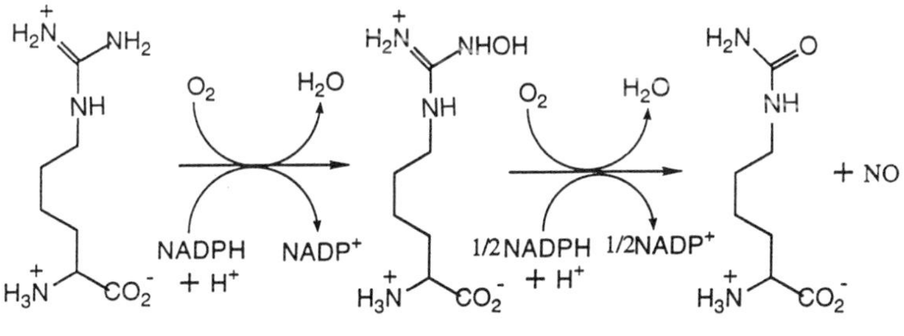

Figure 4.

independent but there is evidence that with these isoforms the calmodulin is very tightly bound, thus they do not require elevated external calcium levels for activation [118]. This is a matter of some complexity and has been discussed in a review [119].

The regulation of NOS has been little explored. NO itself inhibits NOS, presumably because NO interacts with enzyme-bound ferric heme [120]. This could constitute a feedback mechanism preventing overproduction of NO. Pregnancy and sex hormones have a substantial effect on NOSs in guinea pigs. Both eNOS and nNOS are subject to regulation by oestrogen and this could explain some changes occurring during pregnancy as well as some gender differences [121]. Lipopolysaccharides and γ-interferon suppress eNOS activity while activating iNOS. Inflammation could be the trigger for vascular endothelial cells to switch from eNOS to iNOS activity [122]. NOS in lung tissue is stimulated by paraquat and injury resulting from paraquat inhalation can be reduced by the use of NOS inhibitors [123].

A large number of compounds related to arginine are inhibitors of NOS (Figure 5). Among the most commonly used are *N*-monomethyl-L-arginine (LNMMA) and the methyl ester of *N*-nitro-L-arginine (L-NAME) [124]. Both are competitive inhibitors. Aminoguanidine (AG) selectively inhibits iNOS, whereas LNMMA affects both eNOS and iNOS [125]. On prolonged incubation of nNOS with LNMMA the latter becomes a substrate for NOS and gives rise to NO and citrulline [126]. *N*-Methyl-*N*-hydroxyarginine inhibits NOS [127], as does chloropromazine [128], the latter because it is a calcium antagonist.

The mechanism of the first step in the biosynthesis of NO from arginine, a hydroxylation, is fairly well understood [129]. Since NOS is a P-450 type hemo-

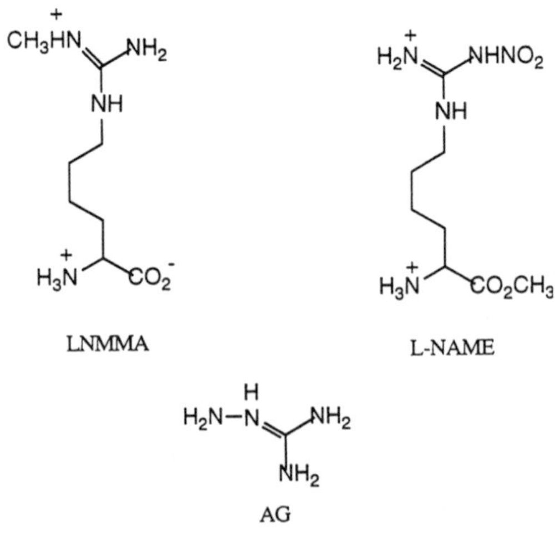

CH_3HN^+ NH_2 NH H_3N^+ CO_2^-

LNMMA

H_2N^+ $NHNO_2$ NH H_3N^+ CO_2CH_3

L-NAME

H H_2N-N NH_2 NH_2

AG

Figure 5.

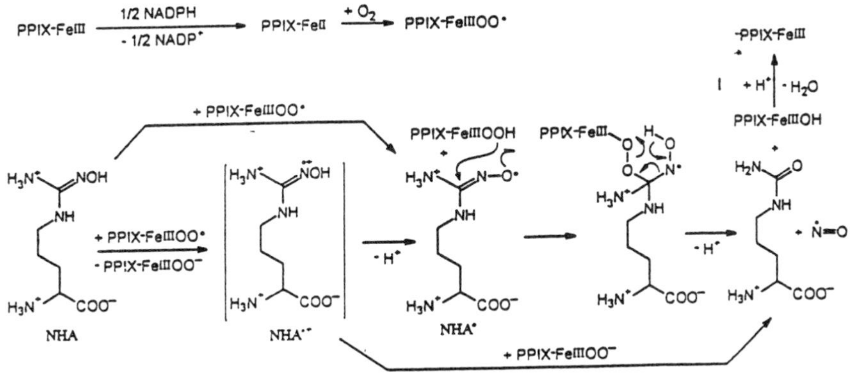

Figure 6.

protein it is probable that this process is similar to other P450 catalyzed *N*-hydroxylation reactions and involves an initial one-electron oxidation of the guanidine nitrogen. It is unlikely that there is direct involvement of tetrahydrobiopterin [130].

The second step, the conversion of *N*-hydroxyarginine into citrulline and NO, is far less clear. The problems of notation have been addressed and a mechanistic pathway, allowing for the variable oxidation state of the heme iron of the enzyme and the radical nature of the departing molecule, has been proposed [131] (Figure 6).

12. MEDICAL USES OF NO GAS

Initially NO in breath was demonstrated indirectly by oxidation and formation of nitrosothioproline and analysis of a derivative by gas chromatography–mass spectrometry (GC–MS). This has been applied to validate chemiluminescence as a method for analysis of NO in breath [13]. More recently there has been direct GC–MS confirmation of the presence of NO in exhaled human breath [133], and this provides the first direct and absolute demonstration of endogenous NO production in humans. This technique demands differentiation between $^{15}N_2$ (m/z 30.00022, parts per thousand) and NO (m/z 29.99799, parts per billion), and is complex and not practical for routine use; its value has been in confirmation of the specificity of more simple analytical techniques that can be used for routine and multiple analyses.

Reaction between nitric oxide and ozone leads to generation of light,

$$NO + O_3 \rightarrow NO_2^* + O_2$$

$$NO_2^* \rightarrow NO_2 + \text{light}$$

and detection of light produced in this way was first applied to measurement of NO as an atmospheric pollutant [134]. A variety of chemiluminescence analyzers is now available for quantitation of NO, and such equipment can be adapted readily for routine measurement of endogenous NO excreted in expired air [132, 135]. The technique is highly sensitive and reproducible, and most investigators report detection limits for NO of 2–5 × 10^{-9} M [136].

The response time of chemiluminescence equipment for the analysis of NO in breath is now rapid (< 0.5), and exhaled NO levels can be determined in a single breath. The method has been used to detect changes in NO production that seem to occur in periods of exercise and with asthma, though the significance of such changes is unclear at present [135].

The technique is unsuitable for assessment of patients who do not have good control over their respiratory cycle. This is because a slow (4 rpm) regular breathing pattern with prolonged exhalation time (>5), is necessary to create the NO plateaux required for standardized measurements. Breathing with normal or low tidal volumes (<1 L) and normal or high frequency (>8 rpm) does not give rise to a stable end-expiratory concentration of NO. Therefore, assessment of NO in the expirate of subjects with chronic dyspnoea or neuromuscular disorders would be difficult by these means.

Further work is required to establish the exact origins of the NO that contributes to the NO profiles recorded with this technology. Both the lower and the upper respiratory tracts release NO into the expirate [135, 137]. Whether or not useful representation of systemic NO production could be obtained from breath analyzers is still to be determined. At the present time it seems likely that the measurement of NO in breath will continue to provide a useful tool for investigation of the respiratory system, but is unlikely to become a widely applicable index for systemic NO generation, particularly under physiological conditions.

The most valuable application of chemiluminescence in this area has been to monitor the exogenous administration of NO in an intensive care unit. Administration of NO as a respiratory gas would have seemed absurd to most anesthetists just a decade ago. Now that the endogenous presence of the molecule has been clearly established in the exhalate, a precedent is set that allows additional administration to be accepted as a comfortable concept. Inhalation of NO gas has been shown to cause vascular and bronchial smooth muscle relaxation in the lungs [138]. The principal mode of action is stimulation of soluble guanylate cyclase with consequent rises in intracellular cGMP concentrations and protein phosphorylation. Inhalation of NO has been used as an effective therapy in pulmonary hypertension of the newborn, and in treatment of adult respiratory distress syndrome (ARDS), a severe and often fatal condition associated with pulmonary hypertension, pulmonary edema, and poor right ventricular function [139]. A number of infants have received inhaled NO for critical pulmonary artery hypertension after operations for congenital heart defects [140].

The mode of delivery in these treatments—a gas to the respiratory system—has a number of important benefits that have been long-awaited. Traditional approaches to the treatment of pulmonary hypertension have involved use of pulmonary vasodilators that affect the systemic circulation causing unwanted general systemic vasodilation and reduced systemic arterial pressure. Such general vasodilators are additionally disadvantageous because they tend to dilate the entire pulmonary vascular bed and augment blood flow to nonventilated and poorly ventilated lung regions. This increases pulmonary ventilation–perfusion mismatch and lowers the arterial partial pressure of oxygen. By contrast, NO inhaled as a gas has no adverse systemic dilator effects, and is delivered only to those regions of the lung that are functional and oxygenated. In this way, ventilation–perfusion mismatch is avoided and systemic oxygen saturation levels are maintained. Prolonged administration of NO in ARDS, followed by withdrawal, has been shown to give sustained improved lung function, suggesting that improved blood flow and oxygen exchange has been achieved during therapy and these have, in turn, allowed improved recovery of damaged tissue. This is an exciting development in the treatment of lung disease. The full potential of inhaled NO in the treatment of various causes of pulmonary hypertension will be realized over the next decade, after full clinical trials have run their course. General application of NO gas in the intensive care unit in the U.K. has been slow to gain acceptance. This is related in part to fear of side effects, such as NO_2 toxicity and methemoglobinemia. In fact, when delivered at between 10 and 80 ppm, these potential problems appear to be insignificant. Concern remains as to how best to monitor for such side effects. A further problem at present is that chemiluminescence equipment is not readily available to monitor accurate delivery of NO to the patient. It is almost certain that there will be considerable developments in the clinical use of NO gas in the future.

REFERENCES

[1] Butler, A. R. and Williams, D. L. H., Chem. Soc. Rev., 22 (1993) 233.
[2] Moncada, S., Palmer, R. M. J. and Higgs, E. A., Pharmacol. Rev., 43 (1991) 109.
[3] Wink, D. A., Darbyshire, J. F., Nims, R. W., Saavedra, J. E. and Ford, P. C., Chem. Res. Toxicology, 6 (1993) 23.
[4] Kharitonov, V. G., Sundquist, A. R. and Sharma, V. S., J. Biol. Chem., 269 (1994) 5881.
[5] Awad, H. H. and Stanbury, D. M., Int. J. Chem. Kinetics, 25 (1993) 375.
[6] Wennmalm, A., Benthin, G. and Patersson, A. S., Brit. J. Pharmacol., 106 (1992) 507.
[7] Spagnuolo, C., Rinelli, P., Coletta, M., Chiancone, E. and Ascoli, F., Biochim. Biophys. Acta, 911 (1987) 59.
[8] Green, L. C., Wagner, D. A., Glogowski, J., Skipper, P. L., Wishnok, J. S. and Tannenbaum, S. R., Anal. Biochem., 126 (1982) 131.
[9] Leone, A. M., Rhodes, P., Furst, V. and Moncada, S., in Kendall and Hill (eds.), *Methods in Molecular Biology*, Humana Press, 1995, in press.
[10] Huxley, A. F. and Niedergerke, R., Nature, 173 (1954) 971.
[11] Ruegg, J. C., *Calcium in Muscle Activation*, Springer-Verlag, Berlin, 1988.
[12] Blinks, J. R., Rüdel, R. and Taylor, S. R., J. Physiol., 277 (1978) 291.
[13] Itoh, T., Kajiwara, M., Kitamura, K. and Kuriyama, H., J. Physiol., 322 (1982) 107.
[14] Somlyo, A. V. and Somlyo, A. P., J. Pharmacol. Exptl. Ther., 159 (1968) 129.
[15] Sobieszek, A., European J. Biochem., 73 (1977) 477.
[16] Somlyo, A. P., Somlyo, A. V., Shuman, H. and Endo, M., Fed. Proc., 41 (1982) 2883.

[17] Casteels, R. and van Breeman, C., Pflügers Arch., 359 (1975) 197.
[18] Reuter, H., Blaustein, M. P. and Haeusler, G., Phil. Trans. Roy. Soc. (Lond.) B, 265 (1973) 87.
[19] Yoshida, Y., Sun, H-T., Cai, J-Q. and Imai, S., J. Biol. Chem., 266 (1991) 19819.
[20] Popescu, L. M., Panoiu, C., Hinescu, M. and Nutu, O., European J. Pharmacol., 107 (1985) 393.
[21] Furchgott, R. F. and Zawadzki, J. V., Nature, 288 (1980) 373.
[22] Waldman, S. A. and Murad, F., Pharmacol. Rev., 39 (1987) 163.
[23] Furchgott, R.F., in Vanhoutte, P. M. (ed), *Mechanisms of Vasodilation*, Vol. IV, New York, Raven Press, 1987.
[24] Palmer, R. M. J., Ferrige, A. G., and Moncada, S., Nature, 327 (1987) 524.
[25] Ignarro, L. J., Buga, G. M., Wood, K. S., Byrns, R. E., and Chaudhuri, G., Proc. Natl. Acad. Sci. USA, 84 (1987) 9265.
[26] Palmer, R. M. J., Ashton, D. S., and Moncada, S., Nature, 333 (1988) 664.
[27] Myers, P. R., Minor, R. L., Guerra, R., Bates, J. N. and Harrison, D. G., Nature, 345 (1990) 161.
[28] Feelisch, M., te Poel, M., Zamora, R., Deussen, A., and Moncada, S., Nature, 368 (1994) 62.
[29] Brunton, T. L., Lancet, 2 (1867) 97.
[30] Feelisch, M. and Noack, E. A., European J. Pharmacol., 139 (1987) 19.
[31] Feelisch, M, Noack, E. and Schroeder, H., European Heart J., 9 (Suppl. A) (1988) 57.
[32] Abrams, J., Amer. J. Med., 91 (1991) 106; Feelisch, M., European Heart J., 14 (1993) 123.
[33] Butler, A. R. and Glidewell, C., Chem. Soc. Rev., 16 (1987) 361.
[34] Butler, A. R., Glidewell, C., McGinnis, J. and Bisset, W. I. K., Clin. Chem., 33 (1987) 490.
[35] Bisset, W. I. K., Burdon, M. G., Butler, A. R., Glidewell, C. and Reglinski, J., J. Chem. Res., (1981) 299.
[36] Flitney, F. W. and Kennovin, G., J. Physiol., 392 (1987) 43P.
[37] Butler, A. R., Calsy-Harrison, A. M. and Glidewell, C., Polyhedron, 7 (1988) 1197.
[38] Flitney, F. W., Megson, I. L., Flitney, D. E. and Butler, A. R., Brit. J. Pharmacol., 107 (1992) 842.
[39] Field, L., Dilts, R. V., Ravichandran, R., Lenhart, P. G. and Carnahan, G. E., Chem. Comm., (1978) 249.
[40] Bauer, J. A. and Fung, H-L., J. Pharmacol. Exptl. Ther., 256 (1991) 249.
[41] Ignarro, L. J., Lippton, H., Edwards, J. C., Baricos, W. H., Hyman, A. L., Kadowitz, P. J. and Gruetter, C. A., J. Pharmacol. Exptl. Ther., 218 (1981) 739.
[42] McAninly, J., Williams, D. L. H., Askew, S. C., Butler, A. R. and Russell, C., Chem. Comm., (1993) 1758.
[43] Stamler, J. S., Simon, D. I., Osborne, J. A., Mullins, M. E., Jaraki, O., Michel, T., Singel, D. J. and Loscalzo, J., Proc. Natl. Acad. Sci. USA, 89 (1992) 444.
[44] Askew, S. C., Butler, A. R., Flitney, F. W., Kemp, G. D. and Megson, I. L., Bioorg. Med. Chem., 3 (1995) 1.
[45] Kamisaki, Y., Saheki, S., Nakane, M., Palmieri, J. A., Kuno, T., Chang, B. Y., Waldman, S. A. and Murad, F., J. Biol. Chem., 261 (1986) 7236.
[46] Braughler, J.M., Mittal, C.K. and Murad, F., J. Biol. Chem., 254 (1979) 12450.
[47] Craven, P. A. and DeRubertis, F. R., J. Biol. Chem., 253 (1978) 8433.
[48] Gerzer, R., Hofmann, F., Bohme, E., Krassimira, I., Spies, C. and Schultz, G., Adv. Cyclic Nucleotide Res., 14 (1981) 255.
[49] Craven, P. A., DeRubertis, F. R. and Pratt, D. W., J. Biol. Chem., 254 (1979) 8213.
[50] Wolin, M. S., Wood, K. S. and Ignarro, L. J., J. Biol. Chem., 257 (1982) 13312.
[51] Kon, H. and Kataoka, N., Biochemistry, 8 (1969) 4757.
[52] Ignarro, L. J., Biochem. Soc. Trans., 20 (1992) 465.
[53] Stone, J. R. and Marletta, M. A., Biochemistry, 33 (1994) 5636.
[54] Brandwein, H. J., Lewicki, J. A. and Murad, F. J., Biol. Chem., 256 (1981) 2958.
[55] Kamisaki, Y., Waldman, S. A. and Murad, F., Arch. Biochem. Biophys., 251 (1986) 709.
[56] Braughler, J. M., Mittal, C. K. and Murad, F., Proc. Natl. Acad. Sci. USA, 76 (1979) 219.

[57] Wu, X-B., Brüne, B., von Appen, F. and Ullrich, V., Arch. Biochem. Biophys., 294 (1992) 75.
[58] Marczin, N., Ryan, U. S. and Catravas, J. D., J. Pharmacol. Exptl. Ther., 263 (1992) 170.
[59] Moncada, S., Rees, D. D., Schulz, R. and Palmer, R. M. J., Proc. Natl. Acad. Sci., USA, 88 (1991) 2166.
[60] Waldman, S. A., Rapoport, R. M., Ginsburg, R. and Murad, F., Biochem. Pharmacol., 35 (1986) 3525.
[61] Schmidt, K., Graier, W. F., Kostner, G. M., Mayer, B., Bohme, E. and Kukovetz, W. R., J. Cardiovasc. Pharmacol., 17 (1991) S83.
[62] Furchgott, R. F., Martrin, W., Jothianandan, D. and Villani, G. M., in Proceedings of 9th International Congress of Pharmacology, Paton, W., Mitchell, J. and Turner, P. (eds.), 1984, Macmillan, London, pp. 149–157.
[63] Megson, I. L., Flitney, F. W., Bates, J. and Webster, R., Endothelium, 3 (1995) 39.
[64] Venturini, C. M., Palmer, R. M. J. and Moncada, S., J. Pharmacol. Exptl. Ther., 266 (1993) 1497.
[65] Radomski, M. W., Palmer, R. M. J. and Moncada, S., Brit. J. Pharmacol., 92 (1987) 181.
[66] Radomski, M. W., Palmer, R. M. J. and Moncada, S., Brit. J. Pharmacol., 92 (1987) 639.
[67] Radomski, M. W., Palmer, R. M. J. and Moncada, S., Brit. J. Pharmacol., 101 (1990) 325.
[68] Radomski, M. W., Rees, D. D., Dutra, A. and Moncada, S., Brit. J. Pharmacol., 107 (1992) 745.
[69] Nguyen, T., Brunson, D. and Crespi, C. L., Penman, B. W., Wishnok, J. S. and Tannenbaum, S. R., Proc. Natl. Acad. Sci. USA, 89 (1992) 3030.
[70] Farias-Eisner, R., Sherman, M. P., Aeberhard, E. and Chaudhuri, G., Proc. Natl. Acad. Sci., USA, 91 (1994) 9407.
[71] Petros, A., Bennett, D. and Vallance, P., Lancet, 338 (1991) 1557.
[72] Marletta, M. A., Yoon, P. S., Iyengar, R., Leaf, C. D. and Wishnok, J. S., Biochemistry, 27 (1988) 8706; Hibbs, J. B., Taintor, R. R., Vavrin, Z. and Rachlin, E. M., Biochem. Biophys. Res. Commun., 157 (1988) 87; Stuehr, D., Gross, S. S., Sakuma, I., Levi, R. and Nathan, C. F., J. Exp. Med., 169 (1989) 1011.
[73] Hibbs, J. B., Vavrin, Z. and Taintor, R. R., J. Immunol., 138 (1987) 550.
[74] Maraj, S. R., Khan, S., Cui, X-Y., Cammack, R., Joannou, C. L. and Hughes, M. N., *The Analyst*, 120 (1995) 699.
[75] Blough, N. V. and Zafiriou, O. C., Inorg. Chem., 24 (1985) 3502.
[76] Beckman, J. S., Beckman, T. W., Chen, J., Marshall, P. A. and Freeman, B. A., Proc. Natl. Acad. Sci. USA, 87 (1990) 1620.
[77] Assreuy, J., Cunha, F. Q., Epperlein, M., Noronha-Dutra, A., O'Donnell, C. A., Liew, F. Y. and Moncada, S., European J. Immunol., 24 (1994) 672.
[78] Koppenol, W. H., Moreno, J. J., Pryor, W. A., Ischiropoulos, H. and Beckman, J. S., Chem. Res. Toxicol., 5 (1992) 834.
[79] Moro, M. A., Darley-Usmar, V. M., Goodwin, D. A., Read, N. G., Zamora-Pino, R., Feelisch, M., Radomski, M. W. and Moncada, S., Proc. Natl. Acad. Sci. USA, 91 (1994) 6702.
[80] Burdon, M. G., Butler, A. R. and Renton, L., unpublished observation.
[81] Boughton-Smith, N. K., Evans, S. M., Hawkey, C. J., Cole, A. T., Balsitis, M., Whittle, B. J. R. and Moncada, S., Lancet, 342 (1993) 338.
[82] Shibuki, K., Neurosci. Res., 9 (1990) 69.
[83] Wood, J. and Garthwaite, J., Neuropharmacol., 33 (1994) 1235.
[84] Garthwaite, J., The Neurosciences, 5 (1993) 171.
[85] Lefebvre, R. A., Baert, E. and Barbier, A. J., in Moncada, Marletta, Hibbs and Higgs (eds.), *The Biology of Nitric Oxide*, 1992, Vol. 1, 293.
[86] Ward, S. M., Dalziel, H. H., Thornbury, K. D., Westfall, D. P. and Sanders, K. M., in Moncada, Marletta, Hibbs and Higgs (eds.), *The Biology of Nitric Oxide*, Vol. 1, 1992, 295.
[87] Burnett, A. L., Lowenstein, C. J., Bredt, D. S., Chang, T. S. K. and Snyder, S. H., Science, 257 (1992) 401.
[88] Kim, N., Azadzoi, K. M., Goldstein, I. and de Tejada, I. S., J. Clin. Invest., 88 (1991) 112.

[89] Franchi, A. M., Chaud, M., Rettori, V., Suburo, A., McCann, S. M. and Gimeno, M., Proc. Natl. Acad. Sci. USA, 91 (1994) 539.
[90] Garthwaite, J., Charles, S. L. and Chess-Williams, R., Nature, 336 (1988) 385.
[91] Kitto, K. F., Haley, J. E. and Wilcox, G. L., Neurosci. Lett., 148 (1992) 1.
[92] Meller, S. T., Pechman, P. S., Gebhart, G. F. and Maves, T. J., Neuroscience 50 (1992) 7.
[93] Shapoval, L. N., Sagach, V. F. and Pobegailo, L. S., Neurosci. Lett., 132 (1991) 47.
[94] Togashi, H., Sakuma, I., Yoshioka, M., Kobayashi, T., Yasuda, T., Kitabatake, A., Saito, H., Gross, S. S. and Levi, R., J. Pharmacol. Exptl. Ther., 262 (1992) 343.
[95] Kovách, A. G. B., Szabö, C., Benyö, Z., Csáki, C., Greenberg, J. H. and Reivich, M., J. Physiol., 449 (1992) 183.
[96] Iadecola, C., Proc. Natl. Acad. Sci. USA, 89 (1992) 3913.
[97] Boje, K. M. and Arora, P. K., Brain Res., 587 (1992) 250.
[98] Nowicki, J. P., Duval, D., Poignet, H. and Scatton, B., European J. Pharmacol., 204 (1991) 339.
[99] Yamamoto, S., Golanov, E. V., Berger, S. B. and Reis, D. J., J. Cereb. Blood Flow Metab., 12 (1992) 717.
[100] Lipton, S. A., Choi, Y-B., Pan, Z-H., Lei, S. Z., Chen, H-S. V., Sucher, N. J., Loscalzo, J., Singel, D. J. and Stamler, J. S., Nature, 364 (1993) 626.
[101] Williams, D. L. H., *Nitrosation*, Cambridge University Press, Cambridge, 1988.
[102] Barnett, D. J., McAninly, J., Williams, D. L. H., J. Chem. Soc., Perkin Trans. 2 (1994) 1131.
[103] Meyer, D. J., Kramer, H., Özer, N., Coles, B. and Ketterer, B., FEBS Lett., 345 (1994) 177.
[104] Bonner, F. T., Hughes, M. N., Poole, R. K. and Scott, R. I., Biochim. Biophys. Acta, 1056 (1991) 133.
[105] Murphy, M. M. E. and Sies, H., Proc. Natl. Acad. Sci. USA, 88 (1991) 10860.
[106] Palmer, R. M. J. and Moncada, S., Biochem. Biophys. Res. Commun., 158 (1989) 348; Mayer, B., Schmidt, K., Humbert, P. and Böhme, E., Biochem. Biophys. Res. Commun., 164 (1989) 678; Mason, A., Bassaye, E. and Busse, R., Arch. Pharmacol., 340 (1989) 767; Palacios, M., Knowles, R. G., Palmer, R. M. J. and Moncada, S., Biochem. Biophys. Res. Commun., 165 (1989) 802; Stuehr, D. J., Kwon, N. S., Gross, S. S., Thiel, B. A., Levi, R. and Nathan, C. F., Biochem. Biophys. Res. Commun., 161 (1989) 420; Tayeh, M. A. and Marletta, M. A., J. Biol. Chem., 264 (1989) 19654.
[107] Castillo, L., Sánchez, M., Chapman, T. E., Ajami, A., Burke, J. F. and Young, V. R., Proc. Natl. Acad. Sci. USA, 91 (1994) 6393.
[108] Leone, A. M., Palmer, R. M. J., Knowles, R. G., Francis, P. L., Ashton, D. S. and Moncada, S., J. Biol. Chem., 266 (1991) 23790.
[109] Stuehr, D. J., Kwon, N. S., Nathan, C. F., Griffith, O. W., Feldman, P. L. and Wiseman, J., J. Biol. Chem., 266 (1991) 6259; Kwon, N. S., Nathan, C. F., Gilker, C., Griffith, O. W., Matthews, D. E. and Stuehr, D. J., J. Biol. Chem., 265 (1990) 13442; Mayer, B., John, M., Heinzel, B., Werner, E. R., Wachter, H., Schultz, G. and Böhme, E., FEBS Lett., 288 (1991) 187.
[110] Bredt, D. S. and Snyder, S. H., Proc. Natl. Acad. Sci. USA, 87 (1990) 682.
[111] Bredt, D. S., Hwang, P. M., Glatt, C. E., Lowenstein, C., Reed, R. R. and Snyder, S. H., Nature, 351 (1991) 714.
[112] Sheng, H., Schmidt, H. H. H. W., Nakame, M., Mitchell, J. A., Pollock, J. S., Förstermann, J. S. and Murad, F., Brit. J. Pharmacol., 106 (1992) 768.
[113] Nakane, M., Schmidt, H. H. H. W., Pollock, J. S., Förstermann, U. and Murad, F., FEBS Lett., 316 (1993) 175.
[114] Pollock, J. S., Klinghofer, V., Förstermann, U. and Murad, F., FEBS Lett., 309 (1992) 402.
[115] Tonetti, M., Sturla, L., Bistolfi, T., Benatti, U. and De Flora, A., Biochem. Biophys. Res. Commun., 203 (1994) 430.
[116] Robbins, R. A., Barnes, P. J., Springall, D. R., Warren, J. B., Kwon, O. J., Buttery, L. D. K., Wilson, A. J., Geller, D. A. and Polak, J. M., Biochem. Biophys. Res. Commun., 203 (1994) 209.

[117] Knowles, R. G., Merrett, M., Salter, M. and Moncada, S., Biochem. J., 270 (1990) 833.
[118] Cho, H. J., Xie, Q-W., Calaycay, J., Mumford, R. A., Swiderek, K. M., Lee, T. D. and Nathan, C., J. Exp. Med., 176 (1992) 599.
[119] Knowles, R. G. and Moncada, S., Biochem J., 298 (1994) 249.
[120] Griscavage, J. M., Fukuto, J. M., Komori, Y. and Ignarro, L. J., J. Biol. Chem., 269 (1994) 21644.
[121] Weiner, C. P., Lizasoain, I., Baylis, S. A., Knowles, R. G., Charles, I. G. and Moncada, S., Proc. Natl. Acad. Sci. USA, 91 (1994) 5212.
[122] Walter, R., Schaffner, A. and Schoedon, G., Biochem. Biophys. Res. Commun., 202 (1994) 450.
[123] Berisha, H. I., Pakbaz, H., Absood, A. and Said, S. I., Proc. Natl. Acad. Sci. USA, 91 (1994) 7445.
[124] Rees, D. D., Palmer, R. M. J., Schulz, R., Hodson, H. F. and Moncada, S., Brit. J. Pharmacol., 101 (1990) 746.
[125] Joly, G. A., Ayres, M., Chelly, F. and Kilbourn, R. G., Biochem. Biophys. Res. Commun., 199 (1994) 147.
[126] Klatt, P., Schmidt, K., Brunner, F. and Mayer, B., J. Biol. Chem., 269 (1994) 1674.
[127] Feldman, P. L., Griffith, O. W., Hong, H. and Stuehr, D. J., J. Med. Chem., 36 (1993) 491; Moynihan, H. A., Roberts, S. M., Weldon, H., Allcock, G. H., Ånggård, E. E. and Warner, T. D., J. Chem. Soc., Perkin Trans. 1, (1994) 769.
[128] Palacios, M., Padron, J., Glaria, L., Rojas, A., Delgado, R., Knowles, R. and Moncada, S., Biochem. Biophys, Res. Commun., 196 (1993) 280.
[129] Marletta, M. A., J. Biol. Chem., 268 (1993) 12231.
[130] Hevel, J. M. and Marletta, M. A., Biochemistry, 31 (1992) 7160.
[131] Korth, H-G., Sustmann, R., Thater, C., Butler, A. R. and Ingold, K. U., J. Biol. Chem., 269 (1994) 17776.
[132] Gustafsson, L. E., Leone, A. M., Persson, M. G., Wiklund, N. P. and Moncada, S., Biochem. Biophys. Res. Commun., 181 (1991) 852.
[133] Leone, A. M., Gustafsson, L. E., Francis, P. L., Persson, M. G., Wiklund, N. P. and Moncada, S., Biochem. Biophys. Res. Commun., 201 (1994) 883.
[134] Zafiriou, O. C. and McFarland, M., Anal. Chem., 52 (1980) 1662.
[135] Persson, M. G., Wiklund, P. and Gustafsson, L. E., Amer. Rev. Respir. Dis., 148 (1993) 1210.
[136] Archer, S. L. and Cowan, N. J., Circ. Res., 68 (1991) 1569; Brien, J., McLaughlin, B., Nakatsu, K. and Marks, G., J. Pharmacol. Methods, 25 (1991) 19; Menon, N., Wolf, A., Zehetgruber, M. and Bing, R., Proc. Soc. Exptl. Biol. Med., 191 (1989) 316.
[137] Gerlach, H., Rossaint, R., Pappert, D., Knorr, M. and Falke, K. J., Lancet, 343 (1994) 518.
[138] Pepke-Zaba, J., Higenbottam, T. W., Dinh-Xuan, A. T., Stone, D. and Wallwork, J., Lancet, 338 (1991) 1173.
[139] Roberts, J. D., Polaner, D. M., Lang, P. and Zapol, W. M., Lancet, 340 (1992) 818.
[140] Jowinois, D., Pouard, P., Mauriat, P., Malhere, T., Vouhe, P. and Safran, D. J., Thor. Cardiovasc. Surg., 107 (1994) 1129.

INDEX

Acetohydroxamate–Al, speciation curves, 230
Activation of acids, 121–122
  amines, 121
  enolization, 122
  epimerization, 122
  oximes, 121
  water, 121
Activation of leaving groups, 119–121
  alkoxy oxygens, 119
  amides, 120
  hydroxide, 119
  oxime, 119
  sulfur, 121
Acyl carrier protein, 40, 77
  description of, 40
  X-ray structural studies of, 40
Adenosine 5′-phosphates, stability constants with Al, 221
Adenylate kinase, 16–18
  action of, 71–72
  domains of, 18
  mechanism of action, 71
  reactions of, 17
Aluminum
  absorption, 233–237
  abundance, 201
  analytical methods, 202–204
  and Alzheimer's disease, 238
  and osteomalacia, 238
  atomic spectroscopy, 203
  binding of albumin, 217
  binding of amino acids, 214–218
  binding of aminopolycarboxylates, 215
  binding of aspartate, 215, 216
  binding of carboxylic acids, 211–212
  binding of citrate, 215
  binding of hydroxycarboxylic acids, 212–214
  binding of inorganic ligands, 207
  binding of oligopeptides, 217
  binding of peptides, 214–218
  binding of phosphates, 218–220
  binding of proteins, 214–218
  binding of Schiff bases, 216
  binding on DNA, 222
  binding on nucleotides, 220–223
  bioinorganic chemistry of, 199–250
  bone disease, 238
  catecholate binding, 223–224
  chemical methods of determination, 202

CO_2-hydroxide system, 211
coordination with biomolecules, 206–229
daily adult intake of, 201
determination of, 202
distribution of in humans, 236
effect on behavior of phospholipids, 220
elimination of, 237
F systems, 211
hydroxamate binding, 224–225
hydroxide system, 207
in human biological fluids, 235
in human blood, 235
metabolism, 233–237
neurological toxicity, 237–238
neutron activation analysis, 203
pharmaceutical uses of, 240–242
phosphate systems, 209–211
phosphorylated proteins, 220
porphyrin binding, 228–229
pyridinone binding, 225–228
pyrone binding, 225–228
salicylate binding, 223–224
scope of analytical methods, 204
speciation curves with acetohydroxamate, 230
speciation curves with ATP, 222, 233
speciation methods, 204–206
speciation of, 239–240
stability constants for various compounds, 229
stability constants of hydroxide system, 208
stability constants of myo-inositol-triphosphates, 219
stability constants of pyridone complexes, 226
stability constants of pyrone complexes, 226
stability constants with adenosine 5′-phosphates, 221
sulfate system, 211
synthetic inositol derivative binding, 219
ternary complexes, 229–233
toxic effects of, 237–239
toxicology of, 239–240
triamino-inositol derivative binding, 219
X-ray fluorescence, 203
Albumin, binding of Al, 217
Alkaline phosphatase, 59–60
active site of, 59
and metal binding, 59
structure of, 59, 82
Animal physiology, role of nitric oxide, 251–277
Artificial metalloenzymes, 141–146
design of, 141, 144
formation of, 141
macrocyclic, 143
reactions of, 142, 144, 145
selective recognition, 143
structures of, 141, 142
Ascorbate oxidase, 151–197
amino acid alignment of, 168
anaerobic reduction of, 173, 174
and nitric oxide, 177
and stability of copper-depleted form, 182
binding site of ascorbate, 188–189
biological function, 154
catalytic mechanism of, 186, 187
crystalline characteristics, 157
crystallization of, 157–158
Cu site diagram, 163
Cu site geometries, 163–167
diagram of azide form, 184
diagram of peroxide form, 181
diagram of reduced form, 180
electron transfer processes, 187–192
electron transfer within the trinuclear copper site, 192

electronic structural studies, 165, 166
functional derivatives of, 176–185
hydrogen bonding, 160
hydrogen bonding scheme, 161–162
interatomic distances, 164
intramolecular electron transfer, 190–192
kinetic properties of, 171–176
molecular properties, 155–157
occurrence, 154
oxidation–reduction potentials, 170–171
preparation of azide form, 183
preparation of peroxide form, 180
preparation of reduced form, 179
reoxidation of, 175, 176
schematic representation of, 158, 159
secondary structure of, 160–163
sequences, 154
spectroscopic properties, 155–157
spectroscopic studies of azide form, 185
steady-state kinetics, 172
structure of, 158
structure of peroxide form, 182
symmetry of, 158
tetramer contact surface areas, 160–163
topology/packing diagram, 160
trinuclear Cu site, 165
X-ray structure of azide form, 183–185
X-ray structure of Cu depleted enzyme, 177–179
X-ray structure of peroxide form, 180–183
X-ray structure of reduced form, 179–180
X-ray structure studies, 157–167

Ascorbic acid, Al complexes, 227, 228
ATP, speciation curves with Al, 222, 233

Bestatin, 50
Binuclear metal ions, catalysis, 129
Biomolecules, Mn-containing, 1–113
Blockade of inhibitory reverse paths, 122–123
enzymatic, 122
hydrolysis of 3-carboxyaspirin, 122
Blue oxidases, 153
inhibition of, 176–177
redox potentials, 171
spectroscopic properties of Cu ion, 156

cAMP-dependent protein kinase, 72
action of, 72
structure of, 72
Carboxypeptidase, 136–141
cooperation, 139
crystallographic studies, 138
description of, 136
inhibitors, 137, 139
mechanism of, 136, 137
model studies, 140
thermodynamic studies, 138, 139
Catalase, 40–41
action of, 40
and iron protoporphyrin IX prosthetic group, 40
Mn containing, 40
properties of, 41
spectroscopic studies of, 77
structural studies of, 41
Catalysis by binuclear metal ions, 129
Catalysis by metal-bound hydroxide ion, 127–129
alkene hydration, 128
amide hydrolysis, 127
and phosphate esters, 128

general base, 128–129
model studies, 127
nucleophilic attack, 127–128
Catalysis by metal-bound water, 125–127
amide hydrolysis, 126
general acid, 126–127
lactonization, 126
nucleophilic, 125
Catalytic efficiency, 129, 133–136
and hydrolysis of phosphate diesters, 134
and ligand effects, 133, 135
and metal size, 133
Catecholates, Al binding, 223
Ceruloplasmin, 153
amino acid alignment of, 167, 168
catalytic mechanism, 185–187
Cu centers of, 156
molecular properties of, 155
redox potentials, 170
spectroscopic properties of, 156
X-ray study, 168
Chelation therapy in Al detoxification, 241
Concanavalin A, 48
Creatine kinase, 16–18
action of, 71–72
active site of, 17
reactions of 17

Deoxyribonuclease (DNase), 43–44
action of, 43
conformational studies of, 44
structure studies of, 44, 78
Dioxygen, 152
Dioxygenase, 13, 14
DNA, Al binding, 222
DNA polymerase, 41–43
action of, 41
active sites of, 43
and metal sites, 42
mutation studies of, 43
proteolysis studies of, 42
structure of, 42
structure of, 77–78
DNase (*see* Deoxyribonuclease)

EDRF (*see* Endothelium-derived relaxing factor)
Effective molarities, 123
Elongation factor Tu, 14, 15
structural studies of, 70
Endothelium-derived relaxing factor (EDRF), 255–256
identification of, 255–256
Enolase, 60–61
action of, 60
crystal structure of, 61
heavy metal derivatives of, 61
role of metal ions, 60, 61
structure of, 83
Ester hydrolysis, 117

FPK (*see* 6-Phosphofructokinase)
Fructose-1,6-bisphosphatase, 44–45, 78
action of, 44
crystallographic studies of, 45
EPR studies of, 44
metal binding sites on, 44
metal sites on, 78
structure of, 78
Fungal laccases, 167–170
amino acid alignment of, 167, 168
redox potentials, 171

G-Actin, 13
Galactosyltransferases, 19
Gastrointestinal tract, and nitric oxide, 264
GC (*see* Guanylate cyclase)
Glutamine synthetase, 45–46, 79
action of, 45
crystal structure studies, 46
metal binding studies of, 79

metal sites on, 45, 46
occurrence of, 45
regulation of, 45
structure of, 45, 79
thermodynamics of binding of, 46
Glyceryl trinitrate, 256
Granulosa virus, 16
Guanylate cyclase (GC), activation of by NO, 258–261
Guanylate kinase, 16–18
and binding sites, 18
domains of, 18
occurrence of, 17
reactions of, 17
studies of, 18

Hexokinase, 72
active site of, 72
spectroscopic studies of, 72
Hydroxamate, Al binding, 224–225
Hydroxypyridones, Al complexes, 227

Immune system, and nitric oxide, 262–264
Inorganic pyrophosphatase, 64–68
action of, 64
binding site, 68
comparison of amino acid sequences, 66–67
metal ion requirements of, 65
structural studies, 65, 84
types of, 64
Iron protoporphyrin IX prosthetic group, 40
Isocitrate dehydrogenase, 15, 16
structural studies of, 71

Laccase, 22, 153
anaerobic reduction of, 173, 174
and azide binding, 184
and nitric oxide, 177
catalytic mechanism of, 185–187
Cu centers of, 156
fully reduced, 183
functional derivatives of, 176–185
inhibition of, 177
kinetic properties of, 171–176
molecular properties of, 155
reoxidation of, 175, 176
spectroscopic properties of, 156
steady-state kinetics, 172
α-Lactalbumin, 19–20, 72
and metal binding, 19, 72
and Mn binding, 19
structural studies of, 19, 72
Lactoferrin, 20
Lectin, 46–49
action of, 47
and conformational studies, 47
characterization of in plants, 47
chemistry of, 79
cleft, 48
diagram of binding sites, 47
metal binding coordination spheres, 48
metal sites on, 79
metal-free protein studies, 47
Leucine aminopeptidase, 49–50
action of, 49
and bestatin, 50
and metal binding sites, 49
metal site on, 80
specificity of, 79
structural characterization of, 49
Lewis acid catalysts, 115–149
Lignin, 22
Lignin compounds, 24

Maltol, Al complex of, 225
Mandelate racemase, 20–21, 72–73
and Eu, 20
and Mg, 20
and Mn, 20
as catalyst, 20
metal site on, 73
structure of, 21, 72

Manganese-dependent peroxidase, 22–24
Masking of anions, 123
Metal ion catalysis, 117–125
  activation of electrophiles, 117
  amide hydrolysis, 117, 118
  carbonyl reductions, 117
  decarboxylation, 117
  ester hydrolysis, 117, 118
  nitrile hydration, 117, 118
  phosphoryl complexation, 119
Metal ions and organic functional groups, 129–132
  and proximity, 130
  carboxypeptidase, A, 131
  hydrolysis, 132
Metal ions as Lewis acid catalysts, 115–149
Metalloenzymes, 136–141
Mn chemistry, 3
  coordination chemistry of, 3
  in biological systems, 3
Mn geometry, 3
  importance of in biology systems, 3–10
Mn in enzymes, 5, 6
  structure determination of, 9
  tabular listing of known structures, 9
  tabular lists, 6–8
Mn in humans, 4
Mn in living organisms, 4, 5
Mn in plants, 4, 5
Mn-containing biomolecules, 1–113
Mn-dependent peroxidase, 73–74
  spectroscopic studies on, 73
  structure of, 73
cis, cis-Muconate cycloisomerase, 24
  catalytic action, 24
  structure of, 74
  structure studies of, 25
Multiple catalytic repertories, 132–133
  amide hydrolysis, 133
  hydrolysis of phosphinate esters, 133
  template effect, 132
Myo-inositol-triphosphates, Al binding, 219

Nervous system, and nitric oxide, 264–267
Nitric oxide (NO)
  and inflammatory response, 264
  and neurodegeneration, 266
  and platelet aggregation, 261–262
  and the immune system, 262–264
  and the nervous system, 264–267
  and vascular smooth muscle relaxation, 253–255, 265
  anion, 268
  chemiluminescence, 272
  chemistry of, 267–268
  guanylate cyclase activation, 261
  in animal physiology, 251–277
  in the gastrointestinal tract, 264
  medical uses of, 271–273
  neuronally derived, 266
  reaction with ozone, 271
  solution chemistry of, 252–253
Nitric oxide synthase (NOS), 268–271
  inhibitors of, 270
  mechanism, 268, 269
  regulation of, 270
  substrates for, 268
Nitrosonium ion (NO^+), chemistry of, 267–268
Nitrovasodilators, 256–258
NO (*see* Nitric oxide)
NO^+ (*see* Nitrosonium ion)
NOS (*see* Nitric oxide synthase)
Nucleotides, Al binding, 220–223

Osteomalacia, induced by Al, 238
Ovalbumin, 25

and Mn, 25
description of, 25

Parvalbumin, 62
and binding sites, 62
and metal ions, 62
description of, 62
mutant studies of, 83
occurrence, 62
structure of, 83
Phosphatases, 10–13
enzyme hydrolysis, 10
Phosphate esters, stability toward hydrolysis, 123
6-Phosphofructokinase (FPK), 50–52
action of, 50
active site of, 51
three-dimensional structure of, 50
3-Phosphoglycerate kinase, 52–53
action of, 52
and metal binding sites, 53
Phosphoglycerate kinase, metal ion binding, 80
3-Phosphoglycerate kinase
spectroscopic properties of, 52
structural properties of, 52, 53
Phosphoglycerate kinase, structure of, 80
Phospholipids, behavior in presence of Al, 220
Phosphorylated proteins, Al binding, 220
Ping-pong di Theorell–Chance mechanism, 172
Platelet aggregation, and nitric oxide, 261–262
Porphyrin, Al binding, 228–229
Proteins with 1 bound metal, 13–39
Proteins with 2 bound metals, 40–59
Proteins with 3 bound metals, 59–64
Proteins with 4 bound metals, 64
Provision of productive conformations, 123–125
and metalloenzymes, 123
Purple acid phosphatase, 11–13
and Mn, 11
di-iron form of, 70
properties of, 12
Pyridinones, Al binding, 225–228
Pyrones, Al binding, 225–228
Pyruvate kinase, 62–64
action of, 63
binding affinities of, 84
function of, 62
nucleotide specificity of, 63
spectroscopic studies, 84
structure of, 62, 64

ras p21 protein, 25–27
binding sites of, 26
crystal structure of, 26
description of, 25
metal sites of, 74
mutations of, 74
structures of, 74
Reactive oxygen intermediates, 262
Recognition of anions, 123
in metalloenzymes, 123
Ribonuclease H, 80
structure of, 80
Ribonuclease, structure of, 80
Ribonucleotide reductase, 54–56
action of, 54
characterization of, 54
diagram of active site, 55
metal site on, 54, 81
Mn activation of, 55
Ribulose 1,5-bisphosphate carboxylase (RuBisCo), 27–28
action of, 27
active site, 28
active site characterization, 75
as part of a multienzyme complex, 75
site-directed mutagenesis, 27
x-ray studies on, 75

Roussin's Black Salt, 257
RuBisCo, *see* Ribulose 1,5-bisphosphate carboxylase)

Salicylates, Al binding, 223
Serine/Threonine protein phosphatase 2A, 10–11
  and Mn sensitivity, 10
Shibuki probe, 264
SOD (*see* Superoxide dismutase)
Staphylococcal nuclease, 29–31, 75–76
  action of, 76
  and cleavage site, 29
  and metal site environment, 30
  diagram of metal binding site, 31
  mutant forms of, 75
  mutant studies, 30
  spectroscopic studies of, 75
  structure of inhibited enzyme, 30
Subarachnoid hemorrhage, 266
Superoxide dismutase (SOD), 31–39, 76–77
  active site of, 38
  and metal ion specificity, 32, 37
  and Mn bond distances, 38
  comparison of amino acid sequences, tabular listing of, 34–36
  homology, 33
  magnetic properties of, 32
  metal sites in, 77
  occurrence of, 76
  primary structures of, 33
  reaction mechanism of, 33
  spectral properties of, 32
  structure of, 76
  types, 31
  with Fe cofactors, 32
  with Mn cofactors, 32
  X-ray structural studies of, 37
Synaptic plasticity, 265, 266

Transketolase, 39–40
  action of, 39
  active site, 39
  and Mn replacement of Mg, 40
  structure of, 77
Triamino-inositol derivatives, Al binding, 219
tRNA, 29
  and Mn, 29
  structural studies, 29

Vascular smooth muscle relaxation, 253–255
Veratryl alcohol, 23

D-Xylose isomerase, 56–59
  action of, 56
  and metal binding sites, 57
  and metal ion specificity of, 57
  and Mn, 58
  metal dissociation constants of, 56
  mutants of, 82
  occurrence of, 56
  pH dependence of, 57
  site-directed mutagenic studies of, 57
  site-directed substitution on, 81
  spectroscopic studies of, 58
  structural studies, 82
  structure of, 56, 57

J

A

I

P

R

E

S

S

Advances in Metal-Organic Chemistry

Edited by **Lanny S. Liebeskind,**
Department of Chemistry, Emory University

Organometallic chemistry is having a major impact on modern day organic chemistry in industry and academia. Within the last ten years, the use of transition metal based chemistry to perform reactions of significant potential in organic synthesis has come of age. *Advances in Metal-Organic Chemistry* contains in-depth accounts of newly emerging synthetic organic methods and of important concepts that highlight the unique attributes of organometallic chemistry applied to problems in organic synthesis. Particular emphasis will be given to transition metal organometallics. Each issue contains six to eight articles written by leading investigators in the field. Emphasis is placed on giving the reader a true feeling of the particular strengths and weaknesses of the new chemistry with ample experimental details for typical procedures. Contributors have been urged to write in an informal style in order to make the material accessible to interested readers who are not experts in the field.

REVIEW: *Advances in Metal-Organic Chemistry* is an attractive volume that will be a worthwhile addition to private collections for active practitioners in germane areas. It should be found in all technical libraries as a useful reference book."

— *Journal of Medicinal Chemistry*

Volume 1, 1989, 393 pp. $109.50
ISBN 0-89232-863-0

CONTENTS: Introduction to the Series: An Editor's Foreword, *Albert Padwa.* Preface. *Lanny S. Liebeskind.* Recent Developments in the Synthetic Applications of Organoiron and Organomolybdenum Chemistry, *Anthony J. Pearson.* New Carbonylations by Means of Transition Metal Catalysts, *Iwao Ojima.* Chiral Arene Chromium Carbonyl Complexes in Asymmetric Synthesis, *Arlette Solladie-Cavallo.* Metal Mediated Additions to Conjugated Dienes, *Jan-E. Backvall.* Metal-Organic Approach to Stereoselective Synthesis of Exocyclic Alkenes, *Ei-ichi Negishi.* Transition Metal Carbene Complexes in Organic Synthesis, *William D. Wulff.*

Volume 2, 1991, 300 pp. $109.50
ISBN 0-89232-948-3

J A I P R E S S

Advances in Metal and Semiconductor Clusters

Edited by **Michael A. Duncan,** *Department of Chemistry, University of Georgia*

Volume 1, Spectroscopy and Dynamics
1993, 314 pp. $109.50
ISBN 1-55938-171-X

CONTENTS: Preface, *Michael A. Duncan.* Photodissociation Kinetics and Optical Spectroscopy of Metal Cluster Ions, *Urmi Ray and Martin Jarrold.* Electronic Properties of Gas Phase Metal Clusters, *Karl-Heinz Meiwes-Broer.* Chemical Bonding in the Late Transition Metals: The Nickel and Copper Group Dimers, *Michael D. Morse.* Spectroscopy and Photochemistry in Clusters and Organometallic Complexes of Silver, *Michael A. Duncan.* The Unique Complexation and Oxidation of Metal Based Clusters, *James L. Gole.* Spectroscopy of Neutral Semiconductor Clusters, *Mary L. Mandich.* Photophysical Studies of Bare and Metal-Containing Silicon Clusters, *Steven M. Beck.* Cluster Ion Photodissociation Spectroscopy, *Daniel E. Lessen, R. L. Asher and Phillip J. Brucat.* Index.

Volume 2, Cluster Reactions
1994, 213 pp. $109.50
ISBN 1-55938-704-1

CONTENTS: Preface, *Michael A. Duncan.* Collision-Induced Dissociation of Transition Metal Cluster Ions, *Peter Armentrout.* Ion-Molecule Reactions of Metal and Semiconductor Clusters: Ionization Potentials and Electron Affinities, *John Eyler.* Theoretical Studies of Clustering Reactions: Sequential Reactions of SiH_n (n=0-3) with Silane, *Krishnan Raghavachari.* Formation, Stability, and Reactivity of Gas Phase Bimetallic Clusters, *Koji Kaya.* Potential Energy Surfaces for the Insertion of the Third Row Transition Metal Atoms (Hf-Pt) into H_2, *Krishnan Balasubramanian.* Metallo-Carbohedrenes: A New Class of Molecular Clusters, *A.W. Castleman, Jr..* The Calculation of Accurate Metal Ligand Bond Dissociation Energies, *Charles Bauschlicher.* Index.

Volume 3, Spectroscopy and Structure
1995, 231 pp. $109.50
ISBN 1-55938-788-2

CONTENTS: Preface, *Michael Duncan.* Metal Atom-Rare Gas van der Waals Complexes, *W. H. Breckenridge, Christophe Jouvet, and Benoit Soep.* Spectroscopic Studies of Large-Amplitude Motion in Small Clusters, *Eric A. Rohlfing.* Study of Small Carbon and Silicon Clusters Using Negative Ion Photodetachment Techniques, *Caroline C. Arnold* and *Daniel M. Neumark.* CRLAS: A New Analytical Technique for Cluster Science, *J. J. Scherer, J. B. Paul, A. O'Keefe, and R. J. Saykally.* Metal-Carbon Clusters: The Construction of Cages and Crystals, *J. S. Pilgrim and M. A. Duncan.* Index.